Rivers of the Sultan

Rivers of the Sultan

The Tigris and Euphrates in the Ottoman Empire

FAISAL H. HUSAIN

Oxford University Press is a department of the University of Oxford. It furthers
the University's objective of excellence in research, scholarship, and education
by publishing worldwide. Oxford is a registered trade mark of Oxford University
Press in the UK and certain other countries.

Published in the United States of America by Oxford University Press
198 Madison Avenue, New York, NY 10016, United States of America.

© Oxford University Press 2021

All rights reserved. No part of this publication may be reproduced, stored in
a retrieval system, or transmitted, in any form or by any means, without the
prior permission in writing of Oxford University Press, or as expressly permitted
by law, by license, or under terms agreed with the appropriate reproduction
rights organization. Inquiries concerning reproduction outside the scope of the
above should be sent to the Rights Department, Oxford University Press, at the
address above.

You must not circulate this work in any other form
and you must impose this same condition on any acquirer.

Library of Congress Cataloging-in-Publication Data
Names: Husain, Faisal, author.
Title: Rivers of the Sultan : the Tigris and Euphrates in the Ottoman Empire / Faisal Husain.
Description: New York : Oxford University Press, 2021. |
Includes bibliographical references and index.
Identifiers: LCCN 2020047015 (print) | LCCN 2020047016 (ebook) |
ISBN 9780197547274 (hardback) | ISBN 9780197547298 (epub) | ISBN 9780197547304
Subjects: LCSH: Water-supply—Political aspects—Euphrates River Watershed. |
Water-supply—Political aspects—Tigris River Watershed. |
Water resources development—Political aspects—Euphrates River Watershed. |
Water resources development—Political aspects—Tigris River Watershed. |
Turkey—Foreign relations—Iraq. | Iraq—Foreign relations—Turkey. |
Turkey—History—1683–1829.
Classification: LCC HD1691.H87 2021 (print) |
LCC HD1691 (ebook) | DDC 333.91/6209560903—dc23
LC record available at https://lccn.loc.gov/2020047015
LC ebook record available at https://lccn.loc.gov/2020047016

DOI: 10.1093/oso/9780197547274.001.0001

Some of this material appeared earlier in The Journal of Interdisciplinary History,
XLVII (2016), 1–25. It is included herein with the permission of the editors of The Journal
of Interdisciplinary History and The MIT Press, Cambridge, Massachusetts.
© 2014 by the Massachusetts Institute of Technology and The Journal of
Interdisciplinary History, Inc

To John R. McNeill

Contents

Acknowledgments ix

Introduction 1

PART I. *The Amphibious State*

1. Fortresses 23
2. Shipyards 40

PART II. *The Water Wide Web*

3. Arable Lands 61
4. Grasslands 79
5. Wetlands 95

PART III. *The Rumblings of Nature*

6. Havoc 111
7. After the Flood 127

Conclusion 145

Appendixes 151
Notes 163
Bibliography 221
Index 255

Acknowledgments

COUNTLESS PEOPLE HAVE given me the most generous help at every stage of writing this book. It is impossible to express my gratitude to all of them by name here. A few, however, must have special mention. First among them is John McNeill. Without his mentorship, this book might never have been written. Susan Ferber and Alan Mikhail offered me their unwavering support and encouragement throughout the completion of this project. To the three of them I am eternally grateful.

During my research, I accumulated an enormous debt to Gábor Ágoston, M. Fatih Çalışır, M. Talha Çiçek, Samuel Dolbee, Şengül Karaloğlu, and Dina Khoury. Different chapters of the book benefited greatly from kind invitations by Hümeyra Bostan to İstanbul Şehir University; by McGuire Gibson to the Neubauer Collegium for Culture and Society at the University of Chicago; by Giacomo Parrinello and G. Mathias Kondolf to UC Berkeley; by Stephanie Rost to the Institute for the Study of the Ancient World at New York University; and by James C. Scott to the Program in Agrarian Studies at Yale University. Christopher Morris and Marc Van De Mieroop readily shared their expertise with me whenever I needed it. Unless otherwise noted, all maps and illustrations were prepared by the talented artist Meredith Sadler. They all have my appreciation and admiration.

The leisure to write this book was made possible by a residential fellowship at the Notre Dame Institute for Advanced Study. Its staff—Meghan Sullivan, Donald Stelluto, Carolyn Sherman, Kristian Olsen, and Paul Blaschko—made my writing sojourn a most pleasant experience. I was fortunate to meet and learn from them.

For research and writing, I have received additional support from Penn State University, Georgetown University, the Institute of Turkish Studies, the Fulbright-Hays Doctoral Dissertation Research Abroad Fellowship, and the Mellon/ACLS Dissertation Completion Fellowship. I am pleased to acknowledge and thank each of them.

Finally, my sincere thanks go out to my department head Michael Kulikowski and to the Penn State University community as a whole—staff, students, and faculty. I am truly privileged to be surrounded by them on a daily basis.

Rivers of the Sultan

Introduction

> Who looks upon a river in a meditative hour, and is not reminded of the flux of all things?
> —Ralph Waldo Emerson, *Nature* (1836)

IN MAY 1534, two decades of tension between the Ottoman and Safavid empires finally boiled over. After securing a truce with the Habsburgs the previous year, Sultan Süleyman I left the Ottoman capital Istanbul and marched east toward the Caucasus, the region where he intended to strike his archenemy, Shah Tahmasp I. After the sultan's departure, conjectures about what could happen following the military confrontation were many and varied. Maybe, people in his entourage thought, Samarkand, Khurasan, Turan, or Baghdad would be next.[1] No one could have predicted that, as the Safavid army retreated deeper into Persia and winter approached, Süleyman would abruptly end the chase and set his sights on Baghdad, which he would triumphantly enter before year's end. The Ottoman camp had even less inkling that the impending campaign would eventually unify the entire Tigris and Euphrates rivers under Istanbul's control. Without forethought, the Ottomans at the end of 1534 joined other natural and biological forces governing, and being governed by, the two longest rivers in West Asia.

Iron and Fire

The Tigris and Euphrates are twin rivers bound by a common geography. They emerge together from the same rocky womb of the Taurus Mountains, and they perish together in the same shallow tomb of the Persian Gulf.[2] In between, the rivers carve diverse landscapes united by a common geographical drudgery—siphoning off the rainfall and snowmelt they capture inland to a single aquatic outlet.

The natural boundaries of this drainage basin rarely dovetailed with the webs of political power that regularly formed and fractured throughout history. Ottoman predecessors had no shortage of political ambition, but they often lacked the capacity to control such a broad geographical unit, roughly the size of Bolivia and half again as large as Texas. Political possibilities changed with the spread of iron metallurgy in the first millennium BC. Abundant and widespread in Earth's crust, the metal allowed states to equip more foot soldiers with iron armor and weapons than had been possible when bronze was king. Aided by the cheaper armaments and larger armies of the Iron Age, the Assyrian and Achaemenid empires were among the first powers in history to ever claim sovereignty over the entirety of the Tigris-Euphrates basin.[3]

Like iron metallurgy 2,500 years earlier, gunpowder technology in the fifteenth and sixteenth centuries opened new possibilities to align the

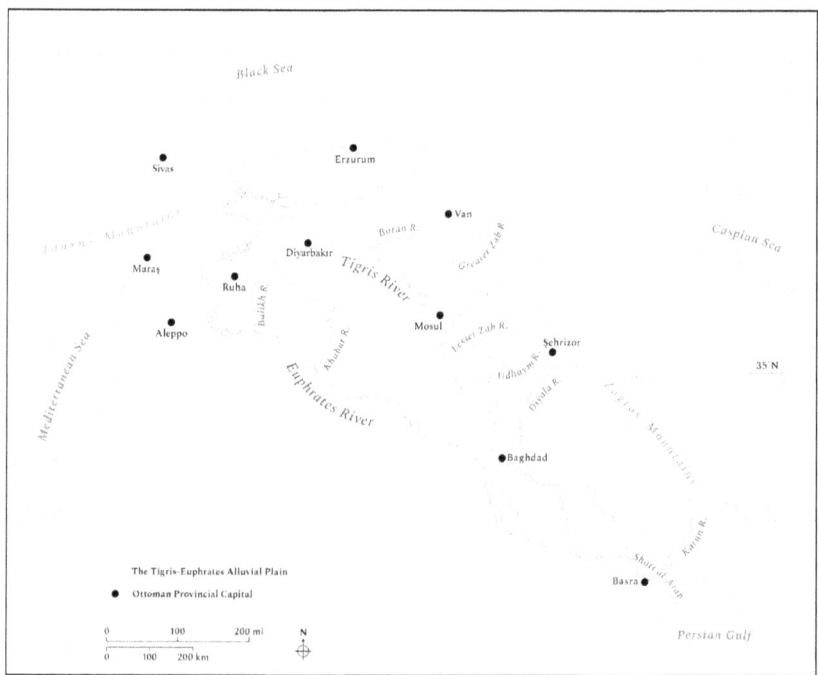

FIGURE I.I. The Tigris-Euphrates River Basin.

political and fluvial landscapes of West Asia. A mixture of saltpeter, sulfur, and charcoal, gunpowder introduced a novel chemical form of energy to the battlefield, overcoming the limits of medieval weaponry operated by human and animal muscle power. From the late fourteenth century, the Ottoman Empire joined the global arms race to harness the power of the new chemical explosive, gradually developing the domestic capacity to manufacture guns and powder. Self-sufficient in quality arms and ammunition, Istanbul achieved military prominence abroad while strengthening its authority at home. It was in this context that the Ottomans could inaugurate a rare period of political unification in the Tigris-Euphrates basin. Firepower strengthened Istanbul's governing hand in this far-flung region, allowing it to maintain a robust bureaucratic apparatus, intimidate local chieftains, and confront rivals successfully for nearly 400 years. Human history and water history in the drainage basin became unified under a single imperial order that was more stable and durable than anything that came before it, enforced by field cannon, artillery fortresses, firearms, and gunboats.[4]

The Waterwheel

The Tigris and Euphrates' life journey on earth, more than a thousand miles in length, forms the continental arc of a planetary circle through which water cycles between the ocean and the atmosphere. "The rivers flow, the oceans perform their slow and elegant gyrations, the clouds congeal and weep," as a scientist pithily describes the process.[5] Aşık Mehmed, a sixteenth-century Ottoman geographer from Trabzon, compared river flow in the context of the hydrologic cycle to a spinning waterwheel.[6]

From the sixteenth century, Istanbul put this natural waterwheel to work for the benefit of its imperial project. It relied on the natural flow of water between mountain and ocean to deal with a chronic ecological imbalance along its eastern borderland. Within the escarpment between Mosul and Baghdad, the thirty-five parallel runs. Above it, mountainous terrain captures higher precipitation, and distance from the equator lowers evaporation from the soil and transpiration from plants. Ottoman provinces in this upstream zone, as a result, enjoyed a moist climate, supportive of dry farming and forest growth. The lowlands south of latitude 35°N stood in sharp contrast to the situation in the uplands. Here, aridity thins out the vegetation cover and makes irrigation necessary for food production. The natural distribution of ore deposits,

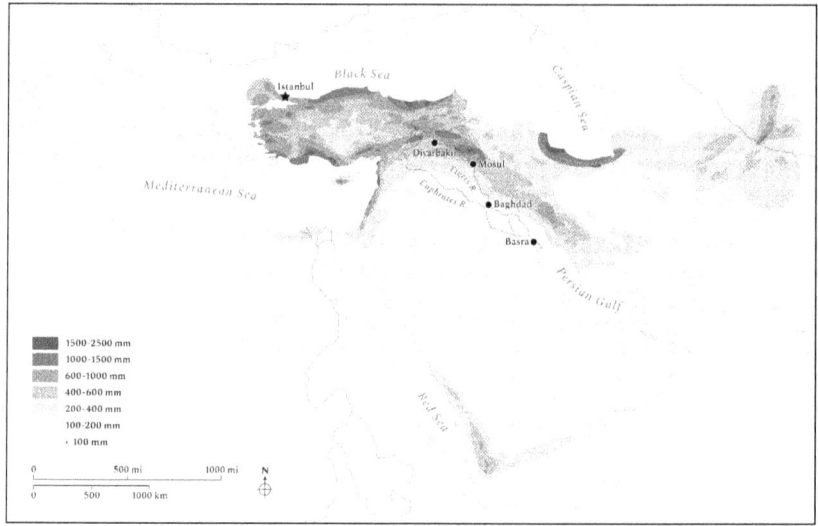

FIGURE I.2. Mean Annual Rainfall in West Asia. After *Tübinger Atlas des Vorderen Orients* (Wiesbaden: L. Reichert, 1977–1993), A IV 4.

abundant in the north and scarce in the south, further widened the mineral wealth gap between the two zones.

The unification of the Tigris and Euphrates allowed Istanbul to rebalance this natural resource disparity along its eastern frontier. It organized regular shipments of grain, metal, and timber from upstream areas of surplus in Anatolia and the Jazira (the stony plateau between the Tigris and Euphrates north of Baghdad) to downstream areas of need in Iraq. Northern grain in the form of wheat and barley supported larger garrisons in the south; metal in the form of armaments equipped the garrisons with gunpowder weapons; and timber in the form of rivercraft delivered both grains and arms. Some river rafts were even broken up for their timber to be redeployed in the south. For some 200 years, this imperial system of waterborne communication anchored the Ottoman presence in Iraq, enabling Istanbul to fend off foreign rivals, suppress domestic revolts, and exploit the organic wealth of the Tigris-Euphrates alluvial plain.

At the turn of the eighteenth century, the confluence of natural and human disasters forced a realignment of the Ottoman Empire's relationship to the twin rivers on its eastern frontier. As war and rebellion raged in the background, drought, plague, and an abrupt shift in the Euphrates' channel plunged Iraq into chaos. Initial attempts to restore order failed to yield lasting success, and the case for devising new mechanisms to reintegrate Iraq into the empire became inescapable. Istanbul settled for a compromise based on sharing power with a military oligarchy based in Baghdad to stabilize the region. Over the course of the eighteenth century, the Pashalik of Baghdad—as this provincial oligarchy came to be known—expanded its zone of influence, enclosing the most critical stretches of the Tigris and Euphrates. Through its extensive networks, the Pashalik provisioned Iraq with the resources that it had long lacked while keeping the Ottoman central government at arm's length. By the late eighteenth century, Baghdad had emerged as an alternative administrative center to Istanbul in the Tigris-Euphrates basin, in control of the waterways most amenable to navigation and irrigation.

The East Is Blue

Writing a history of the Tigris and Euphrates is an attempt to piece back together a jigsaw that time has torn apart. Even though natural scientists take the physical and biological unity of river systems as an article of faith, the twin rivers appear in most historical works as dismembered bodies. In Ottoman historiography, for example, different portions of the drainage basin feature in national and provincial monographs in isolation from each other. Studies framed around administrative units illuminated the adaptability of imperial governance in different localities but rendered invisible the total fluvial system. The Tigris and Euphrates, as a result, have leaked through the cracks of monographs about Anatolia, Syria, and Iraq as well as their cities and provinces.[7]

One aim of this book, therefore, is to overcome the historiographical dams that have divided the Tigris and Euphrates into artificial basins and to demonstrate the utility of adopting a hydro-scale that considers the fluvial system as a continuous whole. Unified under Ottoman hegemony, the natural drainage pattern of the twin rivers fostered intimate bonds between upstream and downstream provinces, transporting not only water and sediment but also boatloads of men, guns, and grain that cemented the Ottoman presence in the east. The hydro-scale clarifies the magnitude and significance of these movements. It reveals, moreover, how the Tigris and Euphrates could expand the reach of natural and political disturbances happening anywhere in their basin. For instance, drought in the highlands of Sivas and Diyarbakır could trigger floods in the lowlands of Baghdad, and security anxiety in the Persian Gulf could spur the construction of riverboats upstream in Birecik.[8]

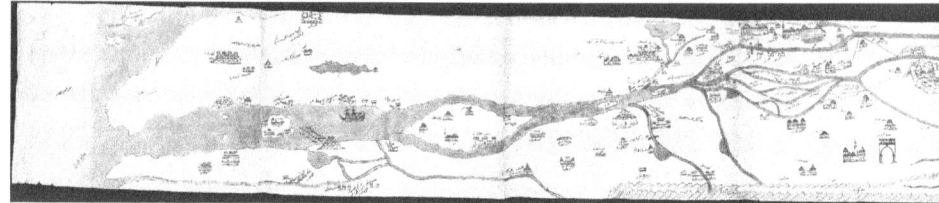

FIGURE I.3. An anonymous Ottoman cartographer drew this map on eight double folio sheets around 1650. The map offers a panorama of the Tigris-Euphrates system, from the Taurus Mountains (*right*) to the Persian Gulf (*left*). Courtesy of the Qatar National Library.

Long before the "river basin" came into vogue as a concept during the twentieth century, human perceptions and institutions presupposed its natural unity.[9] Ottoman officials considered the Euphrates to be an interconnected environmental system when, for example, they floated timber downriver, leaving it to its fate, knowing that flow could carry it over 800 miles to their downstream partners. Ottoman geographers, furthermore, gave expression to the spatial unity of the Tigris-Euphrates basin. One of the earliest Ottoman panoramas of the river system is a remarkable eleven-foot-long map drawn in the middle of the seventeenth century. The map features the entire river system, from the Taurus Mountains to the Persian Gulf, and indicates the major routes, settlements, and holy sites in-between.[10] By the standards of the time, the map is an impressive cartographic achievement. Even travelers flying over the Tigris and Euphrates today cannot see the rivers in their entirety at one time. The magnitude of the landscape, however, could not defy the anonymous cartographer's sense of dimension, which recognized that settlements along the Tigris and Euphrates all belonged to a single fluvial system that could be represented in a single map.

Operating on an unconventional spatial scale invites an unconventional vocabulary. The area of land drained by the Tigris and Euphrates lacks a general historical name. This book will refer to it as the drainage or river basin, a geographical term synonymous with watershed in North American usage and with catchment area in other parts of the world.[11] Within the drainage basin, most irrigation and navigation activities before the age of fossil fuels could not break free from a micro-geography that is referred to here as the alluvial plain. If the Tigris-Euphrates basin before the nineteenth century were a concert hall, the alluvial plain would form the center stage; thus it attracts the lion's share of attention in this book.

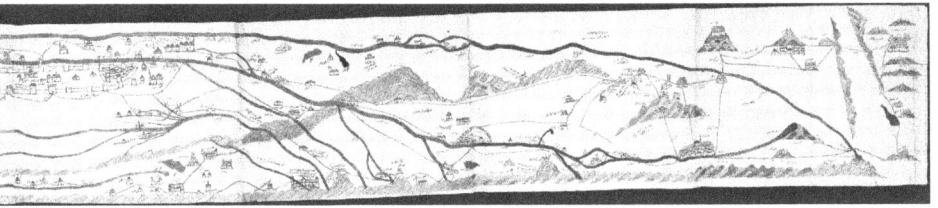

Nature throws the alluvial plain into stark relief. Geologically, it is an extensive depression filled with thick sediment deposits south of Hit on the Euphrates and Tikrit on the Tigris, boxed in by the rocky scarps of the Arabian desert along the southwestern flank, the marshes of Basra and Khuzistan on the southeastern boundary, and the Jabal Hamrin hill range on the northeast. Topographically, the alluvial plain is exceptionally flat, largely sitting at an elevation lower than 165 feet and less than 1 percent gradient.[12] In the middle of the seventeenth century, Ottoman traveler Evliya Çelebi compared it to the Kipchak steppe in the Ukraine.[13] His French contemporary, Jean-Baptiste Tavernier, thought it looked more like the terrain in Holland.[14] Coming from the Mediterranean, travelers could recognize their arrival in the alluvial plain from changes in the vegetation cover. Here the date palm dethrones the olive tree and rules the plant kingdom.[15] Historically, the alluvial plain roughly corresponds to the ancient lands of Sumer and Akkad, the Ottoman provinces of Baghdad and Basra, and Arab Iraq (Irak-i Arab), as the region was referred to in early modern Ottoman literature.[16] To the Tigris and Euphrates, the alluvial plain is what southern Louisiana is to the Mississippi. To avoid prolixity, it will often be referred to as the alluvium or Iraq.[17]

From a purely materialist perspective, the Tigris and Euphrates are modest rivers. If the drainage basin area and the annual discharge volume are used as metrics for global comparison, the twin rivers would rank low, overshadowed by the fluvial heavyweights of the world such as the Amazon and the Congo. Despite their humble geographical standing, they could still play an outsized role in the political affairs of Eurasia when a central administration based in Istanbul coordinated their exploitation with upstream and downstream settlements. The energy of river flow expanded the combat radius of Ottoman

Table I.1. Drainage Basin Area for Major Rivers of the World

Drainage Basin	Area (mi^2)
Amazon River	2,722,000
Congo River	1,429,000
Nile River	1,293,000
Mississippi River	1,255,000
Yangtze River	756,000
Ganges River	626,000
Volga River	533,000
Indus River	450,000
Tigris and Euphrates Rivers	430,000

Source: *The Times Atlas of the World*, 10th ed. (New York: Crown, 1999), 60–61.

Table I.2. Mean Annual Discharge for Major Rivers of the World

Name	m^3/s	Gauging Station	Observation Period
Congo River	41,128.7	Brazzaville, Republic of Congo	1971–1983
Mississippi River	17,701	Vicksburg, USA	1965–1983
Yangtze River	14,583.9	Yichang, China	1980–1983
Danube River	5,456.7	Drobeta-Turnu Severin, Romania	1840–1988
Rhine River	2,287.5	Rees, Germany	1936–1984
Yellow River	1,214.3	Sanmenxia, China	1976–1979
Nile River	1,203.5	Asyut, Egypt	1976–1979
Euphrates River	1,140.5	Hit, Iraq	1964–1972
Tigris River	1,126.5	Baghdad, Iraq	1968–1972

Source: C. J. Vörösmarty, B. M. Fekete, and B. A. Tucker, *Global River Discharge Database* (*RivDIS* v1.0) (Paris: UNESCO, 1996).

armies in West Asia and supported the stability of the eastern frontier as they fought in Central Europe. The twin rivers in this way helped the Ottoman Empire balance its military engagement between the Asian and European fronts.

Their historical influence disproportionate to their geographical size, the Tigris and Euphrates remind us to appreciate the small things in nature. Small rivers can be as complex and as enchanting as large ones. In the words of Henry David Thoreau, writing in 1852, "A brook need not be large to afford us pleasure by its sands & meanderings and falls & their various accompaniments. It is not so much size that we want as picturesque beauty & harmony. If the sound of its fall fills my ear it is enough."[18]

When Continents Collide

Aşık Mehmed's waterwheel metaphor is useful, but seen solely through the lens of the hydrologic cycle, the Tigris and Euphrates would look frozen in time, flowing in a smooth and predictable manner outside of history and apart from living organisms.[19] Only with a sense of geologic time that transcends the limits of the human experience can we fully grasp their dynamism and flux. Luckily, the Tigris and Euphrates are relatively young rivers; in the sea of deep time, their history lies far from the dark abyss and closer to the surface waters.

The Tigris and Euphrates were born at the tail end of a protracted geologic transformation. About 200 million years ago, the supercontinent Pangaea began to break up into some twenty pieces, which slowly floated on the soft upper layer of Earth like ships at sea. Those ships, large and small, collided with one another over time to form the continents we know today. One of the last major continental collisions occurred some 35 to 20 million years ago, when the Arabian plate plunged beneath the Eurasian landmass, forcing it to fold and raise the Taurus-Zagros belt. The mountains' soaring peaks formed water towers, hoarding moisture from the passing winds and releasing it with the aid of gravity down into the deep trench that emerged between the Arabian and Eurasian plates. As perennial streams eroded the jagged summits, they built up the alluvial plain of Iraq and morphed into the Tigris-Euphrates system.[20] The process was millions of years in the making, during most of which the lower stretches of both rivers formed a single network of interwoven channels. Only after the fourth millennium BC did the river system partition into its two discrete courses, one for the Tigris and another for the Euphrates, following abrupt channel shifts and intensified sedimentation that accompanied the expansion of Sumerian irrigation networks.[21]

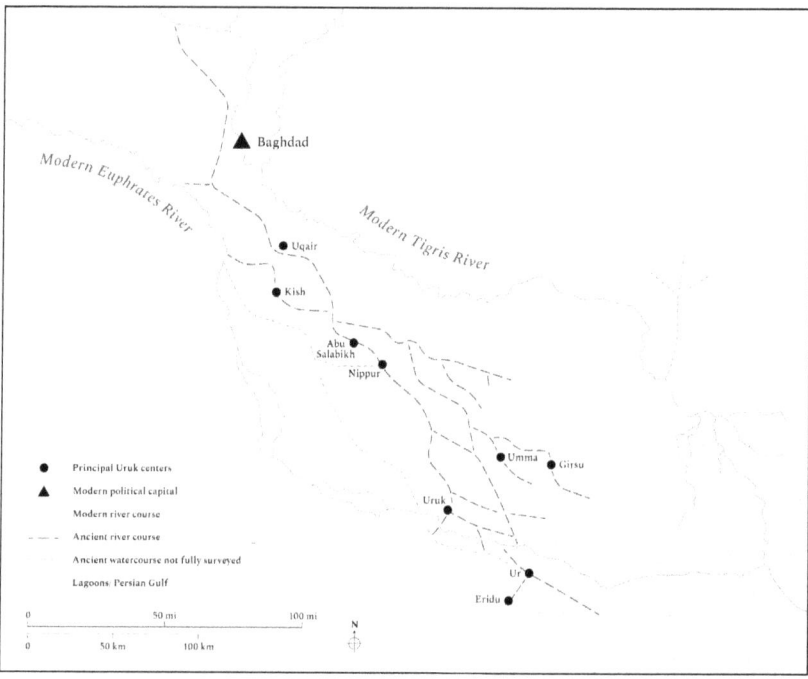

FIGURE I.4. The Ancient Tigris-Euphrates System, c. 4000–3000 BC. After Guillermo Algaze, *Ancient Mesopotamia at the Dawn of Civilization: The Evolution of an Urban Landscape* (Chicago: University of Chicago Press, 2008), 45.

Compared to the snail's pace of tectonic drift and uplift, the tempo of Ottoman history was fast and furious. From the early thirteenth century, waves of Turkish nomads arrived in the mountainous region of western Anatolia as refugees, fleeing the Mongol invasion from the east. In their new home, the Muslim Turks were far enough away to evade the Mongol government in the east and close enough to raid the crumbling Byzantine Empire in the west—for God, gold, and glory. As the power of the Seljuk state in Konya waned, Turkish nomads rallied behind political leaders of different origins in their plunder of Christian settlements in Byzantine territory. Among them, a man named Osman and his band of warriors rose to prominence in the summer of 1301, when they routed a Byzantine mercenary force near Nicaea (modern İznik). This is the first explicit mention of the Ottomans in a contemporary account, written by the Byzantine scholar Pachymeres. The victory marked a major step in the formation of a frontier statelet that would later become the Ottoman Empire, named after its founder Osman.[22]

Expansion characterizes the broad contours of Ottoman political history between the fourteenth and sixteenth centuries. In 1326, the Ottomans captured Bursa from the Byzantines, their first major city and first capital. From east of the Dardanelles Strait in 1352, they crossed into the Balkans and before long occupied Gallipoli, their first foothold in Europe. Shortly afterward they seized and relocated their capital to Adrianople (modern Edirne) in the north, a launchpad for further expansion into Bulgaria, Thessaly, Serbia, Macedonia, and Albania. Flanked on its Balkan and Anatolian sides, Constantinople resembled a Byzantine island in an Ottoman sea, until it was finally engulfed by the armies of Mehmed II in 1453.

Constantinople (hereafter Istanbul) possessed everything that a great power could need—a central location, fresh water supply, wealthy hinterland, natural harbor, and enormous prestige. From this new imperial capital, the Ottoman dynasty was well positioned to further centralize its bureaucratic apparatus, consolidate its territories, and undertake more conquests. A century later, the Ottomans ruled, directly and indirectly, an area roughly the size of modern India, at the juncture of West Asia, North Africa, and Southeast Europe. From a nomadic encampment of some 40,000 tents on the periphery of Constantinople, the Ottomans occupied the throne of the Caesars, at the helm of a world empire that filled contemporaries with awe.

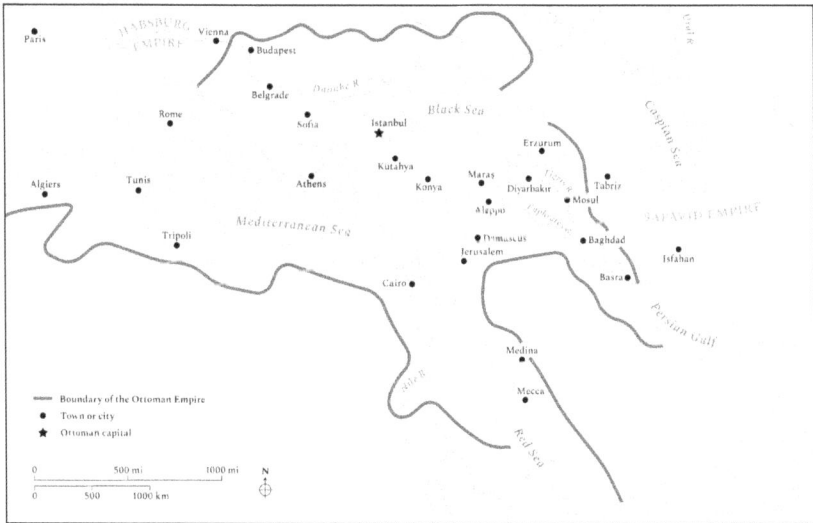

FIGURE 1.5. The Ottoman Empire, c. 1550.

The meteoric rise of the Ottoman Empire required a formidable military machine. Until the middle of the sixteenth century, the bulk of Ottoman armed forces were Turkish cavalry troops, descendants of the dynasty's early backers. Based in the provinces, they reported to active duty with their arms and equipment during the war season and received in return for their service a *timar*, the right to collect tax revenues from an agreed-upon number of fields and villages. The Ottoman dynasty augmented its cavalry force with salaried foot soldiers called the janissaries, for centuries considered to be the mightiest fighting force in the Mediterranean world. Originally, the Ottoman state trained one-fifth of its Christian prisoners of war to serve in the janissary corps as the sultan's elite bodyguard. From the late fourteenth century, however, the growing need for more janissary recruits prompted the state to turn to its own subject population. It forcibly recruited the young sons of Christian peasants mostly from the Balkans, aged anywhere between ten and twenty, through a systematic levy—converting them to Islam, teaching them Turkish, and training them in the military arts. Through the same process, some of the enslaved boys came to occupy the highest positions in Ottoman bureaucracy. For about three centuries, the Ottoman military machine paid for itself; conquest brought land and treasure, which in turn supported more cavalry and infantry soldiers to conquer even more land and seize more treasure.

According to an origin story written down about 250 years later, the Ottoman family's first encounter with the Tigris or Euphrates occurred in the early thirteenth century. Süleymanshah, grandfather of Osman, crossed into northeastern Anatolia with his family, fearful and wary of the Mongols riding in from the east. The patriarch was a shepherd and soon realized that the mountainous terrain of his refuge was inhospitable for his sheep. Determined to go back to Central Asia, he and his family ran into the Euphrates in northern Syria, which they had to ford. "They were the sort of nomad yürüks who didn't know the proper way," according to an early version of the story, "and they pushed on heedlessly into the Euphrates."[23] Süleymanshah drowned in the river along with his horse, but his three sons survived. One of them, Ertoğrul, would return to Anatolia, where he would father Osman, the progenitor of the Ottoman dynasty.

The tragic story of Süleymanshah portrays a vulnerable family in an alien environment—too weak, too naïve, and too nomadic even to cross the Euphrates. Three centuries later, his descendants would return to the river basin not as shepherds but as the charismatic rulers of a major naval power, competing with Venice and Spain for dominance in the Mediterranean and with Portugal in the Indian Ocean. Neither the Euphrates nor the Tigris could drown a family that had mastered the open seas.

Putting in My Oar

The historiographies of the Ottoman Empire and the Tigris-Euphrates basin continue to evolve rapidly. Each generation of scholars has opened new frontiers of research, aided by newly released archival materials and new information about the human past derived from non-textual sources, such as satellite images, sediment cores, and botanical remains.[24] Kenneth Burke, a twentieth-century philosopher of language, once compared this restless world of scholarly exchange to an endless conversation at a parlor. "You come late. When you arrive, others have long preceded you, and they are engaged in a heated discussion, a discussion too heated for them to pause and tell you exactly what it is about.... You listen for a while, until you decide that you have caught the tenor of the argument; then you put in your oar. Someone answers; you answer him; another comes to your defense; another aligns himself against you, to either the embarrassment or gratification of your opponent, depending upon the quality of your ally's assistance.... The hour grows late, you must depart. And you do depart, with the discussion still vigorously in progress."[25]

This book belatedly enters a crowded parlor dominated by numerous conversations. Each chapter is a contribution to one of them, but the book as a whole engages with two general discussions. The first unfolds among historians arguing about the remarkable expansion of the Ottoman Empire; how it successfully and enduringly incorporated vast lands in Asia, Africa, and Europe within its administrative ambit. Halil İnalcık, the late doyen of Ottoman studies, made a seminal contribution to this discussion when he highlighted the systematic recruitment of Balkan and Anatolian noble families into the Ottoman army and bureaucracy. In return for their services to the Ottoman state, the gentry of the newly conquered population retained its privileges and patrimonies; in some areas like Bosnia, it gradually converted from Christianity to Islam.[26] Gábor Ágoston advanced a similar view of Ottoman expansion into frontier regions like Hungary and Iraq, arguing that Ottoman policy refrained from imposing a uniform system of government in favor of incorporating existing local institutions, power-holders, legal customs, and procedures for revenue management.[27]

In the past fifteen years, three major syntheses of Ottoman integration have joined the discussion. Sociologist Karen Barkey emphasized the formation of

Ottoman social capital along the Byzantine-Seljuk frontier during the fourteenth century, a network of relationships brokered across religious and cultural boundaries through war, trade, accommodation, and marriage. At the center of the network, members of the Ottoman household successfully manipulated previously unconnected social groups to further their political project.[28] More recently, the idea of the Ottoman Empire as a brokered enterprise featured in studies of the seventeenth and eighteenth centuries. The development of a monetized market economy in this period, according to Baki Tezcan, empowered the authority of jurists and the janissaries and opened the door for new members of Ottoman society to join the ranks of the ruling elites through financial entrepreneurship. The process, which Tezcan describes as "proto-democratization," weakened royal authority but created a stronger state whose legitimacy was more widely recognized throughout the empire.[29] Following in Tezcan's footsteps, Ali Yaycioglu has argued that the Ottoman state survived challenges to its survival in the age of revolutions (c. 1760–1820) by pursuing reforms based on partnerships and constitutional ties between central and provincial actors.[30] Regardless of the theory, this lively conversation is thick with "pragmatism," "flexibility," and "adaptability" in reference to Ottoman rule in the early modern period.[31]

This book aims to bring a regional approach to the question of Ottoman expansion and integration in the historiography, calling attention to two natural threads that tied the eastern frontier to the metropole—the Tigris and the Euphrates. The construction of an administrative framework around a fluvial geography is crucial to understanding Ottoman state-building in Iraq. Firmly ensconced within the drainage basin, Istanbul could put the energy and fresh water of the rivers in the service of consolidating its downstream territories. River flow supplied a free and renewable source of energy that Ottoman officials regularly tapped to transport men and provisions from north to south, knitting Iraq together with the rest of the Ottoman world in the process. Separated from Iraq by an arduous overland route of more than 1,000 miles, Istanbul would have struggled to hold it without an efficient system of inland water transport. The region repeatedly threatened to fall under the influence of a local warlord, a tribal leader, or the Safavid capital cities of Qazvin and Isfahan, each located less than 500 miles away. In addition to their reliable energy, the Tigris and Euphrates carried the gift of life to the parched deserts of Iraq. River water maximized the food production capacity of an otherwise barren region. Its symbolic and strategic values aside, Iraq became, as a result, economically significant in the eyes of the Ottoman

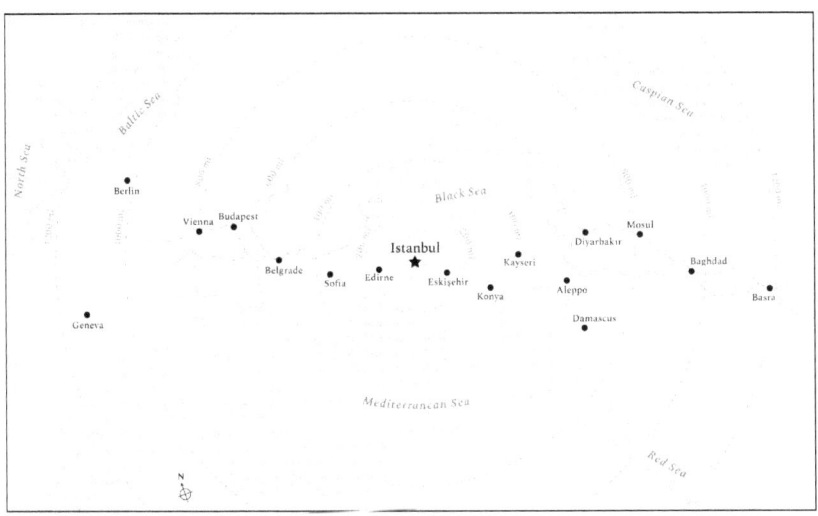

FIGURE I.6. Distance from Istanbul. After Rhoads Murphey, *Ottoman Warfare, 1500–1700* (New Brunswick, NJ: Rutgers University Press, 1999), xiv.

administration, the grazing ground of more than a million goats and sheep, and home to one of the largest date orchards in the world.

The second long-standing historiographical conversation that this book joins is the place of the Tigris and Euphrates in the long history of West Asia. The evolution of irrigation systems dominates this discussion due to the role that irrigation agriculture played in the development of complex societies in an arid setting like Iraq's.[32] From the middle of the seventh millennium BC, the first farmers descended from the northern foothills and colonized the lowlands, where they relied on the natural recession of floods to cultivate the domesticated plants they brought with them and on the rich biomass of marshes and lagoons to supplement their diet.[33] Not until the fourth and third millennium BC did formalized irrigation networks bloom. Under the labor and institutional arrangements of the first cities and states, excavated canals shorter than two miles branched off the main river channels in opposite directions to resemble a herringbone pattern.[34] This epoch witnessed the development of writing and provides the first glimpse into interstate conflicts over water and the internal workings of fairly centralized agricultural bureaus.[35] From Assyria, Babylonia, and then Persia, large territorial empires rose to preeminence in the first millennium BC and reshaped the Tigris and Euphrates at a scale never before seen. Gigantic feeder canals, more than ten miles in length, tapped the twin rivers and cut across the alluvial plain, gradually replacing the Sumerian herringbone system of short diversions.[36] The imperial irrigation configuration reached its zenith under Sasanian rule in the sixth century CE, virtually transforming Iraq into a large settled farm cultivated to its full capacity.[37]

From the ninth century, the colossal transverse canals for which Sasanian Iraq was famous steadily deteriorated. This prompted massive land abandonment and the retreat of cultivated areas to narrow bands along the main river channels. Scholars have offered different explanations for the breakup of the agricultural economy, including unprecedented urban growth at the expense of the countryside, civil wars that left irrigation canals in tatters, the outbreak of lethal epidemics among agricultural labor, and changes in irrigation policies in favor of self-serving landed elites. Each of those factors, among others, likely contributed to the tragic turn of events. Once considered the sole culprits, the thirteenth-century Mongol conquerors have more recently been widely absolved of blame for the agrarian collapse. If anything, they merely administered the final blow to an irrigation system that had already withered away.[38]

From the Mongol conquests until the nineteenth century, precious little is known about the history of the Tigris and Euphrates. The irrigation system of antiquity was a shell of its former glorious self, and for a field preoccupied with settled agriculture, this period thus has held no allure.[39] "Life continued," one survey writes, brushing past the entire last millennium of the rivers' history.[40] A more comprehensive survey published in two volumes relegates the experience of the alluvial plain since the arrival of the Mongols to half a page, a sorrowful coda to the celebrated achievements of ancient civilizations.[41]

This study seeks to fill this lacuna by writing the Tigris and Euphrates, as a coherent unit, into the post-classical history of West Asia. Compared to their ancient radiance, the lights of the twin rivers may have been dim in arable lands, but they continued to shine brightly elsewhere. In shipyards and docks, Ottoman officials transformed the Tigris-Euphrates system into a natural communication network that facilitated the expansion of stable state institutions. And in grasslands and wetlands, local farmers relied on the seasonal rise and fall of the Tigris and Euphrates to raise animals and grow rice, thereby compensating for the deterioration of large-scale irrigation projects that once crisscrossed their landscape. When examined through this multidimensional view, the Tigris and Euphrates assume new significance in the early modern era.

Sources

Ottoman imperial power came with the privilege of methodically recording and preserving the past, both of its own state institutions and of the societies it came to dominate. After conducting a survey in the early twentieth century, the German orientalist Hellmut Ritter estimated that the libraries of the Ottoman capital Istanbul housed about 124,000 Arabic, Persian, and Turkish manuscripts. The figure astonished Ritter. "How could you collect all these books?" he asked Hoca Şerefeddin, whom he identified as "the last great theologian of Turkey." The answer Ritter received was remarkably candid—"with the sword" (*bisseif*).[42]

The same sword that helped to incorporate the Tigris and Euphrates within the Ottoman Empire ensured that Istanbul would regularly receive from the east countless documents that form most of the source base for this study. Among the archives and manuscript libraries that house these documents today, the largest is the Ottoman Archive in Istanbul, now a department within the Turkish Presidency State Archives of the Republic of Turkey (Türkiye Cumhuriyeti Cumhurbaşkanlığı Devlet Arşivleri Başkanlığı). According to the latest published count, the Ottoman Archive boasts about 95 million documents and 400,000 registers.[43]

Because this book concerns the relationship between the Ottoman Empire and the Tigris-Euphrates system, the backbone of its source material originates from Ottoman bureaucracy. Two state departments and their records are especially worth mentioning. The first is the Grand Vizierate (Bab-ı Asafi or Bab-ı Ali), once known throughout Europe as the "Sublime Porte." Its building, a stone's throw from the royal Topkapı Palace, used to be the private dwelling of the Grand Vizier. In the early eighteenth century, the Porte supplanted the Topkapı as home to the central offices of the Ottoman government and as the archival repository for the Imperial Council (Divan-i Hümayun). One of the most important archival sources that the Porte came to possess after the transition were the Registers of Important Affairs (Mühimme Defterleri). Each register in this collection is a compilation of royal edicts issued by the Imperial Council on behalf of the Ottoman sultan to all provinces throughout the empire. Some 425 of those registers have survived, most of which are today in the custody of the Ottoman Archive.[44] This book uses about thirty of them, covering the period between 1552 and 1781.

The second state department germane to the subject of this study is the Imperial Land Registry (Defterhane-i Amire), which operated under the supervision of the chancellor (*nişancı*). Its archive contains land surveys, timar registers, and endowment transactions that guided the decision-making process of the Imperial Council during its meetings. Among its holdings, the cadastral surveys (Tapu Tahrir Defterleri)—the Ottoman equivalent of the English Domesday Book and the Spanish Relaciones Geográficas—are particularly useful. After the early fifteenth century, whenever Ottoman armies conquered a new territory, the Imperial Land Registry instructed deputies to compile a comprehensive survey of the land, primarily its population and revenue sources. When conditions in the region changed for any reason, officials were expected to compile new surveys to update the old ones. Some 3,400 cadastral surveys have passed down to us as a result of this painstaking effort, many of which are kept in the Ottoman Archive in Istanbul, but the majority have been inherited by the General Directorate of Land Registers and Cadasters (Tapu ve Kadastro Genel Müdürlüğü) in the Turkish capital Ankara.[45]

The cadastral surveys gave rise to an entire subfield within Ottoman studies—defterology—that flourished between the 1970s and 1990s but has ever since entered a state of torpor.[46] The unfortunate recent lack of interest in these sources may have to do with natural shifts in scholarly winds as well as the challenges of reading the specialized chancery script in which they were typically written. This book constitutes a modest attempt to revive the subfield through an analysis of about twenty cadastral surveys from Anatolia, Syria, and Iraq. It aims, moreover, to demonstrate how historians can enlist the cadasters, used in the past primarily to chart demographic and economic trends, to write environmental histories of the early modern Ottoman Empire. It offers a systematic examination of the little-studied cadasters of Baghdad and Basra to reconstruct the biomes and subsistence pursuits prevalent in the Tigris-Euphrates alluvium. Reading the land surveys with a fresh, ecological eye contributes to defterology studies by bringing to light the critical role of flora, fauna, and even fungi that have traditionally escaped the consideration of economic historians due to their low value relative to grain crops, such as grass and desert truffle.

By the standards of pre-industrial societies, the Ottoman cadasters are breathtaking in scale, depth, and volume, but like most census data sets, ancient or modern, they have their shortcomings.[47] While the cadasters record valuable quantitative information about crops and livestock, they largely leave out the technical and organizational process of production. How did a palm orchard and grain field coexist in the same area? How did a mammal hypersensitive to heat like the water buffalo endure the sweltering months of summer? And what did a waterwheel on the Tigris and Euphrates look like?

Most often, the cadasters do not answer such questions and focus instead on the magnitude and tax revenue potential of productive activities. They reflect, after all, the priorities of the taxman rather than the historian.

To supplement environmental analysis of this fragmentary cadastral evidence, this book uses more recent ethnographic data on traditional subsistence strategies from the same geographical region as a source of analogy for the sixteenth and seventeenth centuries. Referred to as "direct historical analogy" in the archaeological field, this approach shares with Ottoman environmental historians the assumption that systems of land use in West Asia and North Africa experienced long-term continuities until the gradual mechanization of agriculture during the past 150 years.[48] When documented under similar ecological and technological constraints, nineteenth- and twentieth-century farming techniques can therefore provide a relevant and appropriate frame of reference for interpreting the Ottoman cadastral surveys.[49] In fact, based on more liberal assumptions of cultural continuity, many archaeologists conduct ethnographic fieldwork around their excavation sites to assist their interpretation of ancient texts and artifacts.[50] More ambitious scholars assume similarities with cultures far removed from the region of their archaeological investigation to validate their arguments.[51] This book adopts a far more conservative approach to analogical reasoning, the subject and source sides in its analogies being closer in both time and space than many of those employed in the archaeological literature.

Decades of authoritarian rule, war, and rampant corruption have ravaged large parts of the region under study and have taken their toll on their populations and cultural heritage, including precious archival collections. The experience of the modern state of Iraq, the country to which the Tigris and Euphrates have long been of most vital importance, is particularly heart-wrenching.[52] One of this book's goals, therefore, is to suggest strategies to overcome the enormous loss and damage that the Iraqi archives have sustained. Through a judicious combination of archival sources and literary accounts produced by different arms of the Ottoman state and by different authors, historians can still write studies of the Tigris and Euphrates rivers before the nineteenth century.

———

This book has three parts. Part I (Chapters 1 and 2) focuses on the establishment and transformation of the infrastructure of riverine control, primarily fortresses and shipyards, between the sixteenth and seventeenth centuries. Together, fortresses and shipyards turned the Tigris and Euphrates into Ottoman supply lines, regularly delivering food, weaponry, soldiers, and gunboats downstream. This imperial system of river transport provided land-bound Ottoman garrisons with the means to concentrate and project their power along the eastern frontier. Istanbul used its military superiority to cash in on the land-based wealth of the alluvial plain, where the Tigris and Euphrates were most consequential for farming activities. Part II (Chapters 3, 4, and 5) stays within the first 200 years of Ottoman rule in the region, detailing imperial policies to manage the exploitation of arable lands, grasslands, and wetlands. The productivity of these three zones endowed the state with a firm economic and political footing in the alluvium. Food and revenue shortfalls from one zone caused by natural disaster could easily be offset by inflows from elsewhere. The Ottoman presence in Iraq owed its resilience, in part, to a policy that generally pursued accommodation with ecological diversity over unbridled agrarian development.

Part III (Chapters 6 and 7) documents the transformation of the classical Ottoman model to exploit the Tigris and Euphrates throughout the eighteenth century. Between 1687 and 1702, a major channel shift in the Euphrates undermined the power of Istanbul-appointed governors in the alluvium, preparing the ground for the rise of a new localized system of river management based in Baghdad. Relying on textual and proxy evidence, Chapter 6 demonstrates that the channel shift was triggered by a combination of an unsuccessful river diversion and weather anomalies. Chapter 7 tracks the rise of the Pashalik of Baghdad in the aftermath of the Euphrates' relocation. As the fabric of central authority unraveled in the drainage basin, Istanbul became increasingly reliant on this provincial dynasty to keep Iraq within its imperial orbit. By 1780, the Pashalik accumulated enough power to dominate most of the region between Mardin in Southeast Anatolia and Basra on the Persian Gulf. Through its networks with neighboring provinces and foreign powers, the Pashalik assumed the role formerly held by Istanbul in the region, fulfilling Iraq's insatiable demand for natural resources on its own—by river, land, and sea.

Each of the seven chapters tells the local story of at least one particular social group: the raft makers of Diyarbakır, the carpenters of Birecik, the farmers of Rumahiyya, the marsh dwellers of Hasaka, and the slave soldiers of Baghdad. Taken together, their stories will weave an Ottoman tapestry of the Tigris and Euphrates, the central characters pervading every chapter in this book. More than any other natural force in this region, the rivers' majestic flow rendered visible the immutable interdependence of life, water, and power.

PART I

The Amphibious State

Throughout history, large rivers have created opportunities for political expansion into the remote interiors of the continents. From central Italy, imperial Rome used the Rhône to penetrate Gaul and relied on the Rhine and the Danube to guard its settlements in Germania and Pannonia.[1] In predominantly inland regions like West Asia, most of which are distant from the seacoast, navigable rivers gained greater political significance.[2]

The following two chapters explore the role of the Tigris and Euphrates in the eastward thrust of Ottoman expansion during the sixteenth and seventeenth centuries. Riparian (streamside) fortresses and shipyards upheld Ottoman political authority between the Taurus Mountains and the Persian Gulf, and over Iraq in particular. With them, Istanbul could wage warfare that involved the cooperation of two specialized hierarchical organizations—army and navy. If rivers were "organic machines," as historian Richard White described the Columbia, fortresses and shipyards were the central cogs of the early modern Tigris and Euphrates.[3] At the barrel of a gun, the Ottomans controlled the system of waterborne communication and confronted domestic and foreign challenges to their hegemony.

The energy of the Tigris-Euphrates system conferred upon the Ottoman Empire a competitive advantage and a logistical edge over its rivals. Wind and river currents extended the distances over which Istanbul could field heavy artillery, food supplies, and troops along the eastern frontier. One of the reasons that the Safavid Empire—its chief opponent in the east—could not sustain a more formidable war machine was that the Iranian Plateau lacked navigable waterways comparable to the Tigris and Euphrates. Even though its total land area is three times larger than Iraq, Iran enjoys only half of its neighbor's surface flow.[4] This hydrologic deficiency made the business of provisioning a large army with field guns much more difficult, forcing the Safavids to rely almost entirely on overland transport through the Zagros range.[5] Ottoman armies in the east, as a result, tended to be better armed and fed than their Safavid rivals and could keep Iraq, with brief interruptions, under Istanbul's dominance throughout the early modern period.[6]

1
Fortresses

> They assumed that the distance was great and our sublime court too far away!
> —Sultan Selim II, *Mühimme Defterleri* (1571)

IN DECEMBER 1564, Mevlana Yakub, a religious lawyer from Istanbul, was preparing to take up his new post as the imperial judge of Baghdad City. His luggage included a letter furnished by the Ottoman Imperial Council, written in Turkish and addressed to all provincial authorities serving between Istanbul and Baghdad. "When he enters your jurisdiction," the letter read, "provide a sufficient number of people to accompany him in dangerous and frightening roads and help him reach the well-protected city [of Baghdad] safe and sound. You should know this!"[1]

The terse letter does not detail Yakub's itinerary, but he must have followed one of two routes to reach his destination, those charted by the Tigris and the Euphrates. For millennia, the stream of humanity moving between Anatolia and Iraq flowed along these two arteries, which provided travelers with necessary access to water and a degree of security and convenience in their journeys across the inhospitable Jazira desert.[2] In the words of an English traveler in 1603, the river routes allowed merchants to "avoid and shunne the great charge and wearisommnesse of travell through the Desart of Arabia."[3] In the sixteenth century, the communication networks of antiquity carried on, but they had to be reoriented to a new political map. A radical political realignment placed the reins of the entire Tigris and Euphrates in the hands of Istanbul. At the stroke of a pen, Yakub and countless other diplomats, merchants, and artisans gained the right to safe passage from the Mediterranean to the Persian Gulf. Similar decrees secured the movement of armies, along with their food and heavy equipment, between the northern and southern ends of the river basin.

This chapter focuses on the role of the Tigris and Euphrates in the establishment of Ottoman centers of power in Iraq. The Ottoman system of river transport boosted the clout of fortified cities and equipped them with the resources necessary to subordinate the countryside and repel foreign enemies. The twin rivers, in other words, made Ottoman state building in the southeastern frontier possible.

The Ottoman Empire relied on different strategies for border enforcement. In Central Europe, historians have documented the development of a network of fortifications to secure the front line with the Habsburg Empire.[4] On the eastern edge of the empire, on the other hand, they have emphasized the ambiguity of the border with Safavid Persia and Ottoman reliance on tribal groups as instruments of both military defense and expansion.[5] This chapter brings Ottoman frontier fortresses in the east to the fore, calling attention to the importance of runoff as a unifying power between them. The position of fortresses in their drainage basin—near the headwaters, the estuary, and in-between—shaped the role they played in guarding the Ottoman eastern frontier. One of the reasons Baghdad became a regional hub of Ottoman war and diplomacy was Istanbul's capitalization on the natural drainage pattern of the Tigris-Euphrates system, which funneled precipitation across the Taurus and Zagros mountains into two navigable highways bound for the city.[6]

The Tide of Conquest

Dedicated to continuous warfare against the infidel world, early Ottoman rulers cut their military teeth fighting different Christian powers, initially the Byzantines and later the Habsburgs, Venetians, and Russians. From the late fourteenth century on, the Ottomans would open a new front, turning against their Muslim neighbors in the east. To justify this controversial move, they cast their coreligionists as backstabbers who derailed their sacred mission in the west to expand Islam and gain new converts.[7]

Three military waves rolled eastward and brought the Tigris and Euphrates under Istanbul's control.[8] The first crashed on the shores of the fledgling Safavid Empire (1501–1722) based in Tabriz. Unlike the Sunni Ottomans, the Safavid household adopted the minority Shi'i branch of Islam and challenged Ottoman legitimacy within Islamdom. The success and charisma of its founder Ismail I won the allegiance of thousands of Turkic tribesmen in Anatolia, who waged a devastating anti-Ottoman rebellion in the spring of

1511. The hawkish Ottoman prince Selim openly blamed the havoc on his ailing father Bayezid II for his cautious policy that avoided open conflict with the Safavids. Backed by the elite janissary guard, the prince forced his father to relinquish power and secured the Ottoman throne in April 1512. The Grim (Yavuz), as Selim I was later known, needed two years to hunt down and kill his brothers and nephew and exterminate 40,000 Shi'is in Anatolia before defeating the Safavids at Chaldiran in the summer of 1514. Three years after that, the regions east of the Euphrates between Erzurum and Raqqa and along the Tigris between Diyarbakır and Kirkuk fell under the sway of Istanbul.

The second Ottoman wave lapped against the Mamluk Empire based in Cairo. Ottoman expansion into a buffer zone in southern Anatolia soured relations between the two powers. In June 1516, Selim set out for Syria to accuse the Mamluk sultan of harboring fugitive challengers to his throne and of conspiring with the Shi'i Safavids to form an anti-Ottoman alliance. The two met on the northern outskirts of Aleppo in late August. Selim emerged victorious, securing the western Euphrates bend between Divriği and Rumkale and clinching control over the entire upper drainage basin.

The third and last wave, sent by Selim's son Süleyman I, finally reached the southernmost tip of the Tigris-Euphrates system, washing over the alluvial

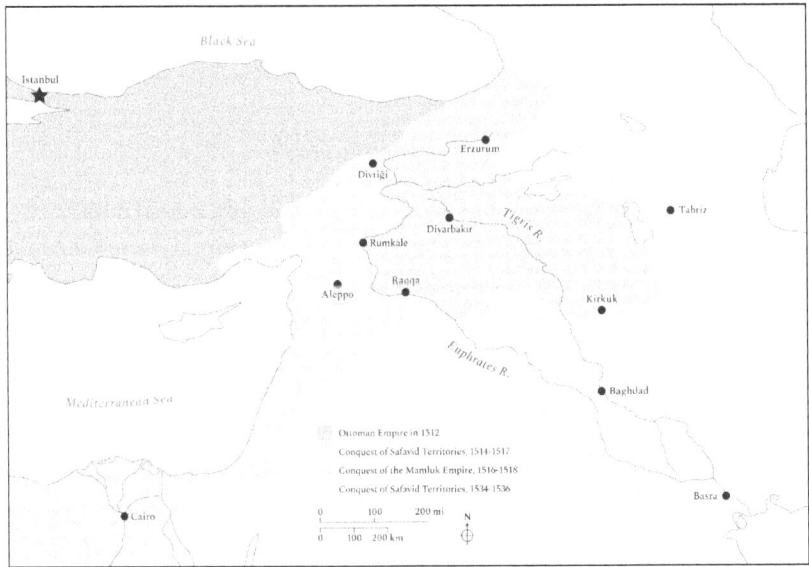

FIGURE 1.1. The Ottoman Conquest of the Tigris-Euphrates Basin. After Donald Edgar Pitcher, *An Historical Geography of the Ottoman Empire from Earliest Times to the End of the Sixteenth Century* (Leiden: E. J. Brill, 1972), Maps XX and XXI.

plain of Iraq. The Ottoman and Safavid empires locked horns once again to settle old and new scores. In 1533, Süleyman dispatched his grand vizier to the Ottoman-Safavid borderland and joined him a year later. In early December 1534, the sultan entered Baghdad unopposed, forcing its Safavid governor to flee for his life to Persia. Local chieftains throughout Iraq, including Basra, flocked to the city to submit to their new imperial masters. In 1546, Baghdad's Ottoman governor Ayas Pasha would reinforce Ottoman authority in Basra by dislodging an unreliable vassal and bringing the city under the direct control of Istanbul.

Between 1512 and 1534, in short, the tide of Ottoman conquests surged eastward, and in three waves it knocked out the Mamluk Empire entirely and pushed the Safavids back into the heart of Persia. Each of the three campaigns was consequential in its own terms, but together they acquired greater importance. The Ottoman Empire's new possessions in Muslim-majority West Asia and North Africa overshadowed its traditional foothold in Christian Europe, both geographically and demographically. West Asia became a platform from which a Mediterranean power based in Istanbul could shape events in the Indian Ocean world. This book examines some of the cumulative environmental consequences of the Ottoman eastern campaigns in the early sixteenth century, in particular, how they inadvertently unified the basin through which the Tigris and Euphrates flowed, and how they created new opportunities and challenges for the Ottoman presence in the region.

Fluvial Infrastructure

Settlements in the Tigris-Euphrates basin existed within a nested hierarchy. By controlling the few in the upper echelons of the pyramid, Istanbul could easily extend its rule to the lower-order towns and villages and the waterways between them. Diyarbakır, the largest on the Tigris, dominated the upper stretches of the river to the borders of Baghdad until 1586, when one of its districts, Mosul, became an independent province. The redrawing of political boundaries, which recurred throughout the sixteenth century, effectively placed the upper Tigris under the jurisdiction of both Diyarbakır and Mosul. Because the Euphrates lacked fortresses comparable in size to those on the Tigris, its upper reaches initially fell under the sway of a city further inland—Aleppo in northwestern Syria. After the formation of the Raqqa province based in Ruha in 1586, the upper Euphrates oscillated between the spheres of Aleppo and Raqqa. As the Tigris and Euphrates flowed closely within Iraq,

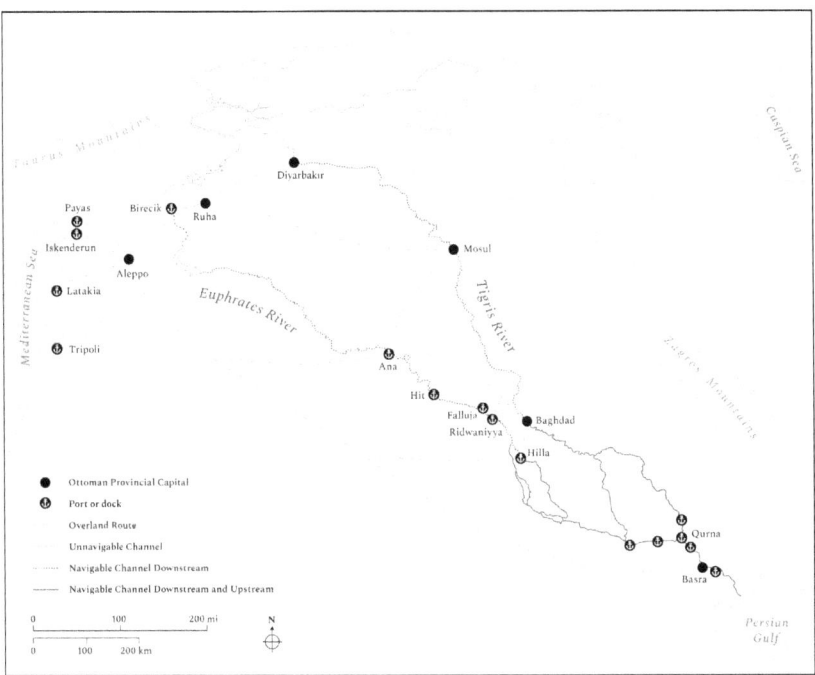

FIGURE 1.2. Communication Routes in the Tigris-Euphrates Basin.

Baghdad in the north and Basra in the south could dominate the remaining segments of both rivers.[9]

What Diyarbakır, Mosul, Aleppo, Baghdad, and Basra shared in common was a strategic location and a sizable taxpaying population, each exceeding 3,000 persons. Because a taxpayer in Ottoman administrative practice often represented the male head of a household, the total population in each fortress (including women, children, and the tax-exempt) was likely well over 15,000.[10] Through different types of investment, Ottoman policy capitalized on the locational and demographic strengths of each large fortress to improve transportation and tax collection between the Mediterranean and the Persian Gulf.

Aleppo was to Istanbul what Palmyra was to Rome, the vital link between the Mediterranean Sea and the Euphrates River, Europe and Asia. Istanbul boosted the intermediary role of Aleppo through two types of investments. Seventy miles northeast of the city, it established the Birecik shipyard. In the west, it developed the northern seaports of the Levant. One of them was Payas at the head of the Gulf of Iskenderun, where Grand Vizier Sokullu Mehmed Pasha (r. 1565–1579) established a large complex comprising a mosque,

Table 1.1. The Taxpaying Populations of
Large Riparian Fortresses

Fortress	Year	Taxpayers
Aleppo	1536	10,079
Amid (Diyarbakır)	1540	4,225
Basra	1590	3,849
Baghdad	1580	3,309
Mosul	1575	3,184

Note: Population numbers represent only those living within fortress walls.

Sources: OA, TT 397, 16-75; OA, TT 200, 17-58; OA, TT 660, 20-54; OA, TT 667, 20-22; TKG.KK, TT 29, ff. 9v-30r; TKG.KK, TT 30, ff. 15v-40r.

bathhouse, elementary school, caravanserai, hospice, and public fountains. Another port that grew in significance in this period was Iskenderun, which Istanbul designated as Aleppo's official port in 1593 in response to the lobbying efforts of European and local merchants.[11]

Proximity and improved access to the Mediterranean made Aleppo from the sixteenth century the residence of European consuls and chartered companies. The city's location at the edge of the Syrian desert at the end of caravan routes also gave it a handy supply of pack animals that travelers could hire or purchase. The encounter between European traders from the sea and animal breeders from the desert in the markets of Ottoman Aleppo had far-reaching economic and even biological consequences. Chief among them was the incursion of Arabian bloodstock among the equine population of the British Isles from the early seventeenth century on.[12] Aleppo officials were in charge of managing Ottoman-sponsored traffic between the Mediterranean and the Tigris-Euphrates basin, purveying the necessary horses, camels, mules, geldings, and oxen. From Aleppo to Iraq, the itinerary could follow one of three major routes from which multiple secondary lanes branched out: the direct desert route to Basra, the Euphrates route through Birecik, and the Tigris route through Diyarbakır or Mosul.[13]

In Diyarbakır and Mosul, Istanbul had to tailor its transportation projects to the notoriously swift current of the upper Tigris, an attribute that gave the river its name, denoting "sharp" and "arrow" in Old Persian.[14] Istanbul resorted to the *kelek*, an ancient raft made of timber and brushwood bundles laid upon inflated goat and sheep skins.[15]

FIGURE 1.3. Kelek Construction in Mosul along the Tigris, c. 1663. From Jean de Thévenot, *Suite du Voyage de Mr. de Thévenot au Levant* (Paris: Angot, 1689), 2:185. Courtesy of the Special Collections Research Center, University of Chicago Library.

Three engineering features made the kelek appeal to imperial officials and travelers alike. First, as a seventeenth-century Parisian traveler realized, the kelek could be assembled without "peg, nail, nor indeed, any bit of iron."[16] Sewn planks supplanted costly and scarce metals and conferred a built-in elasticity in the raft, allowing it to absorb the shocks of hitting rocks and landing on hard shores without breaking up.[17] Second, the kelek's layer of inflated animal skin made travel throughout the year possible, even when the Tigris was shallow during the summer. Rowers largely relied on the natural propulsion of the river current and focused on keeping the kelek on course, away from land, sandbars, and stones.[18] A third engineering feature in the kelek was the ease with which it could be disassembled and recycled, the timber typically sold downstream where it was prized and the skin deflated, dried, and returned upstream for the construction of new rafts.[19]

Ottoman demand for the kelek ratcheted up its production level to an industrial scale in Diyarbakır and Mosul. From the seventeenth century, and

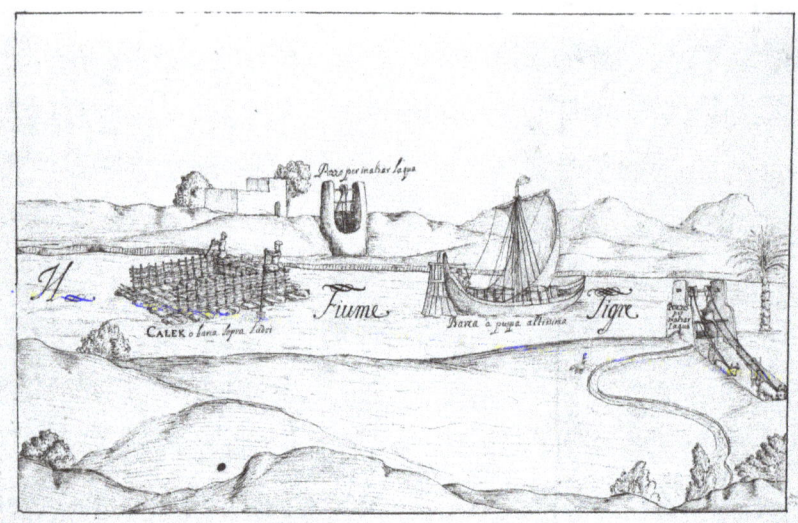

FIGURE 1.4. Kelek Plying the Tigris (*left*), 1673. Source: Ambrosio Bembo, "Viaggio e giornale per parte dell' Asia di quattro anni incicra fatto." Courtesy of the James Ford Bell Library, University of Minnesota.

intensively following the chaos unleashed by Safavid collapse in 1722, Istanbul placed recurrent orders for kelek construction in support of its military forces in the east.[20] In preparations to retake Baghdad from the Safavids in 1637, for example, it ordered the construction of 300 keleks in Diyarbakır.[21] In charge of fulfilling imperial demands were local contractors called *kelekçiyan*. Portrayed in several sources as old and poor, the kelekçiyan worked in the open air on the banks of the Tigris, offering their services to assemble new rafts or serve as skippers.[22] In support of their service for the state, Istanbul augmented their supplies of raw material, primarily watertight animal skins provisioned from eastern Anatolia and Syria. In 1638, for instance, the Ottoman administration ordered Mardin to supply the kelekçiyan of Hasankeyf with as many as 5,000 skins.[23]

Under the insistent pressure of Ottoman demand for their services, raft makers in different towns along the Tigris banded together, strengthened their preexisting ties, and set up their own guilds, referred to in Ottoman documents as the *esnaf-i kelekçiyan*. In Ottoman archival records, the earliest extant evidence of a formal hierarchical structure among raft makers dates to 1638, when the Ottoman state provisioned some 2,000 skins to the headman of the raft makers' guild (*kelekçiyan şeyhi*) in the Anatolian town of Hasankeyf along the Tigris.[24] Raft makers in the Tigris cities of Diyarbakır and Mosul formed their own guilds too, whose leaders could represent and defend their interests in state courts.²⁵ The formation of raft makers' guilds in different nodes along the upper Tigris reflected a broader trend within the Ottoman Empire between the seventeenth and eighteenth centuries, a period when artisan and shopkeeper guilds became widespread, particularly in large cities. In the east, war with Persia fueled this trend, turning raft makers and other artisans into organized contractors for the Ottoman army.²⁶

A cohesive artisan organization offered raft makers numerous advantages. First, it coordinated their efforts to respond to the increasing need for their services more efficiently, as they worked in well-known locations along the Tigris by rotation. Second, working in groups coordinated the dealings of raft makers with other artisan groups, notably the felt makers' and butchers' guilds, from which they procured additional raw materials needed for raft construction. Finally, guild organization granted raft makers official status in society. Ottoman recognition and aid increased the capacity of the kelek construction industry in the upper Tigris, reinforcing the position of Diyarbakır and Mosul as two of the busiest transportation hubs in the drainage basin.²⁷

To control water and land communications in the lower basin, Istanbul poured its resources into Baghdad and Basra. The two fortresses were the largest in Iraq and enjoyed a propitious location. In the south, Basra sat at the crossroads of four water bodies (Shatt al-Arab and the Tigris, Euphrates, and Karun rivers) through which travelers and cargo from all directions passed. North of Basra, Baghdad stood at the point where the Tigris and Euphrates come closest together, separated by a mere twenty-five miles of land.²⁸ As in Basra, locational advantage conferred resilience to Baghdad since its founding in the early Middle Ages. "Massacre, devastation, and oppression have ransacked this city during several hundred years," a British author wrote in the early nineteenth century, "and yet it bears a name, and a certain respectability in the East, solely from the circumstance of its situation being a central depot; or rather, with more propriety, we might call it 'the still important great caravansary of Asia.'"²⁹

Istanbul cashed in on the centrality of Baghdad and Basra in the lower river basin through a network of docks and bridges that served as police stations and toll booths. Before his departure from Baghdad in March 1535, Süleyman I personally ordered officials to do everything necessary to secure the river crossings, a task that later Ottoman administrations regarded as one of the primary responsibilities of Baghdad's governors.³⁰ In Iraq, the pontoon bridge of Baghdad formed the central river crossing.³¹ Depending on the season and water level, the bridge floated on some twenty to fifty boats

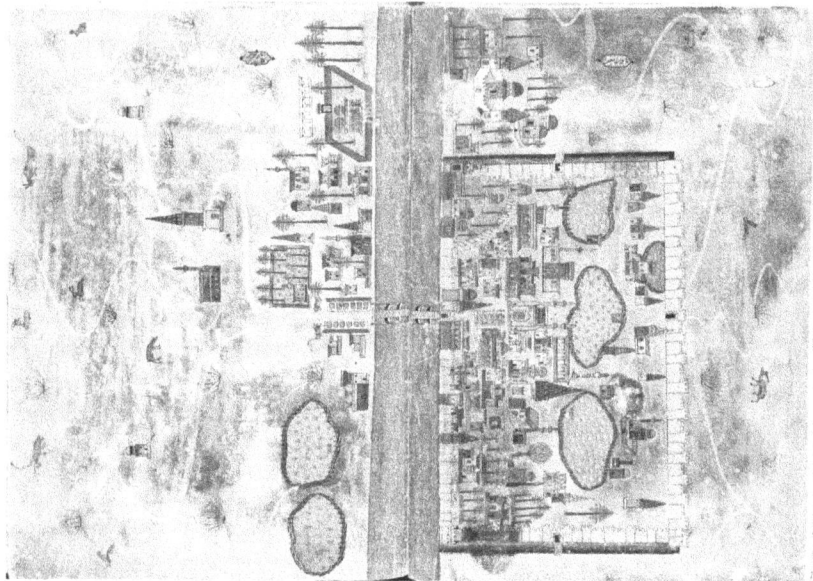

FIGURE 1.5. Baghdad on the Tigris River, c. 1534. Source: Matrakçı Nasuh, "Beyan-ı Menazil-i Sefer-i Irakeyn," İstanbul Üniversitesi Nadir Eserler Kütüphanesi T.5964, ff. 47v–48r. Courtesy of İstanbul Üniversitesi Nadir Eserler Kütüphanesi.

stretching over a river channel that, by a rough German estimate in 1574, was "as broad, as the Rhine is at Strasburg."[32] Heavy iron chains extended from two large anchors buried in the sand on each bank to tie and fasten the boats together. When rafts and vessels had to pass down or up the river, Ottoman traffic police called *köprücüler* temporarily stopped pedestrian movement to open the bridge for passage, unmooring three to six boats in the middle. The traffic police collected different tolls on vessels and pack animals outlined at length in the law code of Baghdad in 1580.[33] For travelers, the Baghdad bridge was a safe and convenient river crossing; for the provincial administration, it was a site of control and source of revenue; for ordinary people, it was a pleasure site where diving competitions took place and coffee houses proliferated.[34]

From Baghdad on the Tigris, Ottoman officials extended their reach and maintained several small docks on the Euphrates. For vessels descending to Iraq, Falluja was the most important transit point during the sixteenth and early seventeenth centuries. Here, "I was searched for money, and the Searcher found all that I had," a London merchant wrote about his encounter with an Ottoman customs agent in 1581.[35] Once cleared, another agent escorted him to Baghdad. During the second half of the seventeenth century, a nearby dock in Ridwaniyya came to replace Falluja as Baghdad's nearest outlet to the Euphrates.[36] For vessels ascending the Euphrates from the south, Falluja and Ridwaniyya were too far upstream and beyond reach. A more convenient dock to them was the city of Hilla, the largest Ottoman fortress on the Euphrates and home to nearly 27,000 people in 1580.[37]

In Basra, Ottoman officials in the second half of the sixteenth century maintained and collected tolls from no less than seven docks within the city itself, on the Shatt al-Arab, and along the lower Tigris and Euphrates. The dock at Qurna stood out due to its strategic location at the confluence of the Tigris and Euphrates, well garrisoned by Ottoman troops and defended by galleys and cannon.[38]

In short, following the Ottoman unification of the Tigris and Euphrates, the largest fortresses along the rivers—Aleppo, Diyarbakır, Mosul, Baghdad, and Basra—received considerable financial and logistical support from Istanbul to improve their communication infrastructure. Imperial investment took different shapes and forms depending on local geography and each fortress's potential. As a result, new seaports on the Mediterranean and a shipyard on the Euphrates came to flank Aleppo, the capacity of the kelek industry in Diyarbakır and Mosul boomed, and docks and bridges in Baghdad and Basra created choke points to capture—and make the most of—increased traffic. Together, these transportation facilities and arrangements resuscitated the river basin's access to the Mediterranean after centuries of relative isolation. For nearly three centuries, the Tigris and Euphrates experienced a slump in traffic as the Mongol-Mamluk conflict and the flourishing of new Ilkhanid and Seljuk centers of power in the north drove a wedge between Iraq and the world around it. Most east-west trade bypassed the lower river valley in favor of a northern route through Anatolia and Persia and a southern route through Egypt and the Red Sea. The regional communication network devised by Istanbul in the sixteenth century revived the status of the Tigris and Euphrates as two of the greatest thoroughfares in Eurasia.[39]

The Roads of Gun and Grain

Through its installations and personnel, the Ottoman administration could intervene to address a deep natural oddity within the river basin. Compared to the catchment area in the north, the southern alluvium had a limited range of natural resources. In terms of cereal crops, the Ottoman Imperial Council complained in 1568 that "those lands are plagued by grain shortage; never have there been [enough] wheat or barley in their granaries since their conquest."[40] Similar complaints appear well into the eighteenth century. "The state's agricultural produce in the districts of the Baghdad province is insufficient," an official report stated in 1749.[41] From the early years of their rule, the Ottomans pursued different policies aimed at dealing with the grain scarcity in Iraq. On some occasions, they even prevented members of the population from pursuing non-agricultural pursuits so locals could remain active in the region's frail agricultural sector.[42] European observers echoed the concern of Ottoman authorities. "The Country from Baghdad to Bussora does not produce Wheat and Barley enough for its own consumption," wrote the resident of the East India Company in Baghdad in 1800.[43] In the unvarnished language of an irritated Italian traveler in 1625, "Bassora hath not sufficient Victuals."[44] The experience of grain shortage in the alluvium, once the breadbasket and tax base of several great powers, was relatively recent and could be traced to the deterioration of the Sasanian irrigation system in the early Middle Ages. The lack of ore deposits, in contrast, was inherent to the tectonic structure of West Asia. The states of ancient Iraq had long struggled to deal with it by importing metals from outside. In the early modern age of gunpowder and standing armies, the alluvium's deficit in ore and grain made it ever more dependent on northern neighbors that enjoyed a surplus of both.[45]

Istanbul relied on river transport to rebalance the resource disparities between the northern and southern ends of the drainage basin. It organized regular shipments of Anatolian and Syrian grain, either raw or cooked as hardtack, to satisfy its garrisons in Iraq nutritionally. On rare occasions, such as 1579, Baghdad received grain originating from as far away as Ottoman Egypt. In addition, Istanbul regularly shipped to its downstream fortresses arms cast out of metals that the alluvium lacked, including guns, mortars, shells, and cannonballs. Along with guns came gunpowder produced in the mills of Istanbul, Thessaloniki, and Gallipoli.[46]

The Tigris and Euphrates formed lifelines for Ottoman defenses along the eastern frontier. Like the Danube in the west, they "made the frontier feel closer" to the imperial center.[47] They channeled natural resources from areas of surplus to areas of need through the treacherous terrain of the Jazira desert. It was through these geographical arteries, carrying vital supplements pumped by the imperial heart, that Iraq stayed within the Ottoman body politic.

Militarization

The Ottoman system of water communication impacted the two ends of the drainage basin differently. In the north, it drove environmental change by adding pressure on livestock for their skins, on arable fields for their harvests, and on Earth's crust for its ore deposits. In the south, the influx of guns and grains transformed Ottoman fortresses into centers for organized violence, exercising the powers of life and death over their hinterlands. As armed confrontations with Safavid Persia broke out throughout the sixteenth and early seventeenth centuries, fortresses along the Tigris, in particular, became highly militarized and home to the largest Ottoman garrisons in the southeastern frontier. From them, military personnel and hardware flowed to smaller outposts. The Tigris fortress of Mosul, for instance, provided support to the Ottoman regiment in Kirkuk near the Zagros foothills. Samawa in the lower Euphrates relied on Baghdad, whose janissaries served in the small town by rotation. At the confluence of the Tigris and Euphrates, Qurna was a military dependency of Basra.[48]

Among its peers upstream and downstream, the Baghdad fortress stood as the most ominous display of Ottoman military might. The Tigris on the west and walls behind a moat on the east enclosed the city within an irregular pentagon dominated by ten towers, each equipped with six to seven pieces of artillery. In the middle of the seventeenth century, the renowned Ottoman traveler Evliya Çelebi counted 1,060 artillery pieces of different types and sizes stationed on the wall towers, behind the gates, and in the citadel. Only in Istanbul, Budapest, and a couple other places in the empire, he wrote, could one find artilleries as impressive as those guarding Baghdad. Day and night, soldiers with smaller firearms manned thousands of crenels spaced between the wall towers. Those unfortunate enough to be discovered by their superiors napping at the crenels during their night shifts were reprimanded with 100 blows from a stick.[49]

The walls of Baghdad were hardy, made of polished mud brick, white stone, and lime and reinforced from inside by arches. During his stay in Baghdad between December 1638 and February 1639, Sultan Murad IV conducted a trial to test the strength of the wall by ordering his troops to open fire at it with cannon. The cannonball pierced the wall without knocking it down, an outcome Evliya Çelebi recounted to demonstrate Baghdad's impressive means of defense. Had the wall been built out of stone, he reasoned, it could have easily crumbled.[50] To European observers, the Ottoman fortifications of Baghdad needed no better vindication than the fact that Nadir Shah (d. 1747), the conqueror of Delhi "whose name shook the east," tried to conquer the city twice to no avail.[51]

In terms of personnel, the Ottoman administration maintained in the city a regular janissary corps equipped with handguns and reinforced it with

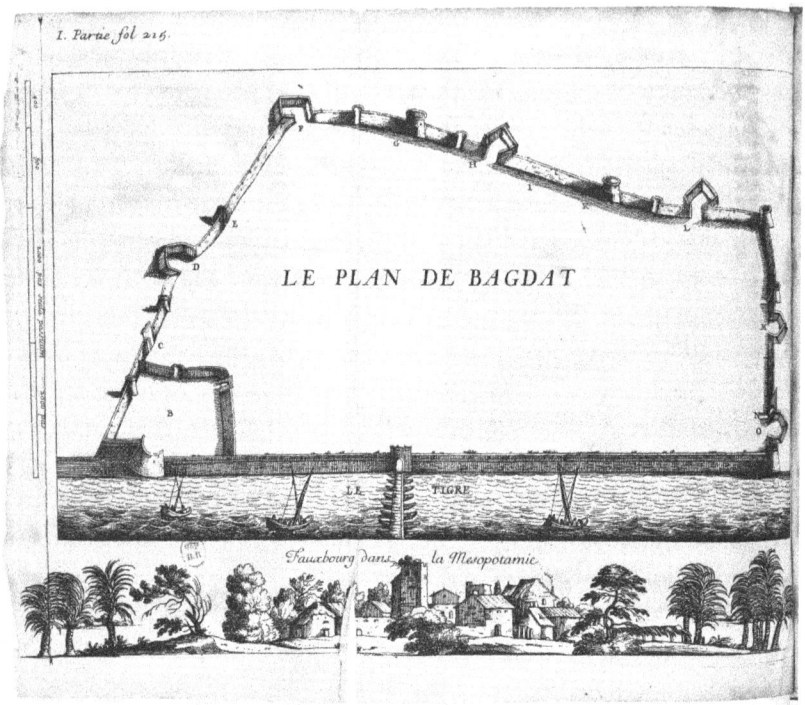

FIGURE 1.6. The Baghdad Fortress, 1652. Source: Jean-Baptiste Tavernier, *Les Six Voyages de Jean Baptiste Tavernier* (Paris: G. Clouzier et C. Barbin, 1676), 1:215. Courtesy of the Bibliothèque Nationale de France.

an artillery corps comprised of artillerymen, artillery carriage drivers, and armorers. Together, the janissary and artillery corps stationed in Baghdad ranged between 1,000 and 10,000 men, depending on the rhythm of military activity in the east.[52] Occasionally, Istanbul deployed additional troops from other provinces and even from foreign countries, including 500 French gunners who were serving in Baghdad in the 1650s.[53] A more renowned unit called up from outside was the provincial military band, whose members were drawn from Egypt. After the nightfall prayers and in celebrations of enthronement, royal birth, and conquest, the band performed on a giant kettledrum known throughout Anatolia and Persia as Baghdad Kösü.[54] Troops recruited by the imperial center fought alongside numerous other local forces and tribesmen who in many cases were far greater in number.[55]

In addition to boosting troop levels, Istanbul invested in the development of military industrial facilities in the Baghdad fortress, including a gunpowder mill and cannon foundry that supported Ottoman war efforts in both the eastern and western fronts.⁵⁶ In the middle of the seventeenth century, the Baghdad gunpowder mill employed about 1,000 powder makers recruited from among those who specialized in pounding coffee beans in the city (*kahve döğücü*). Their humble background working with coffee did not diminish their stature as great masters of their new trade. By Evliya Çelebi's reckoning, the explosive power of the gunpowder they made was second only to that imported from England.⁵⁷ The enlistment of coffee grinders into newly established gunpowder workshops was emblematic of the broader militarization of urban infrastructure and urban society in Baghdad. Popularly known from Abbasid times as the Abode of Peace (Dar as-Salam), the city was transformed by Ottoman policies into the Abode of Jihad (Dar al-Jihad), as it was commonly referred to in imperial correspondence.

Intelligent Control

A crude system of control enforced solely by arms could be unduly wasteful. Ottoman logistical support, therefore, improved the efficiency of riverine fortresses in policing their hinterlands by promoting their cooptive soft power and deep symbolism—sites where Ottoman religious institutions clustered and flourished.⁵⁸ Notables at all hierarchical levels, from the sultan's household to local provincial families, established and subsidized pious endowments in cities to mark their political power and gain divine favor.⁵⁹ The laws, customs, and practicalities governing the relationship between endowed religious structures, on one hand, and their founders, beneficiaries, and revenue sources, on the other, were breathtakingly complex. The Ottoman administration exercised a remarkable degree of oversight over the appointment of staff in charge of religious foundations. A large portion of the correspondence moving between Istanbul and eastern fortresses concerned the appointment of prayer leaders (*imam*), prayer callers (*muezzin*), teachers (*müderris*), sermon deliverers (*hatib*), preachers (*va'iz*), prayer reciters (*dua-goy*), and even floor sweepers (*ferraş*) and cooks (*tabbakh*).⁶⁰ Ottoman judges stationed in each city regularly requested a "noble epistle" (*berat-i şerif*) from the incumbent sultan to confirm new religious appointments and their salaries. When a new sultan ascended the throne, requests from provincial centers streamed into Istanbul seeking the renewal of old epistles.⁶¹ Officials at the Imperial Treasury meticulously recorded the names and salaries of religious appointees in the registers of the Chief Accounting Office (Başmuhasabe Defterleri) and could confirm and dismiss anyone at will.⁶²

Imperial religious patronage and supervision effectively created in major fortresses of the river basin a pro-Ottoman priestly caste whose livelihood depended on its continued subservience to the sultan and his deputies and on its anti-Shiʿi, anti-Safavid missionary work.[63] This priesthood joined forces with the weapons-bearing groups deployed by Istanbul, giving their police supervision the appearance of legitimacy and beneficence. In modern jargon, it represented the propaganda arm of the Ottoman government in the east that tapped into the intangible power resources of ideology and culture rather than the firepower of guns and mortars. In ancient terms, it reprised the alliance of throne and altar that underpinned most political projects in Iraq from the Sumerian city states onward. From the pulpits, prayer rows, and study circles, Ottoman clerics in Diyarbakır, Mosul, Baghdad, and Basra awakened a degree of compliance and trust among the faithful masses and reinforced Ottoman authority without constant resort to the threat of force or to payoffs.

With their buttressed walls and moats, as well as their domed structures and minarets, riverine fortresses in Iraq—to paraphrase architectural critic Lewis Mumford—expressed in concrete terms the magnification of the secular and sacred facets of Ottoman power.[64] They combined the brute coercion of the janissary and artillery corps stationed in the citadels and trenches with the soft power of clerics in houses of worship and learning to exercise intelligent control over their hinterlands and accomplish what each group alone could not.

In control of the entire drainage basin after 1534, the Ottoman Empire retrofitted the Tigris and Euphrates with a network of docks, pontoon bridges, and boat construction facilities that allowed it to manage—and profit from—the movement of peoples and goods between Syria, Anatolia, Persia, and Iraq. More important, the imperial system of waterborne communication created heavily militarized and Sunnified fortresses that stabilized the Ottoman presence in the eastern frontier and held major threats to Ottoman hegemony in check.

Without the Tigris and Euphrates, Ottoman rule in Iraq would have resembled its experience in the Arabian Peninsula or any other inland region that lacked useful transport rivers—far less concrete and contingent on the fickle goodwill of tribal and religious leaders. In fact, the main reason the governors of Baghdad were reluctant to take on the fundamentalist Wahhabi movement in Arabia during the late eighteenth century—despite being repeatedly admonished by Istanbul to do so—was the difficulty of provisioning an army in the middle of the desert. In the words of the British resident in

Baghdad in 1803, "with every thing hostile to him in front and rear, on the right hand and on the left," if the Ottoman governor of Baghdad was able to bring his "broken, dispirited, discontented, disorderly, [and] famished" troops home safely after a foray against the Wahhabis in Arabia, "he will prove himself a great captain."[65] The British resident doubted that he could. In a region crisscrossed by navigable waterways such as Iraq, on the other hand, the Ottoman governor of Baghdad could easily provision his armies with grain and gunpowder to crush seditious movements and confront foreign enemies.

2
Shipyards

> Only those who excel at using fire and water have the
> awesomeness to shake Heaven.
> —He Liangchen, *On Battle Arrays* (c. 16th century)

FIREARMS AND CANNONS fired in unison announcing the departure of 550 Ottoman vessels on the Euphrates River. Spectators in the Birecik shipyard recoiled at the ceremonial shots' piercing sound as the sky darkened with black smoke. The naval mission could finally begin on Friday, July 11, 1567, and embark on the calm summer waters. Carrying a force of some 10,000 janissaries, artillerymen, and irregulars, the squadron had clear instructions from Istanbul—to "purify" the marshes of Iraq from a descendant of the Ulayyan tribe. His crime? He dared to challenge the authority of the newly enthroned sovereign of the realm, Selim II.[1]

After four months of travel, the Ottoman force fought its first major battle in the marsh village of Sadr al-Bahrayn. Soldiers cleared the area at the cost of many lives lost on both sides and started hacking down their enemies' palm and fruit trees with hatchets. "We [will] obey, do not cut our trees!" pleaded the inhabitants, to no avail.[2] Another round of fighting broke out when more tribesmen came on to the scene and confronted the soldiers. Countless locals fell dead alongside their fallen trees.

As it advanced, the expedition turned into a macabre series of atrocities. Captives had their heads chopped off; men, women, and children drowned as they hurriedly fled in boats under the volley fire of the janissaries, who later set fire to abandoned villages, rice fields, and palm groves. Humbled by Ottoman naval power, the rebellious inhabitants surrendered. Their chief's brother came with his men in forty-six canoes to formally tender his allegiance to the country's imperial masters. From the middle of the bushes and trees behind him, villagers slowly emerged and stood in rows to watch their mortified chief's representative kiss the hands of Canbulad Bey, admiral of the Ottoman squadron. "Do not misbehave!" was the admiral's peremptory message to the defeated population before his boat sailed away.[3]

The enormous disparity in manpower and firepower between the two sides could not be starker. Through enormous reed thickets and pools of muddy water, vessels of the Ottoman navy provided the means for thousands of soldiers, along with their arms and provisions, to bring the destructive capacity of their handguns and artillery to the heart of the Iraqi marshes. Aggravating the threat to the local population was the terrifying frequency with which these floating gun carriers were encroaching on its waters. After all, 1567 was not the first year an Ottoman squadron had fought on the Euphrates, nor were members and allies of the Ulayyan tribe its sole targets. It had made an appearance a few times earlier and would remain for many years to come a critical instrument of deadly force against all hostile actors, particularly members of the ill-armed tribal population. To them, cannons of the Ottoman fleet, a sixteenth-century chronicler gloated, caused the tremors of doom announcing the Day of Judgment.[4]

This chapter documents the role of the navy in reinforcing the Ottoman presence in the Tigris-Euphrates basin. Istanbul deployed its gunboats to prey on smaller vessels and riparian villages, carving out a political niche in the east a thousand miles away. Reliance on the gunboat as an assault and transport craft for power projection gave rise to a naval bureaucracy based in two shipyards at both ends of the drainage basin, Birecik in the north and Basra in the south. Collectively, vessels constructed in both locations comprised what the Ottoman administration described as the Shatt al-Arab or Shatt River Fleet (Donanma-i Nehr-i Şatt), one of several fleets throughout the Ottoman Empire. Others operated in the Sea of Marmara and the Aegean Sea and later in the broader Mediterranean, Black Sea, Red Sea, and Indian Ocean—as well as along the Danube River. The Shatt Fleet's founding in the sixteenth century was a local display of a recent global trend in sea warfare that saw the establishment of navies as permanent organizations by increasingly bureaucratized militaries and states.[5] In conjunction with land troops based in fortresses, the Shatt Fleet allowed Istanbul to field a mightier armed force in the eastern borderlands than was previously possible and opened the way for the exercise of Ottoman hegemony over the region.

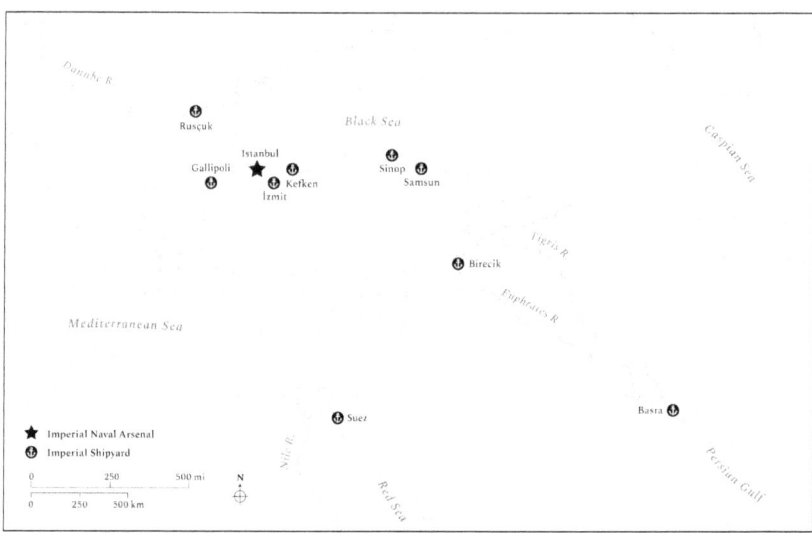

FIGURE 2.1. Major Shipyards of the Ottoman Empire. After İdris Bostan, *Osmanlı Bahriye Teşkilatı: XVII. Yüzyılda Tersane-i Amire* (Ankara: Türk Tarih Kurumu, 1992), 14–29.

Developments in military technology and tactics transformed early modern warfare in a process commonly referred to as the Military Revolution. As gunpowder artillery improved in the fifteenth century, military strategists quickly realized that the new weapon could be critical in deciding the balance of power not only on land but also at sea. From the Dutch East Indies to the Caribbean, empires began converting their oared galleys and sailing ships into floating platforms for heavy artillery. Fighting with guns mounted on ships soon replaced boarding and close combat with infantry weapons as the dominant form of warfare at sea.[6]

What has been missing from this narrative is the pioneering effort of early modern empires to apply a compact version of ship-borne artillery to navigable rivers as a consistent strategy of control. This occurred three centuries before Western Europe's push to conquer inland Asia and Africa with steam-powered gunboats.[7] This chapter recounts this forgotten episode in the Military Revolution and highlights Ottoman innovation in introducing naval artillery to a fluvial landscape in the middle of the sixteenth century. It begins by describing the circumstances that necessitated the establishment and maintenance of a naval infrastructure in West Asia, followed by a closer look at the nuts and bolts of boat construction.

Laying the Foundations

Süleyman I's conquest of Baghdad in 1534 was an entirely overland affair, but it had important naval implications. It brought under Ottoman control the middle Tigris, Iraq's urban heartland and its seat of government since the founding of the Hellenistic city of Seleucia in the fourth century BC. Ottoman authors recognized the political centrality of Baghdad and referred to it as "the seat of government in the lands of Arab Iraq."[8] Ottoman rule in the region rested on this urban pedestal, which provided the foundation for the development and elaboration of government institutions, including a navy. Sources attest to the presence of an imperial squadron (*donanma-i hümayun*) on the Tigris shortly after Baghdad's conquest. In May 1546, officials in Mosul dispatched 120–150 vessels downstream with cannons and provisions to join an expedition to the deep south that resulted in Basra's reincorporation into the Ottoman Empire. The chroniclers who recorded the event do not say who built the squadron, where, or how.[9]

The Ottoman administration chose Birecik (ancient Zeugma) on the upper Euphrates as the site of its first imperial shipyard in the drainage basin. Added to the Ottoman Empire in 1516, Birecik had a dock engaged in the collection of tolls, a modest revenue source that surged to a record high after Süleyman's campaign in Iraq. Little is known of what happened to the Birecik dock until June 1547, when news of its stealthy naval operations reached the Portuguese governor of Hormuz. That month the governor received an Arab merchant from Basra named Hajji Fayat with information about what the Ottomans could do in this "large and well-populated" Euphrates town. "If the Turks had such an evil purpose," he claimed after swearing on his Quran, "they could build in the river Euphrates as many ships as they wanted."[10] Around that time, the Birecik shipyard had a permanent staff of forty-five individuals who received tax exemptions for their services.[11] With support from Aleppo, the naval staff in Birecik received from the Ottoman Imperial Council the first documented order for boat building in July 1552, asking for the construction of 300 vessels.[12] From a modest dockyard in 1516, Birecik by 1552 became a major industrial site.

The stunning impact the Ottoman conquest had on Birecik exemplifies the experience of other riparian settlements. Because of its strategic location amid Anatolia, Syria, and Iraq, Birecik had repeatedly changed hands among the great warring powers of Eurasia—Macedonians and Parthians, Romans and Sasanians, Crusaders and Muslims, Byzantines and Seljuks, Mamluks and Akkoyunlus. Incorporation into the Ottoman Empire offered Birecik a

Table 2.1. Tolls from the Dock of Birecik (*guruş*)

Year	1520	1536	1552	1570
Amount	75,000	180,000	180,000	200,000

Note: Guruş is a large Ottoman silver coin.

Sources: OA, MAD 75, 22v; OA, TT 184, 12; OA, TT 276, 19; OA, TT 501, 20; OA, TT 496, 30.

degree of stability unknown since Roman times, allowing it to become the most prominent shipbuilding facility in the Tigris-Euphrates basin.[13]

The Ottoman Empire chose Birecik as a shipbuilding site because of its unique geo-botanical package. From a steersman's perspective, the town is perfectly situated at the southern edge of the Taurus foothills. Descending in spectacular cataracts from 6,200 feet above sea level in Erzurum, the Euphrates breaks out of the gorges it has cut through Anatolia's high plateaus and establishes a navigable slope by the time it reaches Birecik. Near Birecik's latitude, both the Tigris and Euphrates occupy comparable elevation points (some 1,150 feet), but from there the Euphrates descends gently in wider valleys, while the Tigris falls at a steeper gradient. The upper Euphrates, thus, was hospitable to larger and more sophisticated rivercraft, making Birecik a more attractive base for the imperial navy than, say, Diyarbakır or Cizre on the upper Tigris.[14]

Viewed from the south, Birecik stands out in both its high altitude and its proximity to the Mediterranean. In these respects, the town sharply differs from all downstream settlements, which become more distant from the Syrian coast and drier as the Euphrates descends in a southeasterly direction. Wetter conditions endowed Birecik's adjacent mountain zones with mixed forests of deciduous and coniferous species suitable for ship timber. In his intelligence report to the Portuguese in Hormuz, Hajji Fayat made a direct link between eastern Anatolia's forest reserves and the rise of the Ottoman shipbuilding industry in Birecik. "Near the town of Birecik," he said, "there are great forests, from which comes much fine timber and, in addition, there is also pine-wood from which they [the Turks] could build as many ships as they desired, both large and small, by reason of the abundance of timber to be found there."[15]

Most riparian settlements lacked Birecik's suitable gradient or stocks of timber. Take Basra as an example. Despite its favorable location on the Tigris-Euphrates estuary, Basra lacked a reliable supply of timber. Shipbuilding, as a result, was considerably more expensive for craftsmen there than it was to their counterparts in the north. Portuguese explorer Pedro Teixeira pointed out this deficiency in his description of vessels during his visit to Basra in 1604. "Small as they are," he wrote, "they cost much; for that land has no timber at all, and it is costly to import."[16] Even when Portugal encroached on Ottoman ports in the Persian Gulf in the early sixteenth century, Istanbul used its distant Suez squadron led by former Mamluk naval officers to guard its interests in the region. In the famous campaign of Ottoman sailor Piri Reis from the Red Sea to the Persian Gulf in 1552, Basra simply served as a harbor for his fleet after a failed siege of Hormuz.

From 1559, the Ottoman naval establishment reconsidered its approach to Basra. In that year, it grew concerned that ships owned by the "infidels" came near Ottoman ports in the Persian Gulf, causing fear among Muslim merchants and a decline in customs revenues.[17] Between October 1559 and February 1560, the Ottoman Imperial Council considered the construction of five galliots in Birecik, from where they would be dispatched down to Basra to deal with the Portuguese menace.[18] After an internal investigation, however, the Ottoman vizier Sokullu Mehmed Pasha—who previously served as grand admiral of the Ottoman navy—recommended shipping the necessary timber to Basra and carrying out all construction in the port city, a decision the Imperial Council endorsed.[19] The governor of Basra received the materials necessary to assemble the five galliots in the summer of 1560 and was soon ready to send his new fleet on regular patrol to protect Ottoman ports in the Persian Gulf.[20] Thus, further Portuguese encroachment and several months of deliberation by the Ottoman administration brought about the establishment of a permanent shipyard in Basra, the second in the river basin after Birecik.

From this point, the Birecik and Basra shipyards became part of a vast network of subsidiary shipbuilding facilities throughout the Ottoman Empire, all linked organizationally to the Imperial Naval Arsenal (Tersane-i Amire) in Galata, Istanbul. In addition to the major shipyards, Istanbul built ships in about sixty other smaller locations during the sixteenth and seventeenth centuries.[21] Sultan Bayezid I built the first large shipyard at Gallipoli, through which Ottoman armies crossed from Anatolia to Europe in the middle of the fourteenth century, but by the sixteenth century most of the naval establishment had moved to the shores of the Golden Horn in Istanbul. Selim I and Süleyman I oversaw the transition of the navy's headquarters from Gallipoli to Istanbul and sponsored major expansion projects at the Galata shipyard.[22] The establishment of shipbuilding facilities in Birecik and Basra mirrored the consolidation of the Ottoman naval administration in Istanbul and sheds light on the development of a new link binding the periphery with the metropolis. Just as provincial treasuries and cannon foundries linked the finances and saltpeter industries of the east to Istanbul, the Birecik and Basra shipyards opened the waters of the Tigris and Euphrates to the resources and patrols of the Ottoman navy.

Buildup

Waging war requires massive energy inputs from the environment. The sole aim of the Birecik and Basra shipyards, as military installations, was to exploit the energies of the Tigris and Euphrates to fuel the Ottoman Empire's military operations. Like watermills, the vessels were machines that enhanced humans' ability to capture the kinetic energy of air and water flows. Through them, Birecik and Basra made the rivers a more efficient energy system, moving massive agglomerations of humans, animals, arms, and raw materials. Vessels, moreover, converted the energy of turbulent flow into a rational form. They could transport grain, after careful planning and calculations, between two predetermined points at a forecast time to achieve a desired outcome.[23]

The desired outcome of state-sponsored boat-building activities in Birecik and Basra was to thwart the challenges to Ottoman hegemony in Iraq and the Persian Gulf. Three major players confronted the Ottoman Empire in the east and triggered most gunboat construction orders: one was maritime and European, another overland and Persian, and between them stood an amphibious Arab enemy. Each rival influenced Ottoman naval buildup along the Tigris and Euphrates.

The Ottoman Empire, home to 12 million people in the early sixteenth century, had to contend at sea with the improbable power of Portugal, a nation of some 1.5 million.[24] Relying on guns and sails and motivated by a blend of royal mercantilism and Christian messianism, the Iberian empire burst forth from a promontory of Europe into Asia and established a foothold in the southwestern coast of India before the close of the fifteenth century. To protect Muslim merchant shipping at the turn of the sixteenth century, Sultan Bayezid II became involved in anti-Portuguese maritime operations and supplied arms, experts, and resources to the Mamluk fleet in Suez. The Ottomans inherited the Mamluk navy following their conquest of Egypt in 1517 and enhanced and deployed it to join the competition for honor and profits in the Indian Ocean.[25]

After rounds of fighting on the open seas from Jidda to the Malacca Straits and the signing of an armistice with the Habsburgs in 1545, the Ottomans faced off against the Portuguese in the Persian Gulf, the former based in Iraq and the latter in Hormuz. From Baghdad, Ottoman troops pushed south. They entered Basra in 1546 and conquered Qatif on the Persian Gulf in 1550, prompting Portuguese counterstrikes a year later. The conflict staggered on with tit-for-tat operations and provided the context for Istanbul's decision in 1559–1560 to build its first squadron of five galliots in Basra. Despite reaching an unofficial truce with Lisbon and Goa in 1563, Istanbul continued to beef up its fleet in Basra in anticipation of future engagements. It frequently renovated its ships and ordered the construction of five more galliots in 1571.[26] The naval buildup in Basra proved worthwhile in September 1573, when a Portuguese flotilla raided two Muslim vessels and captured several merchants near Bahrain. An Ottoman squadron of ten galliots in Basra swiftly responded to the incident.[27] Two years later, the Ottoman administration placed an order for eight galleys to be built in Basra in a renewed bid to seize the island of Bahrain from Portuguese control, a plan that never materialized.[28]

The turn of the seventeenth century gave rise to a calmer political landscape in the Persian Gulf. Europe's maritime powers ceased to exert a significant pressure on the buildup of Ottoman naval forces in Basra. An unofficial truce with Portugal remained largely intact until 1622, when an Anglo-Safavid operation expelled it from Hormuz. The Portuguese moved their headquarters to Muscat and ceded to the British and Dutch empires domination of the approaches to the Persian Gulf. Because of their earlier dealings with the Ottomans in the Mediterranean and the benefits they derived from the trading conditions of the capitulations (*ahdname* in Ottoman vocabulary, or treaties of alliance), the British and Dutch were generally on good terms with Istanbul by the time they became prominent players in the world of the Indian Ocean. Their merchants and chartered companies competed with the Ottomans on the open market without imposing the kind of protection racket that characterized Portuguese seaborne trade. Whatever sparring the Ottomans had with Europe's newcomers in the Persian Gulf, it was often settled by diplomats rather than gunboats.[29]

Another challenge to Ottoman armed forces in the east came overland from Persia, home to some 6 million people in the early seventeenth century.[30] From Lahijan, south of the Caspian Sea, in 1499, a twelve-year-old scion of the house of Safi, along with 7,000 of his Turkmen followers, began a conquest campaign that a decade later unified the entire Iranian Plateau, which had been politically fragmented since the collapse of the Sasanian dynasty in the seventh century. In military terms, the resurrected Persian empire was effectively a landlocked power without a navy of its own in the Persian Gulf, largely due to the coast's lack of suitable timber and the difficult terrain that separates it from the dense forests of the Caspian Sea littoral. Unlike the maritime powers of Europe that relied on ships and heavy weaponry, the Safavids tapped into the strength and expertise of mounted Turkmen tribesmen and Caucasian slaves in land warfare. This force was more effective in dealing with Persia's traditional Uzbek enemies in the east than with the Ottomans in the west.[31]

Ottoman and Safavid forces clashed frequently at the Caucasian and Anatolian borderlands, which stood beyond the reach of the Ottoman navy. The two major exceptions were confrontations over Iraq in 1534–1535 and 1623–1638. The first confrontation paved the way for the founding of the Ottoman Shatt Fleet on the Tigris and Euphrates. In 1623–1638, the Shatt Fleet was already in place but found itself in a difficult position. Safavid forces controlled Baghdad on the Tigris and Hilla on the Euphrates and occasionally extended their influence further north. Without access to the most strategic river crossings, Ottoman steersmen lacked the safety of navigation. Shipwrights in Basra, meanwhile, lost access to the necessary raw materials and workforce they typically received from the north. From Baghdad and Hilla, Shah Abbas I dispatched his Qizilbash forces ("the red heads") to different positions along the rivers, where they actively raided Ottoman vessels. Qizilbash raids remained a security concern to Ottoman military logisticians until the forces of Sultan Murad IV drove the Safavids out of Iraq in December 1638.[32]

Still, despite Safavid pressure, the Ottoman army managed to place orders for hundreds of vessels. During a failed attempt to reconquer Baghdad in 1629, Grand Vizier Hüsrev Pasha ordered the construction of 100 vessels in Birecik.[33] A later successful campaign in 1638 involved the construction of 610 vessels for the army and twelve vessels for the entourage of Murad IV to use to cross the Euphrates channel near Malatya.[34] A French traveler witnessed the role of Ottoman riverboats in the 1638 campaign firsthand. "I must confess that in the year 1638," he wrote, "I saw a division of the Grand Signor's army together with ample ammunition of war go down the Euphrates when he went to besiege Babylon."[35] The Ottoman capture of Baghdad and the signing of the Zuhab Treaty in 1639 neutralized the Safavid Empire's threat until its collapse in 1722.[36]

The Ottoman Empire depended on its fleet along the Tigris and Euphrates to fend off its attackers from Persia. While its eastern rival relied entirely on land troops, Istanbul could count on a more complex and effective armed force that included a navy. Despite a brief setback between 1623 and 1638, Ottoman armies ultimately prevailed and maintained their edge over Safavid armies.

The Ottoman state's thorniest challenge in the east were not the empires of Portugal and Persia but rather peoples of more modest means who, along with their herds, made a home for themselves in Iraq. Unlike the episodic encroachment of foreign powers, this domestic menace posed an omnipresent threat to the Ottoman order. What it lacked in technology and resources it made up for in mobility, flexibility, and a peculiar geography. "All the forts and villages are in the water," wrote the Ottoman governor of Basra about one of his riotous districts in 1565. "Capturing them by [conventional] battle and governing them are not feasible."[37] The tribal population of Iraq squared off against the Ottoman state on its own marshy terrain and fought according to its own rules.[38]

With the aid of the Shatt Fleet, the Ottoman army adapted to fighting on this unconventional landscape. Ottoman expeditions against insubordinate tribal groups tended to include a significant naval force. Beyond the needs of the battlefield, the Ottoman fleet was critical for policing and governing the area. In the words of Teixeira in 1604, Ottoman galleys in Basra "are for use in the river and thereabouts, to keep in order the rebellious Arabs, from whom they [the Turks] exact heavy tribute."[39] To crush the Arab tribes at war and exact tribute from them at peace, Istanbul placed numerous small and large construction and renovation orders in Birecik and Basra as well as in Raqqa and Baghdad.[40]

In short, security anxiety and conflicts with Portugal, Persia, and Iraqi tribesmen created incentives for the Ottoman Empire to maximize its naval strength and maintain a provincial squadron in the Tigris-Euphrates basin. The agitated state of affairs was unique to the early modern era, in which the river basin was dominated in its entirety by a Sunni Turkish empire in frequent conflict within and on every side with Christian Europeans, Shi'i Persians, and tribal Arabs of mixed Muslim persuasions. In other historical periods, a naval industry on the scale the Ottomans established was either unnecessary or impossible to set up. It was unnecessary when the river basin was remote from a ruling power's frontier conflicts, as was the case during the Umayyad and Abbasid periods (661–1258). Likewise, the Ottomans' comprehensive naval infrastructure could not have emerged without political cohesion, which the region lacked for most of its history.

Rivercraft

From the fifth millennium BC the Tigris and Euphrates propelled the development of one of the world's most advanced cultures in statecraft, economy, literature, and art. Yet both rivers, with their rapids, shallows, and prevailing northerly wind, arrested nautical development and posed geographical obstacles to the widespread use of large ships. For most of history, the drainage basin's predominant vessels remained small and light, made of local materials such as animal skin, reed, timber, and brushwood.[41] With its immense human, natural, and fiscal resources, the Ottoman Empire broke the innovation bottleneck and introduced a highly developed squadron to the region, based on the more advanced nautical traditions of the Mediterranean that it had inherited from Byzantium and acquired through its rivalries with other maritime powers, particularly the Venetians. Vessels of the Ottoman Shatt Fleet acquired greater offensive lethality by adopting the latest improvements in ship design to tap into the combined capabilities of streamflow energy, human and animal muscle, and heavy ordnance.

In the sixteenth and seventeenth centuries, the galley and galliot formed the backbone of the Ottoman navy on the Tigris and Euphrates and throughout the empire.[42] They constituted a salient feature of the Ottoman order in the drainage basin. The Turk's "chief strength is of gallies," wrote a British merchant passing through Iraq in 1583, "which are about five and twenty or thirty very faire and furnished with goodly ordnance."[43] The galley (*kadırga*) was a large, seagoing, oared vessel developed during the second millennium BC and refined over time by civilizations of the eastern Mediterranean.[44] The Turks became familiar with it following the rise of the first Turkish principalities on the Marmara and Aegean coasts in late medieval times.[45] On the Tigris and Euphrates, the Ottoman galley was armed with one cannon fitted into the prow (*baş topu*), two culverin-type guns (*kolumburina*), and four light battering guns called *darbzen*.[46] The galliot (*kalyata*), a favorite of Mediterranean corsairs, was smaller than the galley and a much more recent innovation. The Ottomans first adopted it in the fifteenth century in their first naval base in Gallipoli, introducing it a century later to the Tigris and Euphrates as well as the Black and Red seas.[47] The Ottoman galliot on the Tigris and Euphrates came in different sizes, the smallest with sixteen and largest with twenty-two oar benches.[48] It could carry nearly as many cannon as the galley did.[49] In action, the firepower of galleys and galliots destroyed local boats and mudbrick forts that had helped the tribes of Iraq remain independent for centuries.[50]

The second half of the seventeenth century marked the Ottoman navy's transition from oar-powered to sail-powered ships, a move that began during the Cretan War with Venice (1645–1669) and was adopted as a policy by Grand Vizier Merzifonlu Kara Mustafa Pasha a year before his death in 1683.[51] Istanbul became convinced of the sailing galleon's advantages over traditional oared galleys, such as its ability to better utilize wind energy, accommodate artillery, and minimize the need for oarsmen and warriors (along with their food and water barrels) on board. Previously used solely for transport purposes, the galleon (*kalyun*) would displace the galley as the empire's main fighting ship by the late eighteenth century.[52] The ripple effect of this transition extended to the Basra shipyard around 1703, when it received the first order for the construction of galleons.[53] A large ship like the galleon, however, was more suitable to the open sea and made little headway in the river basin, where its use was always limited.

More successful on the Tigris and Euphrates than the large galleon was another naval innovation that became prevalent around the same time—the frigate (*firkate*). Powered by oars and sails, the frigate was a long, narrow, and fast ship that could operate successfully on rivers and at sea. The Imperial Naval Arsenal first adopted it in the Mediterranean, in the Black Sea, and on the Danube from 1689 and deployed it in the wars with Europe throughout the eighteenth and nineteenth centuries.[54] A decade after initial adoption of the frigate, the Ottoman administration introduced it to the Euphrates, building sixty of them in the Birecik shipyard.[55] They proved their effectiveness and became thereafter the backbone of the Shatt Fleet.[56] Frigates on the Tigris and Euphrates came in different sizes but on average were sixty-five feet long and ten feet wide and had sixteen oar benches.[57] They could carry a crew of seventy individuals, a number authorities mustered only in major campaigns.[58] The largest frigate was called *baştarda* and normally carried fifty persons. The vast majority were slightly smaller and carried thirty-five: one steersman, two tenders of sails, two gunners, and twenty-eight other persons, perhaps serving as fighters and oarsmen.[59] The Tigris-Euphrates frigate could carry seven cannons: two at the bow and stern, one broadside, and four others with small caliber firing grapeshot (*saçma topu*).[60]

Supplementing the frigate were three riverboats that the Ottoman Empire transplanted from the Danube to the Tigris and Euphrates during the early eighteenth century. The first was called *şayka*, a word derived from Russian (*chayka*) meaning seagull. It was a flat-bottomed, wide, and oared ship built in the shipyards of the Danube from the sixteenth century and used on the Black Sea and Dnieper River as well.[61] The şayka appeared in the

Basra Shipyard in 1709, manned by eight individuals. Around the same time, two other Danubian vessels became part of the Shatt Fleet, the *işkampoye* and *üstüaçık*, both oared and used for communication and the transport of soldiers, carriages, and other heavy loads.[62]

Historically stagnant, the naval industry of the Tigris-Euphrates basin witnessed rapid activity and development under the aegis of the Ottoman Empire. Riverine versions of the Mediterranean galley and galliot arrived first in the sixteenth century and were supplanted by the frigate at the turn of the eighteenth century. With the intervention of the Imperial Naval Arsenal in Istanbul, other vessels from the rivers and seas of the north began to ply the southern waters of the Tigris and Euphrates. None of these technological transfers could have occurred without the free movement of labor across regions once politically divided and now part of a unified empire.

Labor

As major industrial sites, the arsenals of Birecik and Basra needed a large and reliable supply of men to design, build, and operate the boats, something both places lacked. Birecik was a small fort with only four neighborhoods and a taxable population that did not surpass 1,000 for most of the sixteenth century.[63] The Basra fortress's larger population, about four times the size of Birecik's, was handicapped by a low population density.[64] Ottoman officials worried that recruiting locals to Basra's shipyard could reduce the labor force needed to cultivate the land.[65] They could not even find enough carpenters and sawyers among the urban population to work in the shipyard, resorting instead to experts from Anatolia.[66] Basra's population makeup was another problem for shipyard officials. Crewmen from the city were either Arabs or Qizilbash (Safavid supporters of Turkic origin) whom the Ottomans regarded as "neither safe nor reliable" to man the vessels.[67] Istanbul maintained a similar attitude toward Christian crewmen in the Black Sea and Mediterranean, whom it could not trust in battles against Europe's Christian powers.[68]

The local populations of Birecik and Basra alone could not fulfill the demands of an imperial naval arsenal. Ottoman authorities had to adjust the natural population distribution of the land to carry out the sultan's wishes and maintain his navy in the east. They did so by mobilizing laborers from provinces near and far and moving them around like pieces on a chessboard.

Anatolian and Syrian artisans partially fulfilled the demands of the Tigris-Euphrates shipyards for skilled labor. In 1576, for example, Istanbul ordered the governor of Ayntab near Birecik to send fifteen carpenters and sawyers to assist in a construction project in Basra's shipyard.⁶⁹ Likewise, in 1701, Diyarbakır, Maraş, Ayntab, Kilis, Ruha, and Aleppo dispatched dozens of carpenters, caulkers, loggers, and cutlers to work along with local artisans in Birecik's shipyard.⁷⁰

Outside of Anatolia and Syria, Ottoman authorities counted on maritime populations from the far corners of the empire to serve as skilled laborers along the Tigris and Euphrates. Some were hauled from Europe without consent. In 1565, for instance, the governor of Basra petitioned the Imperial Council to send a cooper and an experienced carpenter to his shipyard. Istanbul forwarded the request to the grand admiral of the Mediterranean fleet, ordering him to dispatch the requested workforce from among the "infidel" prisoners he held in custody, likely captured from a European naval force.⁷¹ A less coerced group came from Istanbul. Steward of the Imperial Naval Arsenal recruited missions from the Ottoman capital comprised of a diverse group of artisans, including caulkers, carpenters, blacksmiths, and makers of oars, pulleys, lanterns, and sail cloth. Those recruited hailed from different Istanbul neighborhoods—Eyüp, Tozkoparan, Hacıahmet, Tophane, Kasımpaşa, Samatya, Galata, and Acıçeşme—and found themselves working side by side far away from home in the shipyards of Birecik and Basra.⁷²

The Ottoman administration enlisted shipyard artisans from other coastlines and straits with rich maritime traditions. In 1560, it ordered the governor of the Greek island of Rhodes to dispatch forty to fifty shipwright masters and caulkers to Birecik.⁷³ Officials in Cyprus, Sidon, Tripoli, and Beirut received orders in 1699–1700 to recruit and send an unspecified number of shipwrights, carpenters, oarsmen, caulkers, and augerers (a person who uses an auger to drill holes in timber).⁷⁴ Toponymic nicknames recorded in Ottoman payrolls indicate that artisans from the Black Sea (İnebolu, Çatalzeytin, Amasra, Araklı), Greece (Chios, Kos, Lesbos), and western Anatolia (İzmit, Izmir) occasionally worked in Birecik and Basra.⁷⁵

Among foreign maritime populations arriving in the shipyards of the Tigris and Euphrates, the prominence of Greek artisans stands out, given the great distance they had to travel and their repeated deployments from the sixteenth century. Their vital role in the Naval Arsenal in Istanbul, in missions to Birecik and Basra, and in the maritime history of the Ottoman Empire more generally is comparable to the role they played under the shadow of other empires in control of the Dodecanese islands, including ancient Rome, which actively recruited Hellenized sailors to its imperial fleet.⁷⁶

Artisan wages offer some insight into recruitment practices. Pay changed with workers' position and place of residence. Each artisan corps

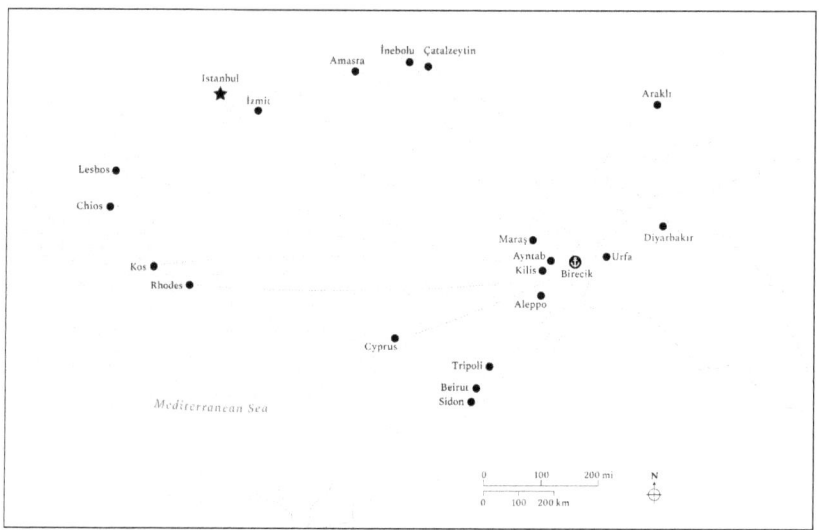

FIGURE 2.2. Deployment of Artisans to the Birecik Shipyard.

(caulkers, carpenters, shipwrights, etc.) had a leader of their own paid about 10 to 30 percent more than those serving under them. Local artisans coming from nearby towns, such as Maraş, Ayntab, and Diyarbakır, were paid about 25 percent less than their peers from more distant locations such as Istanbul or Sidon.[77] Paymasters upheld these standards regardless of workers' faith. The head of the carpenters' corps in 1699–1700 was Greek, yet he was still paid more than the Muslim workers under his supervision.[78] Likewise, in 1700, two Jewish lantern makers received for their work as much as their Muslim co-workers performing comparable tasks, such as pulley making.[79]

Artisans were a minority in the shipyards. The vast majority were crewmen called *levendat*, from the Italian term "Levantino," used by the Venetians to refer to the Levantine population that collaborated with them in the late Middle Ages.[80] In Birecik and Basra, the levendat crewmen worked on a seasonal basis and performed menial tasks, mostly oaring and manning ships. When active, each shipyard employed between 400 and 5,000 crewmen. They were among the lowest-paid of the naval workers, making in a year what a caulker could make in almost two days.[81] Even the meager pay they received was subject to occasional appropriation by higher officials, leading many to run away.[82] A crewman was typically placed on a diet of hardtack, which could easily travel across long distances and maintain its nutritional value to feed large numbers of workers.[83]

Like artisans, crewmen tended to be diverse in their composition. To fill jobs that did not appeal to many, the Imperial Naval Arsenal targeted tribesmen from Anatolia and Syria with no nautical experience.[84] In addition, Istanbul frequently asked officials in Aleppo, Diyarbakır, and Baghdad to sentence convicted criminals in their prisons to forced labor in the galleys.[85] Widespread in the Mediterranean world, the measure was one of the most common forms of punishment Ottoman judges resorted to in all kinds of crimes, from armed robberies and sexual assaults to the sale of adulterated bread. Reliance on both tribesmen and prisoners allowed the Ottoman navy to overcome the perpetual shortage of manpower in the shipyards of Birecik and Basra as well as in Lepanto, Nauplion, Kavala, and Istanbul at an acceptable cost.[86]

The Ottoman administration's ability to assemble all the artisans and crewmen it needed to operate its navy in the Tigris-Euphrates basin was nothing short of extraordinary. The fusion of their knowledge and practical experience, based on the nautical traditions of the Mediterranean, Danube, Black Sea, Tigris, and Euphrates, brought the software needed to generate a hybrid fleet that proved instrumental in uniting the eastern borderlands under Ottoman control. Acquiring the natural resources necessary to assemble the gunboats, the primary hardware of the Ottoman navy, was no less onerous, expensive, and impressive.

Natural Resources

A construction project in the shipyards ended the same way it began—with the sacrifice of two animals to the Divine.[87] They were among a slew of other commodities and valuables that had to be given up for the sake of a gunboat. Above all, boat building in pre-industrial times required massive quantities of timber.[88] Forests on the Marmara and Black Sea coasts fulfilled the Imperial Naval Arsenal's demands due to their accessibility to Istanbul by sea.[89] For Birecik and Basra, the Ottoman administration earmarked landlocked forests in southern Anatolia located within 100 miles of Birecik. Maraş and Malatya supplied most of the timber, but other forested regions around them took part as well.[90] Their abundant pine groves (*çamlık*) attracted Birecik's naval officers and carpenters alike, who refused to trade the pine for alternatives like mulberry.[91]

Species of pine were well suited for shipbuilding. They are on average less dense (thus have a better capacity for floating) and softer (thus easier to work) than hardwoods like oak and beech. Pinewood,

FIGURE 2.3. Forests Earmarked for the Birecik and Basra Shipyards.

moreover, exudes resin, effectively a natural preservative that prolongs its life. Unlike juniper, pine provides tall and straight trunks, ideal for the construction of different boat parts, particularly masts. With pine, Ottoman carpenters tailored the nautical expertise they cultivated along the Mediterranean and Black Sea to the shallow waterways of Anatolia, Syria, and Iraq.[92]

Documents abound that are related to the provisioning of timber to Birecik and Basra. Taken together, they reveal a general pattern. Istanbul typically issued an order to authorities in Maraş and Malatya asking them to mobilize carpenters, loggers, and cutlers and send them in detachments to different mountain zones where trees were to be felled. Even though pine was their chief target, they tended to add smaller quantities of other types, including oak, elm, nut, mulberry, and poplar trees.[93] After the felling, they hired groups of pack animals, several thousands in size and composed of geldings, mules, camels, and oxen to carry the load to different points along the Euphrates' upstream shores.[94] Rafts and floats would pick up the logs from each spot and transport them downstream to Birecik and, if needed, all the way to Basra.[95] At the shipyard, workers sawed the logs into the necessary forms: boards, beams, rudders, keels, masts, oars, and hull floorings, among others.[96] Transportation and labor costs were the main expenses. To cut trees in private groves, the logging detachment had to pay the property's owners and obtain their consent.[97]

Like all organic matter, timber decays, and so too did wooden vessels. Chief among their natural enemies were fire and worms. In Basra, torrid heat was a particularly destructive force. A strategy adopted by shipwrights in southern Iraq since the third millennium BC was to import teak from India.[98] Teak was lighter than oak yet more uniform in structure and akin in toughness and durability. It also has an oily secretion that repels insects and protects iron fastenings from oxidation and its injurious effects on timber.[99] Basra's heavy and costly reliance on Indian timber stood out to foreign visitors.[100] Imported teak served both as a principal building material and, more economically, as a reinforcement for decks and vessels made from Anatolian pine.[101] Another measure to check the natural deterioration of pinewood vessels was frequent maintenance. The Ottoman administration oversaw and financed annual caulking in the Basra shipyard and numerous restoration projects.[102]

Wood was essential not only as a building material but also, together with coal, as a source of heat for the processing of other raw materials, notably iron, used for the manufacture of anchors, nails, hinges, and construction tools such as hammers.[103] The Anatolian towns of Adana and Maraş supplied raw iron delivered by camels to blacksmiths in Birecik, where they smelted it for use locally and in Basra.[104] A document dated to 1700 reveals much about one of Birecik's iron forges, housed in a nearby cave. It contained a cistern for holding tar, coal sacks made of haircloth, two water jars, drainpipes, a furnace, and an anvil.[105]

Caulking consumed a number of resources. The use of sheep and goat hides and oakum was common.[106] Caulkers processed raw tar in copper cauldrons to obtain melted pitch, which they poured on the cracks between boards after inserting the fillers.[107] Early in the sixteenth century, the Ottoman administration secured tar for Birecik from the Greek islands of Rhodes and Kos.[108] The Tigris and Euphrates themselves had their share of natural tar, easily accessible to passing boatmen. On the Euphrates west of Baghdad, Hit was a popular source. "In a fielde neere unto it is a strange thing to see," a London merchant remarked while passing through the town in 1583, "a mouth that doth continually throwe foorth against the ayre boyling pitch with a filthy smoke: which pitch doth runne abroad into a great fielde which is always full thereof. The Moores say that it is the mouth of hell. By reason of the great quantitie of it, the men of that countrey doe pitch their boates two or three inches thicke on the out side, so that no water doth enter into them."[109] On the Tigris north of Baghdad, a Carmelite missionary encountered other sources of tar in 1656. Near the riverbanks, he wrote, "there were some ponds of very hot water, which, boiling out of the earth, brought a great amount of tar to the surface. The tar was gathered from there and brought to the workers by the river, who, after having washed it, stored it to send it elsewhere."[110] Despite the great benefits of the rivers' tar deposits, pack animals struggled to wade through them, sometimes fatally. "Divers Camels have fallen into these Springs," one traveler regretted in 1581, "but none of them could be saved."[111]

On several occasions, the construction superintendent or one of his deputies did not obtain any of the necessary items from their original source. Be it nails, pitch, awning, hawser, coal, or sail cloth, they instead purchased them from nearby cities at market price. From the sixteenth century, Aleppo had been the primary destination for shipyard brokers and occupied the role of the Ottoman navy's marketplace in the east.[112]

Through a shifting combination of imperial commands and trust in the marketplace, Ottoman officials mobilized the natural resources of broad regions to keep the Tigris-Euphrates shipyards supplied with material necessities required for the sustained projection of power in the drainage basin. In its central features, the system for supplying timber, pitch, and other materials was similar to the system in use for supplying carpenters and crewmen. Gunboats of the Shatt Fleet, in many respects, embodied the ecological and human diversity of the Ottoman Empire.[113]

The Ottoman gunboat was the largest and most complex machine in the early modern Tigris-Euphrates basin.[114] Putting it together was a capital-intensive enterprise that depended on the sophistication of a political administration, the goodwill of a subject population, and the bounties of nature. In tandem with the Ottoman shipyards' rise of influence, elaborate systems of finance, labor recruitment, and resource extraction emerged to meet the onerous demands and expectations of a Mediterranean power. Developing and sustaining production capabilities represented a major investment only a power like Istanbul could make, from which it received substantial returns. Gunboats creatively exploited the energy of wind and water flows to facilitate a more intensive marshaling of resources and to bring the explosive power of gunpowder to bear on rivals along the far-flung eastern borderlands.

PART II

The Water Wide Web

The Iraqi alluvium owed much of its physical character and living content to the work of the Tigris and Euphrates. On this lowland the river channels were easier to manipulate and divide into small irrigation canals to meet the needs of crops and forage. The rolling configuration of the northern plains, on the other hand, made very difficult the construction of cross-country canals, an arduous endeavor only a few cruel and high-handed kings could achieve. Scanty and extremely variable rainfall, falling below the 200 mm line, forced farmers of the alluvium to depend more heavily on the Tigris and Euphrates, the wholesale distributors of water in the region.[1] In the north, meanwhile, farming did not require river waters, thanks to abundant winter rains. Increasing in a northeasterly direction, annual precipitation reaches over 500 mm east of Mosul and offers a generous and reliable basis for dry farming. Finally, only in the alluvium does the Tigris-Euphrates flood factor come into the equation. Once the rivers reach this flat zone, they break free from the steep cliffs that confine them upstream and become prone to flooding and movement, scattering around most of the gravel and silt picked up in the north. The annual flood and deposition of sediment have sorted the land surface into marshes, levees, basins, and pastures and created the entire alluvial landscape on which the history of Iraq unfolded. Endowed with a suitable gradient for canal construction, without reliable rainfall, and liable to recurrent flooding, the alluvium over a span of more than six millennia was the focal point for most water management activity. Only in the twentieth century did society acquire the means necessary to intensively exploit the headwaters in Anatolia and Syria, aided by the diesel pump and fossil fuel subsidies.[2]

Enclosed by fortresses and shipyards, the Tigris-Euphrates alluvium could be milked of its organic wealth by the Ottoman state. The following three chapters focus on three major ecological zones arranged by fluvial action: arable lands, grasslands, and wetlands. Each zone harbored a unique community of plants and animals as permanent residents and welcomed visitors that stayed for only part of the year. From a bird's-eye view, the rivers created in the alluvium a *water wide web*, a hydraulic network connecting farmers, pastoralists, and marsh dwellers through which they shared resources and information and enhanced the nutritional aggregate available to them. The web spread risks by expanding the spectrum of subsistence options and conferred greater capacity to adapt to unexpected perturbations in the political and natural environments than possible under a mono-cultural system. As a framework, the web concept breaks free from treating the alluvium as a uniform expanse of desert waste, highlighting instead ecological diversity and the success of human opportunism and ingenuity in integrating arid and wet zones along with arable areas into systems of production.[3]

For most of the sixteenth and seventeenth centuries, Ottoman rule arrested the fragmentation of the alluvium into the private domains of autonomous tribal groups and curbed their individual interests. All food producers came under the protection of a single state authority that promised equal treatment under the law. Despite their agrarian preferences, the Ottomans did not seek to fundamentally alter the balance of power between crop farmers and animal herders with large-scale irrigation works. Instead, they focused on profiting from the alluvium's organic endowment by digging deeper into the resources of every biome while offering entrepreneurs incentives to expand cultivation in small increments. Allocating investments among various ecological zones allowed Istanbul to manage risk and weather the volatility of the eastern frontier. The negative performance of arable lands in one year due to an extraordinary flood or pastoral raid, for instance, could be neutralized by the performance of wetlands and grasslands that come through the same perils unscathed. In some respects, the water wide web that helped hold Ottoman rule in Iraq resembled the "vertical archipelago" of the Andes. Farming at different elevations in the mountain system produced complementary crops that provided Andean kingdoms like the Lupaca with insurance against disaster.[4]

Different hydrologic conditions within the alluvium created different opportunities and challenges. As far as the sources permit, these chapters will elucidate those most prevalent in every biome and the strategies that the local inhabitants developed to cope with them. Disease, for example, features in the chapter on the wetlands, where it was a rampant menace. Likewise, hydraulic engineering, a prerequisite for crop cultivation in basin landforms, is a major topic for arable lands. Their differences notwithstanding, alluvial ecologies—arable, dry, and wet—formed a seamless tapestry, stitched into each other through the seasonal rise and fall of the Tigris and Euphrates.

3
Arable Lands

> The date palms of Baghdad lift their crowns to the sky as
> if they are reaching to the Milky Way.
> —Evliya Çelebi, *Seyahatname* (1656)

FAR FROM THE tempering influence of oceans or mountains, the Tigris and Euphrates maintained near monopoly power over one of life's most precious commodities—water—and were the ultimate brokers of virtually every agrarian enterprise.[1] Those who wished to grow crops and settle in Iraq could not sustain their way of life without river water, which came at a high price in a closed marketplace. The price was greater vulnerability and fewer protections from the Tigris and Euphrates, which could inflict late spring floods, random sediment deposition, and unstable river channels. Without alternative sources of water, grain farmers had to succumb to the mischief of both rivers, find other means of earning a living, or leave.

Under Ottoman rule, tens of thousands of farmers stayed and battled on. This chapter pieces together the agrarian order that Istanbul established in the Tigris-Euphrates alluvium. The political and environmental peculiarities of Ottoman Iraq encouraged the development of a composite irrigation landscape. Small-scale irrigation initiatives run by village-based organizations predominated while a handful of large hydraulic enterprises fell under the purview of the Ottoman state. An intuitive approach could interpret the devolution of most irrigation works to local farmers as yet another sign of Ottoman pluralism, flexibility, and accommodation in the early modern era.[2] No less plausible as an explanation is Istanbul's political realism. Ottoman policymakers may have realized that there were diminishing returns to a more interventionist approach to irrigation agriculture in a frontier zone like Iraq; micromanagement could simply make the system overly costly and cumbersome.

Whether dealing with China, India, Mesoamerica, or West Asia, the historiography of irrigation has revolved around questions about scale and management style. Were large hydraulic projects necessary to exploit major rivers in arid environments for irrigation agriculture? Did irrigation networks require top-down bureaucratic control?[3] Before the middle of the eighteenth century in early modern Egypt, small-scale irrigation projects predominated. The Ottoman state devolved authority over their operation to Egyptian peasants, who exercised great influence on the Ottoman management of water resources.[4] This chapter argues that the requirements of perennial irrigation in Iraq, in contrast to the basin system prevalent in Egypt, placed greater demands on Ottoman bureaucratic and financial resources. Different factors shaped the degree of both state and local involvement in the management of the Tigris and Euphrates. Together, cultural, political, and environmental considerations forged a hybrid irrigation landscape, small in scale and local in character, but with bulky elements that conformed to centralized imperial control.

Imperial Imperatives

The Ottoman Empire had to reconcile three imperatives to determine the degree to which it managed Iraq's hydrologic commons. The first was a cultural attitude that it shared with other premodern polities worldwide, dubbed "traditionalism" in the Ottoman context.[5] The Ottoman state fashioned itself as the guardian of the traditional wisdom of society—tried and true laws, values, and procedures. A traditional posture called for Ottoman intervention in water management to follow the example of ancient rulers and thus preserve the "natural" order of things. Süleyman I, for instance, considered the construction of waterworks "a caliphal prerogative alluding to his image as the second Solomon," according to architectural historian Gürlu Necipoğlu. "His name not only commanded the jinns and animals, but also the winds and water."[6] More broadly, Ottoman legal codes frequently invoked the doctrine of precedent when issues of irrigation were under consideration, prefaced by conventional phrases such as "from old times" (*kadim-ül-eyyamdan*), "in the times of former sultans" (*salatin-i maziye zamanlarında*), and "according to ancient custom" (*adet-i kadime üzere*).[7] A comment by Ottoman traveler Evliya Çelebi during his visit to Baghdad in 1656 encapsulates the traditionalist sentiment. A canal project recently undertaken by the provincial governor, he wrote, had made "the land of Iraq more prosperous than it was in the age of the caliphs."[8] Like mosques and shrines, irrigation works buttressed Istanbul's legitimacy in the region by creating a smooth lineage between the glorious past and the Ottoman present.[9]

In a diligent effort to preserve the legacy of ancient Iraq in water management, the Ottoman administration recruited two long-standing corps of hydraulic engineers, rationalizing its decision by highlighting their deep roots in society and service to ancient rulers. Members of the first corps held the title of *sekkar* and were deployed in the Diyala region northeast of Baghdad. Their job was to excavate and clear canals three to four times a year and to repair dams and breaches in riverbanks. In return for their labor, the Ottoman finance director of Baghdad gave corps members a document called *temessük* that entitled them to a percentage of the harvest in the district where they served. Members of the second corps of hydraulic engineers, called *karikh*, specialized in gravity-flow irrigation and were deployed more widely throughout the Baghdad province, reportedly in every single village. They received tax exemptions on the harvest they cultivated for their service.[10]

Esteem for the status quo, therefore, steered the Ottoman Empire toward an active approach to water management, but it restrained ventures to entirely remodel the irrigation infrastructure of Iraq. The desire to preserve preexisting facilities and arrangements took precedence over the potential rewards of transforming them. To implement the traditionalist agenda in the region, Süleyman I, the conqueror of Iraq, commissioned a judge named Kadızade Sheikh Mehmed to compile a report detailing "the laws and rules governing the population in the past during the reign of former sultans known for exercising justice."[11] Based on the report's findings, Süleyman formulated his first law code for Baghdad in 1537, in which he declared his intentions to uphold the rules and customs that governed the province before the arrival of the Shi'i Safavids.

The second imperative for Ottoman hydraulic management in the alluvium was what Ottoman writer Kınalızade Ali Çelebi (d. 1572) first called the "Circle of Justice" (*daire-i adliyye*), a concept of state comparable to the Chinese Mandate of Heaven that had provided a model for good government in West Asia since ancient times. For the Ottoman Empire in particular, the Circle of Justice was central to its political ideology. A shorthand version of it states: "No power without troops, no troops without money, no money without prosperity, and no prosperity without justice and good administration."[12] Seen within this circle, irrigation agriculture was a major source of wealth whose management had a higher moral purpose than personal enrichment. Above all, it empowered the Muslim sovereign to protect his realm, dispense justice, and legitimate his rule.[13]

The Ottoman sultan was therefore ultimately responsible for the welfare of the peasants and their irrigation needs. This sense of royal duty was on full display during the reign of Süleyman I, who portrayed himself as the restorer of justice and prosperity after what he decried as a Safavid dark age in Iraq.[14] One of his most important reforms pertained to lands owned by the state, leaving the small percentage of privately owned properties in the hands of their owners. In a proclamation issued around 1540, he wanted subjects to know that Iraq was open for agrarian enterprise, stating: "Whoever comes and irrigates desolate and vacant lands to bring them under cultivation will be exempted [from taxation] for three years," an extension of the one-year exemption he had initially offered a few years earlier.[15] The Ottoman Imperial Council asked provincial authorities to search in particular for people willing to reclaim canals that had been abandoned.[16] To those who responded positively, the administration awarded possession (*tasarruf*) and usufruct (*istighlal*) rights but retained ownership (*rakaba*). The bargain, available to farmers in Baghdad and Basra through an arrangement called the *tapulama* (leasing under *tapu*), conferred privileges on landowners that ordinary tenants lacked, including transfer and heritable rights and security of tenure as long as contract conditions were observed. Soon after ascending the throne in 1566, Süleyman's successor Selim II added an important element to his father's reforms by lowering the annual tax on harvests in the Basra province from one-third to one-fifth.[17]

Tax breaks and a favorable legal framework for private landholding made agricultural development an attractive investment. The Basra province gives the clearest signal of growth in arable production during the early phase of Ottoman rule, witnessing between 1550 and 1590 an upsurge in the harvest estimates for date palm, rice, wheat, and barley. In addition, we can infer the positive response to Ottoman policies from forty extant cases of land reclamation throughout Iraq during the second half of the sixteenth century.[18] In all cases, farmers interested in reviving a particular canal or wasteland had first to obtain a title deed (*tapu*) from the local treasury for a lump sum payment. The acquisition occurred either by formal application or through an auction that granted the title deed to the highest bidder. Once the deed was secured, the deed-holder began reclamation work with his own capital and labor and paid the state a percentage of his annual harvest, after which he received a document called *temellük* confirming his ownership. Both the statistical data on crop harvests and the anecdotal evidence of canal and land reclamation attest to the tangible results of Ottoman agrarian reforms.

Finally, Istanbul's commitment to water management in Iraq had to reckon with the novel map of the early modern Middle East. Süleyman I had completed a radical geopolitical reconfiguration that brought the Middle East's three major imperial heartlands together under one administrative roof: western Anatolia and the southeastern Balkans, Egypt, and Iraq. With its Anatolian, Balkan, and Egyptian territories, the Ottoman administration could easily dispense with Iraq, which lacked its rivals' access to the Mediterranean, proximity to the imperial capital, and water resources, be it rainfall in the case of Anatolia and the Balkans or a streamflow in tune with the agricultural cycle in the case of Egypt. This state of affairs sharply contrasted with the privileged status Iraq had enjoyed under Persian tutelage in classical antiquity. The compact wealth of its alluvial land stood out in a polity poorly endowed in agricultural resources, dispersed in smaller concentrations among inland basins, mountain valleys, and the Caspian Sea littoral. No farming region could vie with Iraq for royal attention in the Persian imperial framework. The Ottoman reconfiguration of the Middle East alienated irrigation agriculture in Iraq by dragging it into an unpropitious competition with far more attractive sites for agricultural investment.[19]

Beyond economic interests, Ottoman water management in Iraq was a balancing act between cultural, ideological, and political considerations. Fidelity to the precedent of ancient states and the desire to fulfill its role as a just

Table 3.1. Production Estimates for Major Crops in Basra, 1552–1590 (*tağar*)

Crop	1552	1590
Wheat	1,455	3,101
Barley	4,064	6,896
Date Palm	10,557	23,435
Rice	4,388	9,524

Note: *Tağar* is a load of grain. Its volume and weight varied greatly depending on the locality, time period, and grain being measured. Around 1540, the Ottoman law code interpreted one tağar in Baghdad as the equivalent of ten *kile* (a more common imperial unit), roughly 55 pounds in sixteenth-century Istanbul. See OA, TT 1028, 13; Halil İnalcık, "Weights and Measures," in *An Economic and Social History of the Ottoman Empire, 1300–1914*, vol. 1, *1300–1600*, ed. Halil İnalcık and Donald Quataert (New York: Cambridge University Press, 1994), xl.
Sources: OA, TT 282; TKG.KK, TT 30.s

government inspired Istanbul to be continuously engaged in matters related to agricultural development and irrigation. On the other hand, caution about change or innovation and the availability of more promising opportunities for agricultural investment around the Mediterranean basin curbed whatever commitments the Ottoman administration could make to Iraq. The calculus would considerably change during the nineteenth century, when Western models came in vogue as sources of inspiration in competition with those from the ancient past and the loss of Egypt and most of the Balkans from the Ottoman Empire, giving greater prominence to Iraq in the eyes of Istanbul.

The Farmer's Imperatives

The local farmer's overriding concerns revolved around water and land, the primary factors of arable production. If Egypt was the gift of the Nile, Iraq was a precarious adaptation to the Tigris and Euphrates. Both rivers gradually rise with the beginning of the rainy season in late fall, until they peak when the spring snowmelt reaches Iraq in April and May. The hydrologic climax occurs when the winter crops need water the least—well after the sowing season and around harvest time—and leaves them most vulnerable to the raging torrents. On the other extreme, the Tigris and Euphrates diminish from June and hit rock bottom in September and October, when the summer crops should be irrigated and the need for water was paramount. For the arable farmer, the flow cycle of the Tigris and Euphrates was intrinsically twisted counterclockwise.[20]

Organisms evolve adaptations to survive and exploit different flow regimes.[21] Humans are no exception. To extract fresh water from the peculiar flow of the Tigris and Euphrates, states and farmers had to work harder than their Egyptian counterparts and engineer a more sophisticated and costly irrigation infrastructure. The overall goal of their individual efforts was creating artificial peaks coinciding with the sowing rather than harvest seasons—synchronization, in other words. The endeavor may seem an overly ambitious modification of natural flow patterns for a pre-industrial society to undertake, but it could be achieved through water-lifting devices and canal diversions, which conducted water to fields when it was most in demand.

Whether to grow wheat or pomegranate, irrigate by canal or jug, or seek the aid of the Ottoman state or shun it, Iraqi farmers had first to contend with the conditions of the soil under their feet. As their slopes flatten out south of the Tikrit-Hit line, the Tigris and Euphrates lose velocity and the power to transport their sediment load, most of which they jettison over Iraq en route

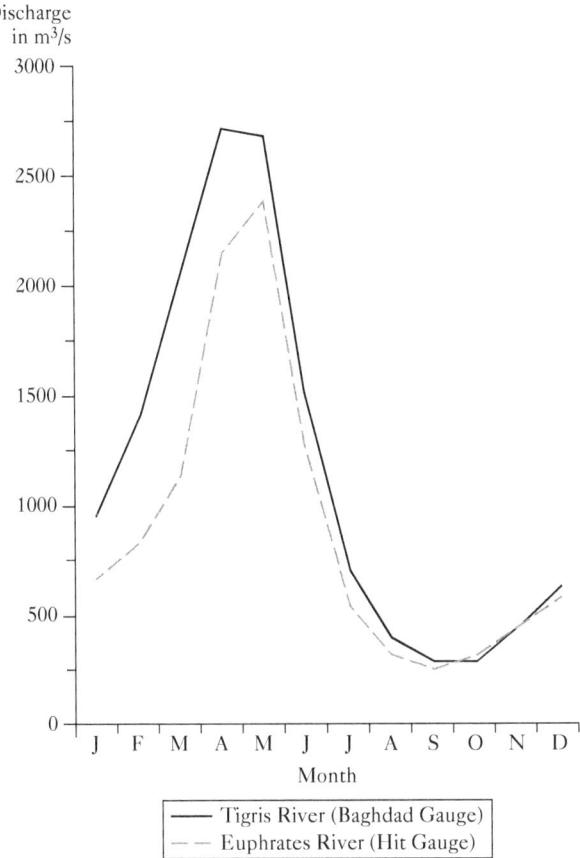

FIGURE 3.1. Mean Monthly Discharge of the Tigris and Euphrates Rivers. After *Tübinger Atlas des Vorderen Orients* (Wiesbaden: L. Reichert, 1977–1993), A V 4.

to the Persian Gulf. By one rough calculation, the rivers deposit 47.5 million metric tons of sediment material over the region every year. If spread evenly over the landscape, this volume would add a sedimentary layer of about 0.82 inch per century.[22] Like all rivers, however, the Tigris and Euphrates never deposit their sediments evenly. They create instead two prominent landforms within the alluvium—the levee and the basin. Each landform had a different soil composition, elevation, and distance from the rivers. The variations between both were subtle, imperceptible to the eye of a casual observer but consequential to the interaction between the farmer and the state.

The Tigris and Euphrates deposit their largest and heaviest sediment particles first adjacent to their channels, building up narrow ridges called levees on both sides of each river. The river levees rise to a height of as much

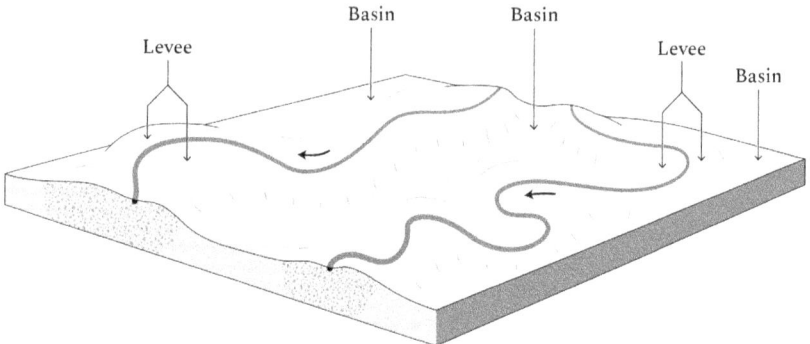

FIGURE 3.2. Diagram of River Levees and River Basins. After Piet Buringh, *Soil and Soil Conditions in Iraq* (Baghdad: Ministry of Agriculture, 1960), 144.

as ten feet above the land surface, making their irrigation difficult without the use of a mechanical device to lift water from the rivers running below. Two ancient lifting devices dominated the irrigation landscape of early modern Iraq. One was the waterwheel, referred to in Ottoman documents as *dolab* or *natule* (from *noria*). They tended to concentrate in the northern river stretches of the alluvium to take the advantage of a stronger current, better capable of turning the wheel. The other dominant lifting device was made out of pulleys and powered by human and animal muscle power. Ottoman officials called it *öküz dolabı*, *charad*, and *dalya*.[23] The device stood out to an English mission to Persia that passed through Iraq in 1599. Farmers on the Tigris and Euphrates, the travelers wrote, "have four bulls yoked together, and a device with a wheel set hard to the river side, with two great ropes; and, at the end of either rope, two long buckets made of the hide of a buffalo, and as the one cometh up the other goeth down, which bringeth up the water, that runneth in little trenches, and watereth the ground, in some places ten miles, some more, some less."[24]

As for crop preferences, farmers dedicated the levees to the lofty date palm, where the tree could keep "its head in the fire, its feet in the water," according to the local saying.[25] Fire came in the form of an intense and sustained summer heat under a cloudless blue sky. Beyond the northern edge of the alluvium, colder winters and higher precipitation levels prevent the fertilization of the tree flowers and destroy the ripening fruits. Its crown bathed in optimal sunshine in the south, the date palm could immerse its roots in water year-round through the porous subsoil of the Tigris-Euphrates levee crests, which keep the tree well drained and safely distant from the saline groundwater.[26]

FIGURE 3.3. Waterwheel on the Euphrates. After Hans H. Boesch, "El-'Iraq," *Economic Geography* 15, no. 4 (1939): 351.

Farmers gave the date palm the levee soil and, in return, obtained dietary and practical value from virtually every part of it. The main prize was in the date fruit, a berry rich in sugar (over 80 percent) and made up of dietary fibers and some protein. Today the fruit is primarily consumed as a sweetmeat, but back then it was also a staple article of diet comparable to bread and potato. For the poor masses, it was their sole food item. Societies near and far held the date of Iraq in high esteem. Ottoman traveler Evliya Çelebi estimated in the middle of the seventeenth century that Iraq had seventy varieties of dates, exported every year in hundreds of ships to India and by hundreds of thousands of camels overland to Hamadan and Isfahan in Safavid Iran.[27] Aside from the berry itself, grinding the date stone produced oil and left behind paste that was used to feed livestock. Locals used the tree's feather-like leaves to build their house roofs and make baskets.[28]

Above all, at the sedimentary throne of the river levees, the date palm served the farmer in its role as a keystone species, on which many other species depended. It could play this special role due to its elevated platform and graceful height. "I have not seen a palm this tall and excellent, neither in Egypt, the Sudan, nor in

FIGURE 3.4. Lift Irrigation by Animal Power (*Öküz Dolabı*). From Verney Lovett Cameron, *Our Future Highway* (London: Macmillan and Co., 1880), 2:250. Courtesy of the Newberry Library.

Tlemcen," Evliya Çelebi wrote of the tree.[29] It grew to a height of sixty to eighty feet, with leaves at the top ranging between ten and twenty feet in length.[30] The local population had long realized that when strategically planted, the giant date palm could lower the surface and air temperatures by providing a natural shade canopy and by releasing the water it absorbs from the ground into the air as water vapor. The date grove, in this respect, served as an artificially designed cooling amenity and heat sink for towns and fields.[31] It allowed farmers to establish multi-story gardens on the levees, comprised of the date palms themselves at the top, smaller fruit trees such as pomegranate and citrus in the middle, and plots of cereals, vegetables, and legumes at the bottom.[32] A British traveler took note of the interplanting of crops when he passed through Basra in the late eighteenth century. "The date-trees being planted about ten feet from each other, and full of leaves at top, afford a very good shade; and the people are enabled to cultivate the ground during the whole day, without suffering much in convenience from the heat of the sun, which out of the shade, and in the middle of the day, is at this season not to be endured," he wrote.[33]

Hand watering the date palms and other crops on the elevated levees did not require complex social organization. Local farmers managed their waterwheels, pulleys, and draft animals without external support. Likewise, the states of ancient Iraq had long eschewed direct involvement in the local management of lift irrigation structures.[34] After their conquest of Iraq in the sixteenth century, the Ottomans came to the same realization. They acknowledged that establishing irrigation projects on landforms raised by the annual flood was impractical. Early Ottoman regulations explicitly stated that such areas were best irrigated by the water lifting devices already in place and run by locals.[35] The Ottoman administration therefore limited its role to monitoring and taxation, imposed on both the irrigation devices and the harvest. In the sixteenth century, for example, it monitored about 150 waterwheels on the Tigris and Euphrates within Iraq and collected an annual charge on their use.[36] As for the monitoring and taxation of levee crops, the Ottoman administration was unusually meticulous in its attention to the date palm. Around 1540, it counted exactly 305,253 palm trees in the Baghdad province alone.[37] If Basra and the Shatt al-Arab region, which had historically boasted a far higher concentration of palm groves naturally irrigated by tidal action, are

FIGURE 3.5. The Hilla Fortress Surrounded by Palm Trees, c. 1534. From Matrakçı Nasuh, "Beyan-ı Menazil-i Sefer-i Irakeyn," İstanbul Üniversitesi Nadir Eserler Kütüphanesi T.5964, ff. 67v–68r. Courtesy of İstanbul Üniversitesi Nadir Eserler Kütüphanesi.

included, date palm numbers across the Tigris-Euphrates alluvium may have run into several millions.[38] The Ottoman administration taxed the annual date harvest by weight and the sale of dates by basket, generating nearly 4 percent of Iraq's total revenue potential in the sixteenth century.[39]

Thus, farmers relied heavily on water-lifting devices to bring under cultivation the levees of the Tigris and Euphrates, zones they largely devoted to date palm horticulture. Levees offered excellent soil and proximity to the rivers, and palms gave in return incalculable economic, nutritional, and practical benefits that no other plant could offer. The tree reshaped life around it by providing an auspicious micro-climate, cooler and more humid than the rest of the country, for the cultivation of other fruits and vegetables. Neither orange nor apricot could have proliferated in Iraq without the aid of palms. The Ottoman administration laid its claim to the levees not through irrigation engineers but through clerks and scribes. They monitored the farmer and collected the tax from the irrigation structures and the harvest. The arrangement suited both parties and contributed to the development of one of the largest date groves in the world, a lasting aesthetic heritage, and a setting for folktales and songs.

After depositing the heaviest sediments nearby to form levees parallel to their channels, the Tigris and Euphrates scattered the lighter fine clay particles farther away, creating throughout the alluvium natural depressions called basins. Here the farmer had to deal with a set of concerns different from the elevation hurdle encountered on the levee crests. More than anything else, distance from the rivers and high levee barriers complicated efforts to irrigate basin landforms and make them productive. Until the nineteenth century, Egyptian farmers faced a similar dilemma but could count on the annual Nile flood, which naturally irrigated their basins on schedule and prepared them for the sowing of winter crops. With the late flood peak of the Tigris and Euphrates, the Egyptian system of basin irrigation was out of the question in Iraq. Instead, farmers relied for millennia on perennial irrigation to expand the frontiers of cultivation into alluvial basins. The system was highly elaborate, dependent on a network of branching canals that deflected the main flow to agricultural units behind the levees. Canal construction entailed the installment of additional water-control devices, a combination of weirs, dams, and sluices that assisted in raising and diverting the rivers into the distributary canal intakes during the low flow period.[40]

The flat alluvial plain formed a great stumbling block to canal excavation. Much of it slopes at very low gradients (between 1:5,000 and 1:10,000), severely constraining the efforts of basin farmers to establish an irrigation canal at a higher gradient (ideally about 1:1,000) to avoid excessive sedimentation. Farmers had two options to overcome this engineering conundrum. The cheaper and more popular one involved channeling water down the levee slopes, which provided the necessary steep gradient for flow. Farmers simply needed to create a crevasse in the levee wall (*kharq* in Ottoman jargon) to divert the flow and a short canal to carefully manage the distribution of water between fields and gardens. Because their flow hinged on the steep levee, the crevasse canals could not travel too far from it, averaging less than two miles in length. Their excavation and subsequent maintenance fell within the capacities and resources of small family farm units.[41]

Farmers, as a result, independently managed their basin fields when these fields were irrigated by short crevasse canals. Mediating their relationship with the Ottoman state were the holders of the tapu title deeds who were ultimately responsible for the maintenance of the irrigation canals and the payment of a negotiated share of the annual harvest to the treasury.[42] Compared to the minimal role they played on levee tops, Ottoman officials were more active along the crevasse canals of basin depressions, every year deploying to them corps of excavators to lend a hand in canal clearance efforts.[43]

The other engineering solution to the alluvium's flat surface was extravagant and less common. It broke away from the geographical limits of the levee by constructing irrigation canals that followed the grade of the alluvium itself. But to achieve a slope sufficiently steep and conductive to flow on the alluvium's gentle gradient, this canal formula had to be long, its tail end far distant from the head gate upstream. Such canals were too large and too complex to be managed at the local level. Agricultural communities along them, therefore, had to seek direct intervention from the Ottoman state, counting on its deep pockets and administrative apparatus to execute the regular maintenance work.

From the sixteenth century, the Ottoman administration assumed direct responsibility for three canals of this class: one on the Tigris and two on the Euphrates.[44] Most important among them was the Dujayl (Little Tigris), an ancient canal derived from the Tigris below Samarra and terminating in Baghdad's northwestern suburbs.[45] Under Ottoman rule, the Dujayl was the primary source of water for an entire district by the same name, the most profitable farming region on the Tigris after Diyala northeast of Baghdad.[46] The two other canals were smaller and on the Euphrates. The first, called the Shahi Canal, served Najaf and was inherited from the Safavids, the rulers of Iraq before the Ottomans.[47] The second, called the Süleymani Canal, served Karbala and was built by Süleyman I after his conquest of the region in 1534.[48]

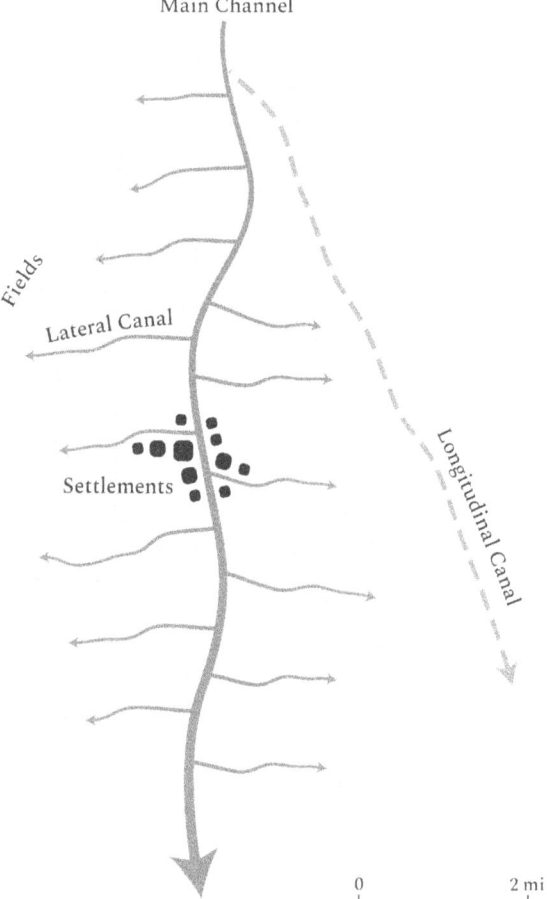

FIGURE 3.6. Layout of Large Longitudinal and Short Lateral Canals. After Tony J. Wilkinson, "Hydraulic Landscapes and Irrigation Systems of Sumer," in *The Sumerian World*, ed. Harriet E. W. Crawford (New York: Routledge, 2013), 44.

The Ottoman state managed both canals primarily to accomplish a religious goal—providing fresh water to the shrines of the Prophet Muhammad's cousin Ali (d. 661) in Najaf and his grandson Husayn (d. 680) in Karbala. But the canals had great agricultural significance as well. Hundreds of rural households relied on them for settlement and cultivation, and they generated considerable revenues for the Ottoman treasury.

In coordination with Istanbul, the provincial bureaucracy in Baghdad went to great lengths to manage these three massive canals. In particular, it had to regularly intervene and dredge the canal beds, which were chronically filled with silt due to their large size and low gradient. The Ottoman

Arable Lands

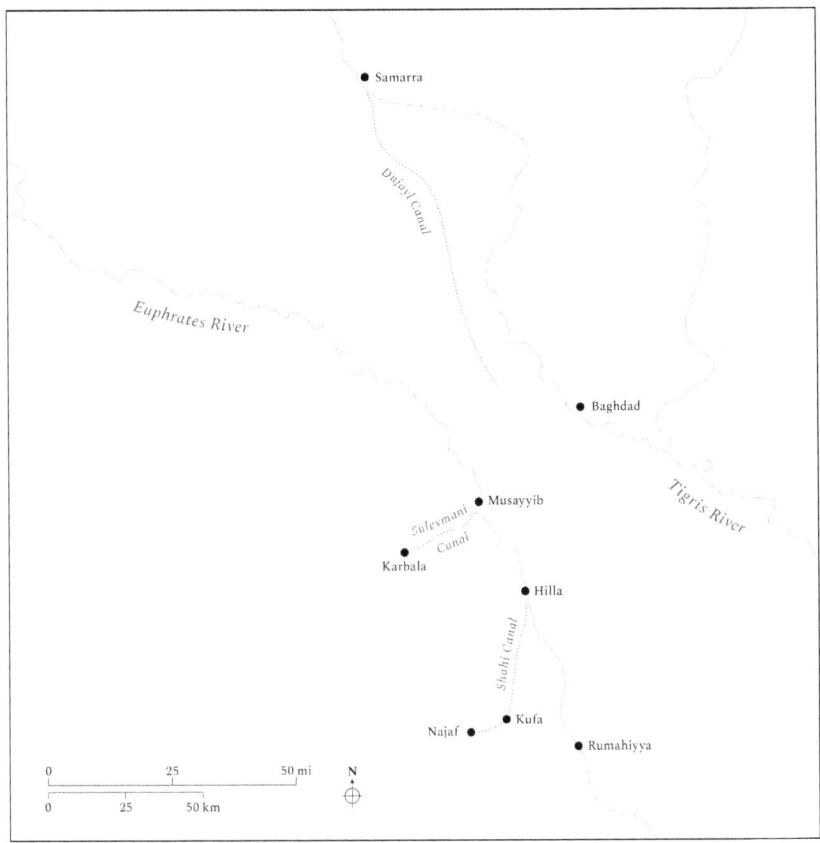

FIGURE 3.7. Major Ottoman Canals in the Tigris-Euphrates Alluvium.

Table 3.2. Large Canals Managed by the Ottoman Administration, c. 1560–1580

	Dujayl Canal	Shahi Canal	Süleymani Canal
Source	Tigris	Euphrates	Euphrates
Serviced Area	Dujayl District	Najaf	Karbala
Length (miles)	60	25	10
Settlements	38	85	8
Households	1,177	1,754	352
Revenues (akçe)	1,446,963	1,164,552	477,229

Notes: Akçe is a small Ottoman silver coin; canal lengths are approximate.
Sources: TKG.KK, TT 29, ff. 119v–140v; TKG.KK, TT 379, ff. 22v–29v; OA, TT 386, 37–85.

administration financed and recruited the necessary manpower to keep the problem in check, carrying out large-scale sediment clearance efforts that a single farming community could not accomplish.[49] On the Dujayl Canal, for example, it recruited 900 men from villages throughout Iraq for an annual forty-day dredging campaign.[50] When specialized expertise was necessary, the Imperial Council in Istanbul intervened to provide it. In 1574 and 1579, for instance, it ordered Diyarbakır's governor to dispatch a sufficient number of master workers in stone (*ustad taşçı*) to Iraq to repair the canals' clogged head gates.[51] Other measures the state directed include appointing superintendents to oversee irrigation works, establishing workshops tasked with the canals' regular maintenance, and constructing tertiary branches to expand the reach of the canals.[52]

Choosing what to grow in basin landforms was a simpler matter than figuring out how to irrigate them. Farmers pushed the cultivation of cereal crops down to the basins because they demanded less water than the levee crops. Like their Sumerian predecessors, early modern farmers focused on the cultivation of wheat and barley. In the sixteenth century, these two crops alone generated for the Ottoman treasury 18 percent of the alluvium's entire estimated revenue. The dry northern part of the alluvium was far more productive in both cereal crops than the marshy south, where date palm and rice stifled the expansion of competing grains. Between the two grains, barley reigned supreme throughout the region. Its high salt tolerance and suitability as a fodder crop particularly appealed to the predominantly mixed farmers in the region.[53]

In their agricultural practices, then, Iraqi farmers configured their agrarian landscape around two fundamental natural forces that had shaped agriculture in the region for millennia—flooding and sedimentation. Their early modern strategies, as a logical result, drew on the proven methods and experiences of earlier generations that had to deal with the same fluvial processes. Sediment deposition by the Tigris and Euphrates divided the alluvial plain into two major agricultural landforms—the levee and the basin. The local inhabitants relied on uplift irrigation to turn levees into flourishing orchards and kitchen gardens and on perennial canals to turn basins into fields of wheat and barley. Geography limited their options to two canal designs—a short one built on the steep slope of the levee, and a large one on the gentle slope of the alluvium. Local farm units took charge of the short levee canals and interacted with the state for additional support and recordkeeping. For the large canals that scarred the surface of the alluvium, farmers forfeited their autonomous mode of irrigation and invited the state to step in. Regardless of the canal formula adopted, farmers devoted basin landforms to the most lucrative cereal crops, particularly wheat and barley. In its revenue-generating capacity, therefore, a river basin was the premodern agrarian equivalent of an oil well. Whether large or small, operated at the local or state level, it fell under the purview of the Ottoman administration that monitored and taxed its harvest.

Be it in taxation, public administration, or justice, Ottoman rule from one province to another exhibited significant differences shaped by local conditions.[54] The same held true for Ottoman irrigation policy in different riverine environments. In Egypt, the Nile's flow regime was in perfect harmony with the needs of the winter crops and required basic management routines that largely operated at the local level. In Iraq, on the other hand, the Ottoman Empire encountered a more artificial system of irrigation that constantly twisted the arms of the Tigris and Euphrates into the needs of the crop farmer. By slowing or impeding the rivers' flow through canal diversions, perennial irrigation in Iraq intensified the accumulation of fluvial silt and clay and required constant intervention, by farmers and by the state.

The Ottoman Empire was committed to maintaining and amplifying age-old arrangements to exploit the Tigris and Euphrates for irrigation agriculture, thereby striking a balance between its economic, cultural, and political priorities. It did not pursue a uniform approach throughout the alluvium. Instead, it tailored its irrigation policy to the needs of different landforms within the region. On levees irrigated by water lifting devices, its intervention was least important. It therefore devolved the duties of irrigation management to local farmers, limiting its role to monitoring and tax collection. In basins irrigated by short canals built on levee slopes, it was more active and provided support to dredge the canal beds. The most centralizing features of Ottoman policy appeared in basins irrigated by long canals built on the alluvium's gradient, which no single farm unit could manage on its own. The Ottoman administration assumed full responsibility for three long canals on the Tigris and Euphrates, directly managing them through appointed officials who oversaw large-scale dredging efforts, the construction of tertiary branches, and the staffing of maintenance facilities.

The inherent vulnerability of perennial irrigation, compared to the basin variety on the Nile, stems from its highly interventionist approach to the natural flow regime. The water diverted by the system's canals, regardless of their size, is never proportional to the diversion of the sediment load carried in from the rocky uplands. Constant human intervention is critical to artificially maintain the necessary balance between the twin supplies of water and sediment in any given canal. In addition, human disturbance of the natural hydrologic cycle with canal diversions inadvertently caused a breakdown in the movement of salt, transferred together with silt by streamflow from the sedimentary rocks in the north. The accumulation of salts in the soil, or salinization, obstructed the absorption of water and key nutrients by crop plants and caused far-reaching damage to agricultural yields.[55]

Because of the deep instability of perennial irrigation, arable production in Iraq, whether directed by state or local effort, always remained "a system of low return and high risk" to the individual farmer.[56] More environmentally sustainable were subsistence strategies that adapted to the natural flow regime instead of seeking to modify it.

4
Grasslands

> In the economy of nature the natural vegetation has its
> essential place.
> —Rachel Carson, *Silent Spring* (1962)

OUTSIDE THE SETTLED, arable clusters watered by the Tigris and Euphrates lay the drier parts of their alluvium, where biological activity was low and the stock of organisms limited. A variegated layer of native grasses and flowers fluctuated seasonally and spread thinly across the open river plain, in sharp contrast to the densely packed stands of cereal crops along the river channels. The diversity, transience, and dispersal of this vegetation cover complicated the monitoring and assessment of the court scribe and defied monopolization and rationing by any central authority. The inedibility of grass compounded its volatility and minimized the rewards of its appropriation.

Rural groups chiefly based on a family nucleus filled the void left by the functionaries of urban bureaucracy. After the last glaciation some 10,000 years ago, farmers gradually devised and honed different strategies designed to chase and exploit this frustratingly intractable flora—notably, through the intermediacy of gregarious, migratory mammalian herbivores.[1] Over millennia, the husbandry and herding of ruminant animals, or pastoralism, evolved as the most viable economic activity in zones lacking irrigation and sufficient rainfall and functioned as a rapid response system, finely tuned to seize hold of grass.[2]

Crisscrossed by two large river systems, Iraq was an attractive destination for livestock in an endless pursuit of grass. When the pastures of neighboring regions burned or shriveled, those watered by the Tigris and Euphrates remained lush and green. This chapter examines the role of the alluvium as a major source of pasture and as a pastoral hub in West Asia. The pastoral economy sustained by river water fell under the purview of the expansive Ottoman Empire following the eastern campaigns of Süleyman I in the 1530s. The Ottoman administration thereafter actively intervened to exercise greater control over those who grazed their livestock along the riverbanks in different times of the year. If Istanbul imprinted arable lands with canals and embankments, its intervention took a less tangible form in the grasslands of the Tigris and Euphrates. It involved the construction of new social structures—namely, large herders' associations—under which Ottoman officials regrouped small and autonomous pastoral tribes. Through this kind of social aggregation, the Ottoman state found a way around the impregnable walls of tribalism, mobility, and dispersal that had traditionally shielded the pastoral economy from state appropriation.[3]

Agricultural systems worldwide have traditionally combined the growing of crops and the raising of livestock, with variations in the emphasis within the mix from one place to another. The field of Ottoman agricultural history, nevertheless, remains uneven. It shows arable lands tilled by peasant households (çift-hane) in the foreground, while obscuring the world of animal husbandry.[4] Even in studies where pastoralists stand front and center, less is written about their contributions to Ottoman agriculture and more about their other roles, as warriors, craftsmen, purveyors of pack animals, or migrant workers.[5] In the words of one historian: "The Ottoman dream of a sedentary paradise with its regular, predictable revenues from pacific farmers had no place for pastoral nomads."[6] This chapter will not question the fact that the Ottoman state did have an agrarian dream. Instead, it shows that Ottoman agrarian predilections could be tempered with realism in a landscape well suited to animal rearing like Iraq, giving a free rein to a pastoral engine of economic development no less important than arable production.

Outside the agricultural cocoon of the Tigris and Euphrates, this story illuminates the political significance of mobile pastoralism as an instrument for frontier expansion in the late medieval and early modern world. Amid the violence and instability prevalent in frontier regions, resilience marked the specialized keepers of livestock. They successfully exploited zones of high risk by keeping their wealth stored in movable assets. With their sheep and cattle, they could move ten miles a day if necessary to get away from approaching armies, while the fields of arable farmers were left vulnerable to seizure and burning by soldiers.[7] Pastoral resilience drew the attention of different agrarian states in need of a reliable ally in the quest for political expansion.

Frontier regions, as a result, became the birthplace of remarkable alliances between urban powers and pastoral groups. After the political fragmentation of Muslim Andalusia in the early eleventh century, for instance, Christian Castile promoted large-scale sheep herding, sometimes with military escorts, in its southward push toward the pasture-rich tableland of inner Iberia.[8] Seventeenth-century New England and the Chesapeake provide a comparable case study. Surrounded by abundant land but short of labor, English farmers adopted a free-range system of animal husbandry that turned into an engine for colonial expansion, whereby roaming domestic livestock established English rights to more territory without regard to Native American claims.[9] Likewise, from eighteenth-century Cape Town, the influence of the Dutch East India Company and the Dutch Reformed Church spread into the South African interior on the heels of white pastoralists (*trekboers*), who acquired their livestock and herding skills from the indigenous Khoikhoi people, only to displace them later.[10] To this global pattern, this chapter adds the story of a forgotten partnership struck between urban bureaucrats and rural shepherds along the Ottoman-Safavid frontier in Iraq.

The Grass Factory

"Men and oxen exchange work," Thoreau wrote in *Walden*.[11] Pastoralism, in this Thoreauvian logic, denotes a relationship of mutual give and take. Under the crook, herbivores curtailed their natural instincts and modified their social ties to form a more manageable herd and push forward the frontiers of food production. In return, herbivores received the attention and devotion of a good shepherd, who had to cater to their numerous behavioral and physiological needs. Pastoral care, Thoreau argued, could be backbreaking, turning men into de facto beasts of burden at their oxen's service. "I am wont to think that men are not so much the keepers of herds as herds are the keepers of men," he quipped; "the former are so much freer."[12]

To fulfill his part of the unequal exchange, the herder had first and foremost to resolve the tension between the seasonality of plant growth and the daily needs of the livestock for fodder. How could he provide a steady supply of forage for his sheep and cattle when forage growth rates varied throughout the year? Modern farming answers this basic, yet fundamental question with forage conservation techniques such as haymaking and ensiling, which harvest and store grass in times of plenty and make it available to livestock when fresh grass is in short supply. This way, Dutch dairy farmers and Texas cattle ranchers can even out variations in forage availability across the seasons and raise their livestock in permanent housing facilities, with ready access to the milking parlor or the slaughterhouse.[13]

The answer offered by modern systems of intensive livestock farming was not available before the mechanization of agriculture and without heavy capital investment in equipment, buildings, and synthetic farm chemicals. Most pastoralists worldwide, instead, had historically opted for mobility as a strategy to meet the nutritional requirements of their livestock, moving between the best and most accessible pastures around the year.[14] Mobility offered pastoralists a continuous supply of grass in its natural, fresh condition but tied them to the whims of temperature and precipitation patterns that govern grass growth. The bargain seemed fair in parts of the world with a high, evenly distributed rainfall. In other regions that experience seasons of intense heat and drought, like the subtropics, it was a more difficult proposition.[15]

West Asia typifies the climatic challenges faced by mobile pastoralists in subtropical latitudes. Finding grazing ground was relatively easy during the mild winter and spring months, when the skies released most of their annual precipitation total and gave rise to abundant and reliable pastures. Pastoralists' gains, nonetheless, quickly evaporated with the onset of the summer. Rising temperature and aridity rolled back the winter-spring range and the boundaries of pastoral activity. Amid an expanse of parched land, herders sought refuge in green pockets that could survive the prolonged dry spell of summer and early fall. The most prominent among those pockets of grass were found in mountains. In the southern coastlands of the Black and Caspian seas, high pastures were spared regional drought thanks to year-round orographic lifting, the mechanism by which mountains force moist air to rise above its condensation level and trigger precipitation. Drier mountain ranges, like the Taurus and Zagros, relied less on rainfall and more on snowpack to conserve their summer pastures, conveniently made available for forage when the ice melted away under soaring temperature.[16]

But highlands were not the sole natural factories of summer grass in West Asia. In the flat plains south of the rocky Perso-Anatolian belt, the Tigris-Euphrates system aptly filled the vacuum left by mountains. Every summer, it generated a verdant range through a flow regime no less impressive than the atmospheric lifting mechanisms that predominated in the north.

The Hydro-Logic of Mobility

On his way to Baghdad, German doctor and botanist Leonhard Rauwolff spent nearly two months in the spring of 1574 sailing down the Euphrates

River. The journey afforded him time to observe the aquatic vegetation as well as life in the desert. "But it is no wonder that the *Arabians* are so restless," he wrote, "for they are full of Want and Nakedness, have not to fill their Belly, nor to cover their Body withal; besides, they have nothing else to do, and are used to idleness from their very infancy, and then because they hate to Work, they are forced to wander like Vagabonds from one place to another."[17] The desert Arabs, according to this working theory of mobile pastoralism, were "so restless" and mobile simply because they were lazy, and given their abject poverty, they had nothing to lose by wasting their time wandering as vagabonds. Coming from an agrarian country with no mobile pastoral traditions and observing at a distance from a moving deck, Dr. Rauwolff was oblivious to the rules and aims underlying movement along the busy sheepwalks of the region.

Not to be confused with wandering, pastoral mobility, along with its associated tents, hearths, and dunghills, was an organized social scheme regulated in Iraq by the hydrologic patterns of the Tigris and Euphrates. The annual winter and spring floods acted as a centrifugal force on pastoralists, forcing them to withdraw their herds to the safety of dry plains, deserts, and foothills around them. Around this time of the year, those dry zones reached their peak in forage production due to recent bursts of rainfall and favorable temperature, making them ready to welcome the herds approaching from the river valley. Toward the west, for instance, the Syrian desert would be carpeted with bulbous bluegrass, wormwood, and gaily colored yarrows. South toward Arabia, thatching grass, medick, and the vine of colocynth developed. Pastoralists heading to the Zagros foothills north and east would encounter on their way Mediterranean hairgrass and needleleaf sedge, dotted by crimson Asian buttercup. Desolate plains between the Tigris and Euphrates harbored more hairgrass as well as camelthorn and milkvetch. In each direction, the distance traversed by pastoralists to graze their livestock depended on the rain— the drier the year, the poorer the pastures, the harder the flocks were pressed to move away from the riverbanks.[18]

When the Tigris-Euphrates flood subsided and most desert vegetation burned in the summer and fall, the rivers beckoned shepherds and flocks from all directions to converge back around the green pastures that their receding waters had left behind. Around the riverbanks, willow, poplar, tamarisk, and lotus sweetjuice proliferated. Knotweed and foxtail pricklegrass occupied recently exposed mud flats. The drying wells of the marshes made dense associations of common reed and southern cattail more accessible to livestock. In saline depressions, salt-resistant and salt-tolerant plants thrived, including alkali sandspurry, Indian walnut, and mamoncillo. Even on the surface of flowing water, colonies of watermoss and water snowflake floated and could land on any of the riverbanks during their journey to the Persian Gulf. This is just a small sample of a much larger mass of riverine plants that formed a safety net for ruminants during the hottest and driest months of the year.[19]

Just like that of settled cultivation, the demanding routine of mobile pastoralism in Iraq was finely tuned to the Tigris-Euphrates flow regime. Nearly two centuries after Dr. Rauwolff's boat trip, a British East India Company servant traveled along the Euphrates on horseback and offered a more sensible view of the pastoral system, highlighting the fluvial force that poured into its vein. Spring is the season, Eyles Irwin wrote in 1781, "when the expected rising of the Euphrates and Tygris, drives the Arabs into the desert, to seek for pasture for their flocks. It is so designed by Providence, that the waste we have travelled over, shall become verdant in many places in another month. When the summer heats have burnt up the grass, the tribes return to the river, which has by this time shrunk to his former bounds." "This is the only variation in their pastoral lives," Irwin continued. "War and bloodshed give a different color to their political ones."[20]

Meadow Transhumance

A pastoral system designed around a river flow regime suited Iraq's flat alluvial plain. With minimal difference in altitude, the landscape could not support transhumance in the classical sense, summarized by the Spanish verse: "The valleys in the winter, the summits in the summer, are whitened by the flocks as if they were covered with snow."[21] West Asian pastoralists in the mountainous north could sing a similar Mediterranean verse as they moved vertically between rain-fed sub-montane plains and the high summits of the Taurus-Zagros belt. In the southern lowlands, on the other hand, the Tigris and Euphrates defined the rhythm of pastoral movement and enclosed herders and flocks within a horizontal universe, a patchwork of sand and grass, wetland and dryland constantly rearranged by the rivers' rise and fall. The river-centered pastoral system is not unique to Iraq.[22] In the Indus basin, a historian called it "circulation with flocks."[23] Geographers working on the Great Hungarian Plain, the *Alföld*, use the term "meadow transhumance" to describe it.[24]

The Tigris and Euphrates have a special knack for supporting meadow transhumance. Unlike Egypt, protected by natural borders within the Nile Valley, the Tigris-Euphrates alluvium is geographically open to intrusion on

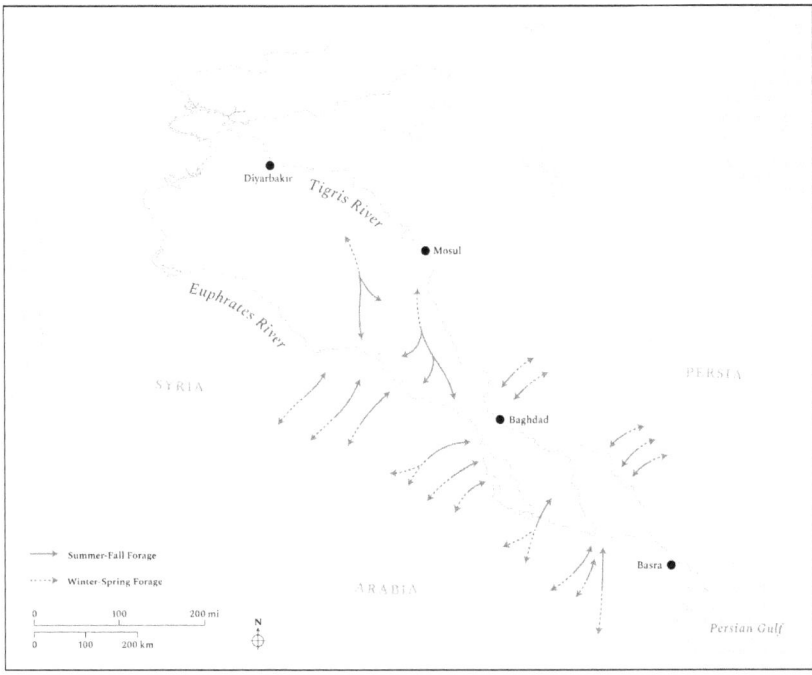

FIGURE 4.1. The Movement of Shepherds in the Tigris-Euphrates Alluvium. After Grahame Williamson, "Iraqi Livestock," *Empire Journal of Experimental Agriculture* 17 (1949): 51.

all sides by mobile pastoralists coming from desert and mountain.[25] During the blistering heat of summer, the critical season when forage was a scarce commodity and nearly depleted in the surrounding deserts, the Nile River swelled and closed Egypt's gates to grazing. The subsiding waters of the Tigris and Euphrates, meanwhile, kept the vital summertime pass to Iraq open and bequeathed to pastoralists verdant meadows, on which their flocks could subsist until winter's first rains. The twin rivers, in a sense, naturally transformed West Asia's distinctive geography and hydrology into meadow transhumance.

Pastoralists who moved around the Tigris and Euphrates brought manpower and capital within the reach of Ottoman provincial centers, but created tensions with those who worked the soil in the same orbit. With the calendar of their annual chores organized by the same fluvial force, specialized growers of crops and animals lived cheek by jowl, in closer proximity than in most other places on earth. Sedentary enclaves existed within pastoral zones without broad natural boundaries separating them. Even by the standards of West Asia, where both groups interacted so regularly, "what makes Iraq's case unique," one historian remarked, "is the intensity of the exposure and the swiftness of the oscillation" between settled and mobile societies.[26] Physical proximity, conflicts of interest, and inveterate prejudices frequently descended into wars. The historical record of Iraq abounds with references to clashes between the two groups, in which the Ottoman state often came to the rescue of the settled, tax-paying peasant. In 1574, according to one early report, an Ottoman commander led thirty units against a rogue group of desert Arabs in the vicinity of Basra, managing to hunt down and behead twenty of them.[27]

Recurrent bloody episodes notwithstanding, the region's fluvial landscape promoted a highly diversified system of agriculture in which the interests of crop and livestock farmers regularly overlapped. If the two groups had to coexist for one reason, it was to share the bounty of arable lands, which occupied a critical place in the forage network of Iraq. Different irrigation structures maintained by crop cultivators inadvertently gave rise to different kinds of plant associations that pastoralists could graze their flocks on. Buttercups, for instance, grew near ponds and small creeks dammed off from river channels, while nutgrass and turkey tangle fogfruit proliferated in wet areas around canals and ditches.[28] Cultivators, moreover, purposely devoted arable areas to forage production. Passing by the Euphrates southwest of Baghdad in 1758, a British naval surgeon noted: "All the country about us is neatly cultivated for pasturage, the river being with great care and judgment admitted, or shut out, agreeable to the necessity of the herdsman."[29] Furthermore, every year by June, when the winter crops had already been harvested, the stubble in fallow and waste lands offered an important source of fodder during the summer and fall seasons. Throughout the year, different grasses among the rows of cultivated crops served as animal feed, such as scarlet pimpernel and yellow sweet clover in winter and bindweed and stinkgrass in summer.[30]

For the debt they owed to the cultivator for using his field for pasture, pastoralists paid in cash and kind. Arrangements commonly involved the payment of rent or a share of the livestock's dairy, meat, and textile products. Other pastoralists could offer their pack animals for the cultivator's transportation needs, share breaking news about the world around them, or guard their host's field from potential attacks.[31] Cultivators undoubtedly took into account the value of a coordinated visit by a flock of sheep. Sheep droppings fertilized their fields, and their grazing organically removed noxious vegetation that competed with crops and trees without the use of herbicides or machinery.[32]

Aside from their economic implications, these reciprocities fostered a rare cultural continuum between growers of crops and keepers of livestock in the region. In many parts of the world, the dividing line between the two groups was drawn along ethnic, linguistic, and religious lines—Europeans and Comanches, Russians and Tatars, Chinese and Mongols. In Iraq, on the other hand, both settled and mobile folks often spoke the same language, professed the same religion, and sometimes even claimed the same tribal affiliation. "One brother often resides in a black tent and moves with the herds for the greater portion of the year while the other occupies a reed or mud-brick house and engages in cultivation," according to the late archaeologist Robert Adams.[33] Every winter, a settled tribe could send a pastoral detachment in charge of grazing flocks in better desert pastures, while the rest remained on irrigated fields engaged in the cultivation and harvest of wheat and barley. On return, flocks would graze within the boundaries of tribal land.

The Tigris-Euphrates system was no less critical to mobile pastoralism than it was to irrigation agriculture. In the flat southern plains, it was the star that shaped the fate of pastoralists traveling in its orbit. Like the mountains of the north, the twin rivers generated ample supplies of wild grass—a vital gift during the second half of the year—and structured the pattern of pastoral movements along horizontal axes, giving rise to the system of meadow transhumance. While supportive of both, the Tigris and Euphrates favored a pastoral way of life over settled cultivation with a natural flow regime better synchronized with the needs of mammalian herbivores than those of winter grain crops. The average Iraqi pastoralist, in other words, lived an easier life and enjoyed greater security from natural disaster than the average Iraqi peasant. This is, perhaps, why Dr. Rauwolff claimed that the desert Arabs "hate to Work," largely because they did not have to toil away as much as those who tilled the land. Common complaints about the "ill-timed" Tigris-Euphrates flood are thus misleading, for they reflect the view of the grain farmer, not the shepherd.

Bleat in the Air

In the desert tracts not reached by the Tigris and Euphrates, between the ruins of ancient cities and villages, only a few deadly animals could survive the fierce competition over scarce and erratic resources—the lion, jackal, osprey, and hyena.[34] Remarkably, the most abundant species to thrive among those titans was sheep. Morphological and behavioral adaptations have given this modest herbivore a head start in extreme habitats, as cold as the snowy hills of Scotland and as hot as the scorching deserts of Iraq. Stuffed with proteins and amino acids, sheep's hydrophilic, woolen coats insulate them against heat and cold through tiny pores that ventilate their bodies and trap moisture between fibers, effectively operating as a built-in thermal regulator. Sheep breeds of arid West Asia, moreover, responded to the limited supplies of food and water by increasing their reserves of body fat. In camels, those reserves come in the shape of humps. Sheep, meanwhile, concentrate their fat deposits around their tails. Fat tails, like fat humps, made their bearers better adapted than cattle to long-distance migration under the heat of the sun and away from rivers and marshes. As social beings, sheep are less disposed to fighting than pigs and easier to manage in large numbers. Making them even more amenable to large-scale management is their fondness of company and highly developed flocking instinct behind a bellwether (often a black goat). Humankind has benefited greatly from such forms of spatial, social, and emotional intelligence in sheep, which ironically have become proverbial for stupidity and lack of individuality in contemporary culture.[35]

Through these morphological and behavioral adaptations, sheep triumphed in the hierarchy of life, becoming the most important mammal in numeric and economic terms for most of Iraq's recorded history.[36] In the second half of the sixteenth century, Ottoman surveyors accounted for some 1.1 million caprids (mostly sheep, with a small proportion of goats) throughout the region, generating 1.4 million akçe. To put these figures into perspective, the population size and revenue potential of caprids far outstripped those of the bovid population combined (cattle and water buffalo). Nearly 97 percent of the recorded sheep and goats was owned by groups based in the drier northern alluvium, where larger herds thrived and enjoyed a greater range of movement than possible in the marshy delta plain of Basra in the south.

Early modern Iraq was a land of sheep enthusiasts. The vast majority of rural households owned a flock of sheep, usually mixed with some goats. Settled households recruited family members to graze their animals daily on fallow fields and young shoots within the village. If local pastures were insufficient, herders had to travel considerable distances. Another option available to small farmers was to entrust a professional shepherd to tend to their flocks when nearby pastures ran low. A large number of Old Babylonian contracts between sheep owners and professional herders attest to the prevalence of this practice in ancient Iraq.[37] Sixteenth-century Ottoman cadasters hint at a similar arrangement. Several specialized shepherds and water buffalo herders appear living among the residents of several neighborhoods in

Table 4.1. Major Herbivores of the Tigris-Euphrates Alluvium, c. 1552–1580

Herbivore	Tax Potential (akçe)	Population Size
Caprid	1,409,277	1,149,597
Water Buffalo	794,974	41,532
Cattle	202,153	25,452

Sources: OA, TT 282; TKG.KK, TT 29.

Baghdad, where they could easily sell their services to their wealthy urban neighbors who owned sheep as a form of capital.[38] If seasonal employment was indeed what made those professional pastoralists appear in the crowded neighborhoods of Baghdad, it would confirm earlier precedent and highlight another aspect of mobile pastoralism in the region—as a form of wealth management service for rich clients.

No mammalian herbivore was as effective as sheep in surviving the harsh environmental conditions of Iraq and in unlocking the energy stored in its green mantle. Despite their agrarian biases, Ottoman tax collectors could not turn a deaf ear in a landscape that resounded with the mournful bleat of sheep. The standards and procedures they followed in a dense field of standing grains, nonetheless, proved ineffective to account for this mammal. The sheep was, needless to say, an entirely different beast.

Royal Herds

Architects of fiscal policy worldwide grappled with the mobility and dispersal of sheep herds. Before the age of aerial surveys, what could they do to keep track of an animal population dotting an expansive landscape? A strategy that gained prominence in the Mediterranean basin involved aggregating small herding groups into larger units with a clear set of duties, privileges, and reporting procedures. In the Extremadura and La Mancha south of Madrid, for example, the Kingdom of Castile maintained the *mesta*, a national guild of sheep owners active between the thirteenth and nineteenth centuries.[39] Between the mountains of the Abruzzi and the plains of Puglia in southeastern Italy, the Kingdom of Naples established a similar, yet smaller, guild known as the *dogana* of Foggia (1447–1806).[40]

In the grasslands of Iraq, the Ottoman Empire followed a comparable Mediterranean strategy of pastoral aggregation and established three

Table 4.2. The Ahşamat

Year	1540	1544	1580
Households	—	9,550	12,134
Sheep Population	—	721,963	728,500
Cattle Population	—	3,135	15,273
Tax Potential (akçe)	284,996	1,160,175	1,337,225

Sources: OA, TT 1028; TKG.KK, TT 228; TKG.KK, TT 29.

herders' associations. It called the largest among them the Ahşamat, an association of Arab, Turkmen, and Kurdish tribes specialized in sheep herding.[41] The group's name is derived from the Persian word *hasham*, meaning attendant, and by extension attendant of livestock or nomad.[42] The Ahşamat association operated as a migratory megapolis with more than 12,000 households in 1580, nearly double the number of households living within the walls of Baghdad and Basra combined. They worked in seventy-three regiments in charge of grazing nearly 700,000 sheep, with an official called the "shepherd leader" (*çoban beği*) and later "Ahşamat commander" (*Ahşamat ağası*) serving as points of contact between state and shepherds. By all indications, the Ahşamat experienced relentless expansion throughout the sixteenth century.

Within the alluvium, Ottoman authorities gave official status to two other groups of sheep herders. The first was the Qara Ulus (Black People or Nation), a predominantly Kurdish confederation based in Anatolia.[43] In the sixteenth century, about 4,500 Qara Ulus households came from the north to pasture some 120,000 sheep in Iraq. Smaller in size was a group based along the Euphrates west of Baghdad called the Qara'ul, a Mongolian word meaning "vanguard" or "scout," indicative of the role member tribes played during the Mongol period.[44] Together, the Ahşamat, Qara Ulus, and Qara'ul constituted a quarter of Iraq's population and were estimated to generate 5 percent of the region's tax revenues during the second half of the sixteenth century.

Joining any of those Ottoman clubs came with benefits for shepherds. One was greater freedom of movement between different provincial jurisdictions within the royal patrimony. This issue particularly mattered to members of the Qara Ulus, who needed to descend southward into Iraq seasonally without major bureaucratic hurdles. Membership in the herders' associations,

Table 4.3. The Qara Ulus

Year	1544
Households	4,516
Sheep Population	121,170
Tax Potential (akçe)	258,727

Sources: TKG.KK, TT 228, fol. 45r.

Table 4.4. The Qara'ul

Year	1540	1544	1580
Households	681	—	—
Cattle Population	500	—	—
Caprid Population	8,173	—	—
Tax Potential (akçe)	62,102	198,475	175,817

Sources: OA, TT 1028, 108; TKG.KK, TT 228, fol. 8r; TKG.KK, TT 29, ff. 36v–40v.

moreover, granted royal protection. If they came under attack, servants of the Ottoman sultan in the provinces had to intervene. In 1678, for example, an autonomous Arab tribe east of the Tigris managed to murder the commander of the Ahşamat. In response, the Ottoman governor in Baghdad dispatched a force of 4,000–5,000 cavalries that exacted revenge on the assailants. The firm response sent a message to the public that the safety of the sultan's shepherds was inviolable.[45] Shepherds in the official associations also enjoyed royal justice. Just as it appointed a "sea judge" who traveled with the grand admiral of the Mediterranean fleet, Istanbul appointed what could be described as a grassland judge, migrating with the Qara Ulus seasonally. The Ottoman judge dispensed justice wherever the shepherds pastured their flocks and served as their point of contact with the central government.[46]

Like the Castilian mesta and Neapolitan dogana, the Ottoman herders' associations in Iraq rationalized the production of a vagrant resource like sheep. By placing a bewildering welter of groups into three uniform administrative containers, the Ottoman state could observe, record, count, and tax a large segment of its herding population.[47] Members of the Ahşamat, Qara Ulus, and Qara'ul surrendered their autonomy to the Ottoman state but balanced this sacrifice with the privileges of royal patronage.

Sheep Supplements

Seldom were the shepherds of Iraq purists about sheep. Instead they oriented their mobility in support of other productive activities. Most often, they cultivated the soil of marginal lands. Even specialized shepherds affiliated with the Ahşamat and Qara'ul associations grew wheat, barley, sesame, cotton, pulses, and millet.[48] In extreme cases, such as mass death among the flocks, some pastoral groups switched completely to settled cultivation.[49]

More than growing crops, cattle made a convenient supplement to sheep herding, especially among the Ahşamat and Qara'ul associations. Differences in anatomy and complementary grazing habits made sheep and cattle a great herbivorous team. Broader muzzles and stiffer lips allow cows to graze taller pastures with their dexterous tongues, leaving behind the shorter grasses for the more selective sheep and goats to nibble at with their smaller mouthparts and cleft upper lips.[50]

To the array of crops and livestock they embraced, shepherds between Baghdad and Damascus added desert truffles, a species of the mushroom family distantly related to the far more prized variety in Europe. Whether black, white, or yellow, truffles belong to the kingdom Fungi and do not photosynthesize. They rather contract out this critical process to the organism that does it best—a green plant. The organic partnership forms when a truffle uses its fine white fibers (mycelium) to envelop the roots of a host plant, sucking up from it sugar and other carbohydrates in return for water, minerals, and protection from pollutants and bacteria. The deep bond between plant and fungus explains why truffle hunting worked in perfect harmony with mobile pastoralism. Spring is the high season for both pastures and truffles in the desert, where they grow in tandem with each other following the first spring rains. Hunting for truffle, therefore, caused minimal disruption to the daily forage. As they pastured their flocks, herders simply looked for a few telltale signs along the sheepwalks, usually in the form of cracked bumps beneath squat rockrose bushes caused by swelling truffles.[51]

Packed with protein, truffle served the role played by tofu in modern American cuisine, as a vegetarian meat substitute. Particularly to those lost in the desert, it provided much needed emergency support. On their way to India around 1758, one Mr. Barton and his servant, robbed by bandits, had to walk bare-footed to Baghdad "without meeting with any other support than the *Truffles* of the *Desert*, that happened then to be in season, and which they found in great plenty."[52] Nutritionally secured travelers used the abundant resource to stage mock fights. In April 1774, a British commercial consul noted that "truffles were found in such plenty that our cameliers and others diverted themselves with pelting each other with them on their journey."[53]

From the sixteenth century, truffle hunting in Iraq was popular enough to be regulated by Ottoman law. State authorities referred to the fungus using both the Turkish name (*domalan mantarı*) and the Arabic one (*fuqa*). They put a price tag on the resource, asking truffle hunters to surrender a quarter of every 13.5 ounce they stockpiled as a levy to the provincial administration.[54]

Even though it tended to center around the husbandry of sheep, mobile pastoralism in the alluvium was a diversified economy that made room for crops, other animal species, and even fungi. Through tribal networks, it integrated grain cultivation in the same marginal lands where sheep were fattened. Other pursuits, such as the tending of cattle and the collection of truffles, were easily incorporated without veering off the paths beaten by sheep and herder.

Outside observers like Dr. Rauwolff thought of mobile pastoralists as a disorderly and dull lot. Pastoral society in this part of the world, however, had quite a rigorous structure. Unlike the pastoral structure prevalent in the west and the north, that of Iraq was programmed to respond to changes in the Tigris-Euphrates flow regime rather than changes in altitude. In winter and spring, pastoralists responded to the swelling of the Tigris and Euphrates by moving their flocks to drylands that teemed with pastures. When the twin rivers calmed and subsided, pastoralists returned, allowing their flocks to feast on the vegetative belts around the river channels. Mobile pastoralists, in this seasonal pattern, danced to the tune of the Tigris and Euphrates.

Animal (particularly sheep and goat) husbandry was an inherently unstable mode of production that demanded maximum flexibility. The tribe, with its network of relationships based on segmentary lineages, offered a sure and firm operational unit, an anchor for households regularly displaced in a world void of a stable territorial framework. Where the agrarian system failed, the triad of pastoralism, mobility, and tribalism triumphed and became an economic force to be reckoned with. They went hand in hand and formed a coherent package able to conquer all but the harshest landscapes.

A renowned historian once wrote that "the Ottoman regime was incompatible with a nomadic economy and tribal customary law."[55] Yet the Ottoman state, grudgingly or otherwise, reconciled its agrarian preferences with those of mobile pastoralists in Iraq for political and economic expediency. Istanbul managed and even augmented the pastoral economy by organizing most shepherds into umbrella associations. The apparatus of provincial government relied on pastoral subsistence strategies to reach the region's least accessible and bio-productive zones and to dip deeper into their vegetative wealth.

The exploitation of grass, it must be remembered, formed only one major area of common interest between state and tribe. Military considerations were part of the calculation as well. Collaboration yielded to Istanbul thousands of battle-hardened tribesmen who could be conscripted into the ranks of its land and naval forces, serving as warriors, oarsmen, and steersmen under the command of governors and admirals. In return for their services, tribal chiefs obtained numerous privileges, including official recognition, land and tax collection grants, and protection that buttressed their legitimacy locally and enabled them to amass power and wealth in competition with their rivals. The relationship could be described as a symbiosis that worked to the advantage of both. The levers of imperial control, however, could sometimes be burdensome, leading shepherds on numerous occasions to visit the Ottoman judge in Baghdad and other towns and complain about the abuse and humiliation they experienced at the long hands of government agents.[56] When access to justice was no longer available in times of upheaval, pastoralists took matters into their own hands.

5

Wetlands

> I enter a swamp as a sacred place, a *sanctum sanctorum*.
> There is the strength, the marrow, of Nature.
> —Henry David Thoreau, "Walking" (1862)

WETLANDS OCCUR IN areas seasonally or permanently saturated with water, a middle world between terrestrial and aquatic ecosystems exhibiting features of both. Like lakes and oceans, their soil lacks oxygen, yet they harbor shrubs, trees, and other large plants found in forests and grasslands. The red maple, for instance, is a deciduous tree that dominates the Great Dismal Swamp in Virginia and North Carolina. The biological composition of wetlands is diverse and abundant, allowing subsistence to be spread across several food webs obtained through crop cultivation, animal raising, fishing, and hunting. The diversity of this ecosystem is reflected in the plethora of terms used to describe it in different parts of the world—muskeg, swale, bayou, turlough, to name only a handful in the English language. In Arabic, the situation is just as complex. According to a British geographer visiting Iraq in the 1830s, "The Arabs are as rich in names for their picturesque marshes, as the Norwegian is for his variously-formed mountains."[1] Rather than get bogged down in the global terminological mire, this chapter will simply call the Tigris-Euphrates wetlands "the marshes," as they are commonly referred to.[2]

Wetlands have been more secretive than most other biomes, shrouding their inner residents in thickets and waters. Adding to their enigma was the long-standing topographic prejudice of agrarian societies, which often scouted out the "city upon a hill" and shunned the supposed squalor of the lowlands. This chapter uses Ottoman bureaucratic materials to reconstruct the early modern history of this forbidding ecosystem at the lower ends of the Tigris and Euphrates. Soon after its conquest of Iraq, the Ottoman state recognized the traditional subsistence value of the marshes and made a scrupulous effort to regulate and profit from their exploitation. It applied the lessons it had learned on arable lands and grasslands to manage the most profitable marsh crops and animals, namely, rice and the water buffalo. Ottoman regulation of water buffalo herding, in particular, closely resembled the aggregating policies used for sheep herding. Through this eclectic approach to farming and a great deal of improvisation, the land-based Ottoman Empire incorporated the expansive sheets of water in Iraq.

In the historiography of post-classical West Asia, a prevalent environmental story narrates the expansion of arid lands under the hooves of mounted nomads charging from the east. As two scholars describe the aftermath of the Mongol conquests in the thirteenth century, "In an astonishingly short period, most parts of Iraq turned into desert."[3] And another account says, "At the beginning of the early modern period, desertification and recession had already occurred on a large scale, perhaps most notably in the alluvial plains of Mesopotamia (including Khuzistan)."[4]

This oft-repeated story is unfortunately based on flimsy evidence. To start with, the causality between nomadic onslaughts and political crises, on one hand, and desertification, on the other, is elusive. Comparable political conditions following the collapse of the Abbasid dynasty in 1258, for instance, fostered the opposite environmental conditions—rewilding the Tigris-Euphrates alluvium with wetlands through unchecked seasonal and extraordinary floods.[5] Other fluvial landscapes witnessed analogous revival of their wetlands during periods of political disintegration, such as northern Italy's Po Valley following the dissolution of the Western Roman Empire in the early Middle Ages.[6] This chapter, therefore, challenges the desertification narrative of the post-Mongol era by bringing to light the prominence of wetlands in the history of early modern Iraq.

Hydrography

Different hydrologic pathways provide wetlands with water, including precipitation, surface runoff, groundwater, and tides. The predominance of one over the others depends on a wetland's local geography and climate. The Iraqi marshes rely for their water inputs on the dispersal and overflow of the lower Tigris and Euphrates. Sapped of their energy by a flat terrain and irrigation projects, the twin rivers overspill and form marshes as they meander toward the Persian Gulf. Passing by the lower Euphrates in the late eighteenth century, the French consul in Baghdad Jean-Baptiste Rousseau poetically described the fluvial process. The landscape was, he wrote, "furrowed by a multitude of small streams which, when detaching themselves from the Euphrates, meander in the plain, cut across, intermingle, and finally return to the river, like lovely children who diverge a little from their mother while playing, and then return to throw themselves into her arms."[7] The marshes were the byproduct of this momentary lapse of fluvial concentration, between the playful branches and their caring mother channel.

Until the late twentieth century, the Iraqi marshes were among the largest in the world, a giant aquatic complex that can be conveniently divided into three parts. The core of the complex centered in the triangle between Kut, Nasiriyya, and Basra.[8] In the footsteps of their medieval Muslim predecessors, Ottoman geographers referred to this core as the Flat Lands (Al-Bata'ih).[9] In administrative circles, a more common term was the Islands (Cevazir or Cezayir). Bureaucrat Feridun Bey (d. 1583) offered one of the earliest Ottoman descriptions of the Islands, portraying them as the black hole of the river system. "There," he wrote, "the water of the Euphrates, the River of Baghdad [Tigris], and the rivers flowing in the environs blend together and splinter left and right like irrigation ditches. As a result, the Islands comprise 360 rivers, the local people say."[10] The two other components of the marsh complex were centered around the Tigris and Euphrates farther north. Along the Tigris, a large swamp formed in the Aqarquf depression west of Baghdad and around sixty smaller ones scattered between Baghdad and Ctesiphon and near Qara Tepe.[11] On the Euphrates, some eighty small marshes existed near Hilla and Rumahiyya.[12]

This network of permanent and seasonal marshes clustered in Iraq, but its significance extended to the entire river system. It acted as a regulator of flow, fluctuating in tandem with the seasonal rise and fall of the Tigris and Euphrates. In the spring, the wetlands brimmed and expanded by absorbing excess runoff and storing it over some 10,900 square miles on the plains, an area the size of Massachusetts. During the remainder of the year, the wetlands gradually released the waters back to the trunk channels and shrank, reaching a minimum of 3,200 square miles in the fall—slightly bigger than Delaware.[13] Through this seasonal cycle of swelling and contraction, the marsh complex reduced the severity of the annual flood during harvest time and eased the pressure on protective works established by local farmers. In addition, marshes provided an insurance against drought and cleansed polluted waters. If an artist were to draw the Tigris-Euphrates system in human form, the Iraqi marshes would constitute the body's kidneys.

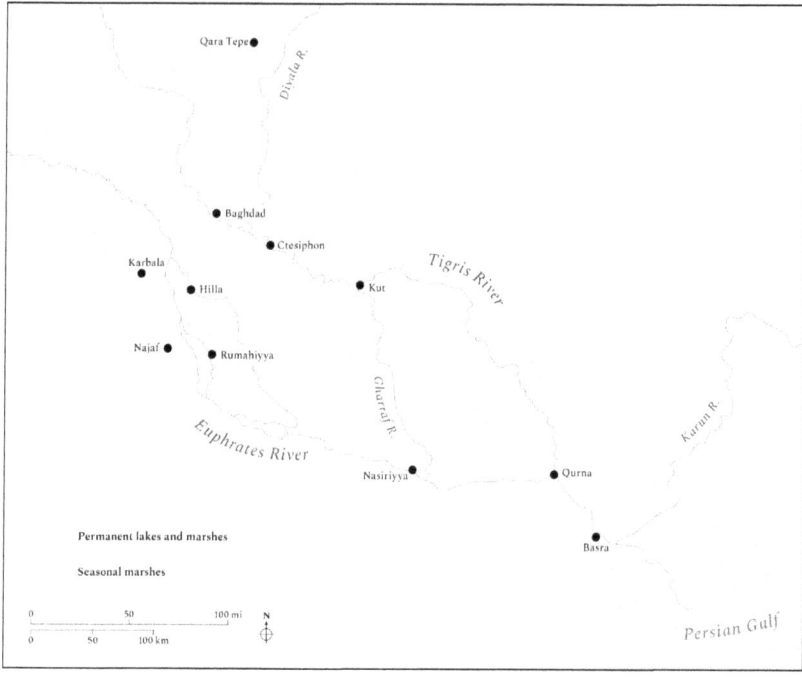

FIGURE 5.1. The Tigris-Euphrates Marshes. After Ahmad Susa, *Atlas al-Iraq al-Hadith* (Baghdad: Mudiriyyat al-Masaha al-Amma, 1953), Map 11; TKG.KK, TT 29.

Settlement

In wetlands, terrestrial animals, especially humans, face an acute housing shortage. Dwellers of the marshes coped with this by clinging to bits of earth that rose above the water. Some of the islets they settled were natural hillocks, others manmade, vernacularly called *jibasha* or *dibin*. A family staked out its artificial islet by fencing the perimeter of the desired location with reed bundles, within which alternating layers of soil and rushes were amassed. Once the heap of earth and greenery rose above the water, it was trampled down tightly and made into a compact, solid mound suitable for settlement. In the flood season, a family could raise the surface of its islet floor by adding fresh rushes as needed. Mobility in this engineered archipelago relied on canoes about six feet long built out of local and imported timber.[14]

Population estimates for the marshes in the early modern era are difficult to come by. Ottoman cadasters do not explicitly state whether a registered tribe or village lived in a marshland or not. The only extant information is about subsistence modes, which often hint at the environment where a population lived. As a rule, the water buffalo is a wetland's biological signature, the only habitat where it could survive and remain productive in Iraq. In ecological jargon, the water buffalo is more "habitat-specific" than more versatile species such as sheep or budgerigar. The giant mammal is among a number of species, such as the fire salamander, that like to hang out in pools of calm water, steering clear of fast-flowing streams. In Iraq, the marshes afforded the buffalo those oases of serenity.[15]

Water buffalo colonies recorded in the Ottoman registers, therefore, strongly indicate the presence of a wetland ecosystem nearby. In the late sixteenth century, roughly 12,000 taxpayers (or close to 70,000 people in total if their families are counted) relied primarily on water buffalo for subsistence and likely either lived or made a living in a wetland environment. The actual figure must be higher, for in 1552, Ottoman scribes conducting the cadastral survey of the Basra province complained that many marsh dwellers were in a state of rebellion and could not be registered.[16] Even the conservative figure of 70,000 is a significant number by the standards of the day—about 18 percent of Iraq's entire population then.[17]

For animals fond of dry land, snug homes were all too rare in the marshes. Humans had to build them from scratch. Every family initiated the process by building an artificial islet before it put down roots in the water. The settlement scheme proved popular, attracting tens of thousands of people. But to sustain their way of life over the long term, marsh dwellers had to overcome the environmental challenges of a perilous habitat.

Disease

Until recently, outsiders considered the marshes to be a diseased organ of nature. The air was impure and the water stagnant, they often grumbled.[18] In the spring of 1560, the Ottoman governor of Basra penned his own concerns about public health in his marshy province. "Heavy air" had ravaged the Ottoman garrison in the region, he wrote to his superiors in Istanbul. Some troops had died, while others were left bedridden.[19] Across the globe, written sources of this kind have given currency to the portrayal of wetlands as unhealthy ecosystems, more fit for mosquitoes and other dangerous creatures than humans.[20] In the words of historian Fernand Braudel, on the flat plains of the Mediterranean basin, "water is synonymous with death."[21] The Iraqi marshes offer a counterexample to this stereotype and demonstrate how still water on a flat plain could create life and community. Imperial records bear witness to a large human population in Iraq that relied on wetlands for settlement and subsistence. It survived the health hazards due to a fortuitous defense system, partly natural and partly human, that degraded the lethality of diseases endemic in fresh or brackish bodies of water, particularly malaria.

The ecology of Iraq formed the first line of defense against malaria and reduced the efficiency of its transmission in the marshes. It did so by breaking up the nexus between heat, humidity, and rain that sustains the malaria parasite. During the warmer months of the year, when mosquito hosts (female anopheles) grow and multiply, Iraq experiences intense desiccation that minimizes the water level of the marshes, leaving vector populations with minimal hydrologic support for their breeding. When winter rainfall and spring snowmelt bring high water to the marshes, average temperatures fall below those favorable to mosquito breeding. This balance between hot and cold, wet and dry conspired against the malaria parasite and made the completion of its life cycle more difficult in Iraq than in other tropical and subtropical countries. The dominant mosquito species that could endure the frustrating ecology of the marshes was *Anopheles pulcherrimus*, a poor carrier of malaria.[22]

Human biology possibly created another defense line against the proliferation of malaria in the marshes. Three archaeologists have recently hypothesized that the local population of ancient Iraq may have developed genetic shields after many generations of exposure to malaria and other infections since the early development of Sumerian agriculture in the Neolithic period.[23] Heritable traits aside, people could acquire resistance by surviving repeated outbreaks of the deadly disease.[24] Wilfred Thesiger, a British explorer who operated as a roving doctor in the marshes during the 1950s, was surprised to find among locals only a few cases of malaria, most of which, he believed, were contracted outside the region. He concluded that most marsh dwellers must have acquired some immunity to the endemic diseases in their environment.[25] "It is a case of the survival of the fittest," a British military officer and his wife wrote of life in the marshes during the early twentieth century, "for the infant mortality is appalling; but those who survive to maturity have hardened in the process."[26]

Even though they were ignorant at the time of the vectors behind malaria transmission, marsh dwellers took general preventive measures against biting flies that inadvertently restricted the spread of malaria. A basic step involved the construction of huts in the open water away from reed beds, where insects converged in great numbers. A more sophisticated plan of action deployed buffalo dung, easily and cheaply available to every household. When lit and allowed to smolder, the manure produced thick billows of acrid smoke that kept mosquitoes at bay.[27]

Ecology, immunity, and human decisions reduced the threat of malaria but failed to prevent other risks to health and safety. Sanitary conditions were poor, particularly during the spring, when rising waters mixed drinking supplies with excrement. Dysentery and bilharzia, among other complaints, were not uncommon. No less threatening to human settlement was the constant threat of giant wild boar. Shunned by a Muslim population that did not consume its meat, the swine thrived among the reed beds, devastating rice crops and goring anyone who happened to stand in the way. No human genes or dung fumes that restricted malaria transmission could repel the veritable scourge of wild boar, nor were clubs and spears effective in containing its threat on a long-term basis. Only the proliferation of firearms and ammunition since World War I could give locals the means to bring the animal to the brink of extermination.[28]

Marsh life had its risks and inconveniences, but none of them deterred human settlement. Disease, in particular, merely curbed population growth in the marshes, the same way it operated in crowded cities around the early modern world. The confluence of natural processes and social initiatives allowed nearly a fifth of Iraq's population in the late sixteenth century to call the marshes home. Once settled, marsh dwellers had to put together their own food guide.

Soil

Wetlands form stumbling blocks to agrarian development. Their soil exerts a major stress on the productivity and development of living organisms. Saturated with water, their soil pores lack oxygen and are incapable of meeting the demands of most rooted plants for respiration. Until the late twentieth century, most agrarian states devised incentives rewarding the drainage and destruction of wetlands as a crude remedy for their stubborn hydric soil. The Ottoman Empire was no exception. After its conquest of Iraq in the sixteenth century, it rewarded those who drained the marshes and put them under the plow with title deeds, tax exemptions, and rights to capital gain. In 1552, for instance, one Khawaja Fayyad paid the Ottoman administration in Basra 16,000 akçe to acquire the title deed of an abandoned marsh island, increasing its value threefold after five years of hard reclamation work. In another case from 1578, an Ottoman mercenary from Herzegovina named Mehmed brought under cultivation a marshland in the Khalis district northeast of Baghdad City. The Baghdad Treasury awarded him a title deed for a payment of 1,000 akçe and a reduced annual tax rate of one-eighth on his harvest.[29]

Absent enterprising individuals like Khawaja Fayyad and Mehmed the Herzegovinian, the Ottoman administration had to come to terms with the marshes and all the limitations they imposed on arable farming. It brought under its purview and thereby taxed the produce of two traditional forms of cultivation that could occur in marshy regions. The first form occurred on the edges of perennial marshes void of vegetation and deployed extensive farming techniques that obtained a relatively low yield from large areas. The agricultural operation required minimal inputs of labor, for the overflow and retreat of marsh water did most of the work to fertilize those edges and turn them into a moist surface of loam. Farmers simply scattered seeds into furrows naturally created after the soil dried out. If not picked up by birds, the seeds took root. In smooth areas without natural cracks, farmers took the additional step of plowing the soil with a crooked stick dragged by two draft animals. Here, both winter and summer crops, including wheat and barley, could grow in modest amounts. Like tilling the soil, irrigation on marsh edges was comparatively easy. In their natural rise and fall, the Tigris and Euphrates took charge of the process. Farmers simply maintained embankments to retain the waters and inundate agricultural plots as needed. Because of its reliance on the chances of flood and water recession, cultivation was a riskier, less predictable, and less profitable venture on marshland than elsewhere.[30]

The Ottoman administration treated a marsh where this extensive form of cultivation occurred as a taxable agricultural unit called *hor*. Located in deep basin depressions, hors were usually uninhabited or only temporarily settled, serving as a satellite exploitation area for pastoral groups and nearby settlements based on more elevated grounds. As marginal lands, the hors of Iraq are comparable to the *gastinae* in medieval Europe and *mezraʿa* throughout the Ottoman Empire. Provincial authorities took pains to squeeze as much money as possible out of hor cultivation and auctioned off abandoned marshes to make sure no land with a capacity for production remained idle.[31]

The other form of cultivation that the marshes supported was far more intensive and more important to the Ottoman state. It centered around rice, a crop that coped with low oxygen levels in marsh soil through its unique roots, equipped with a porous tissue full of air spaces. Like snorkels, the rice roots facilitate the diffusion of oxygen from the aerial shoot into the waterlogged, buried root. Farmers concentrated the cultivation of rice in marshy regions, where the thirsty crop could find large and assured supplies of water during the summer months. From April, farmers soaked the rice seeds in water and laid them under the sun until germination. They cleared the ground and made low-earth dams called berms to divide fields into blocks. By June, when the rivers began to subside, farmers transplanted the young shoots into the rich silt deposited to the fields by the river floods. The crops required constant watering through controlled inundations until the harvest season, which occurred anytime between June and October depending on crop quality and local conditions.[32]

Rice cultivation was backbreaking but well worth the effort. From the farmer's perspective, rice was primarily a major boost to human nutrition. Even after milling, a process that significantly reduces its vitamin and iron content, rice alone could meet the daily needs of laboring adults for carbohydrate and protein.[33] For the Ottoman state, rice was its most lucrative summer crop in the entire alluvium. Similar to its policy in Anatolia and the Balkans, the central treasury in Istanbul controlled the vast majority of land devoted to rice cultivation in Iraq, generating some 6 percent of the region's revenue potential while recognizing freehold status for a small number of fields.[34] Like dates, rice's long shelf life made it an ideal foodstuff for armies at the eastern extremity of the empire.[35] For both the farmer and the state, it stood out from other crops for its high yield. "The rice grows extraordinarily," the French consul noted while passing the paddies of the Euphrates southwest of Baghdad in the late eighteenth century; "a single grain produces up to six or seven stems, and each stem carries about fifty grains."[36]

The Ottoman Empire advanced its agrarian agenda into the marshes primarily by laying claim to two traditional systems of crop cultivation that could circumvent or tolerate waterlogged soil. The first was extensive and involved the cultivation of winter crops such as wheat and barley on marsh margins. The second was intensive rice cultivation and relied on copious water supplies in the marshes for irrigation during the summer. Even without a large-scale drainage effort, the Ottoman state could still promote arable cultivation in the marshes, while preserving the habitat required by vulnerable wildlife.

Refuge

Dolphins, seals, and walrus are aquatic mammals, but none of them call the marshes home. Among other things, shallow waters and dense thickets are too restricting for their movement. According to an Ottoman provincial official in the seventeenth century, so impenetrable were the marshes that neither an elephant nor a snake could pass through them.[37] Only a couple of large mammals with a thick skin (literally and figuratively) could defy such natural barriers and force their way into a wetland to live. The largest is the hippopotamus, which dominates the wetlands of sub-Saharan Africa. In the Iraqi marshes, the star is the water buffalo. Narrow waterways choked with vegetation could not intimidate these powerful swimmers. In the sixteenth century, the Ottoman administration could tax some 41,000 water buffalo in the alluvium—a high number, not far behind the 52,000 figure recorded for the modern Iraqi state during the early twentieth century.[38] If the French explorer Jean-Baptiste Tavernier, who visited Iraq in 1652, is to be believed, the region's buffalo population was as high as 144,000.[39]

In the marshes, the water buffalo population lived on the edge of the possible. The winters of northern regions were too cold for it, sometimes causing mass mortality among its ranks. Most water buffalo between Samsun and Tokat in Anatolia, for instance, perished in December 1743 due to freezing temperatures.[40] In the south, the animal faced the opposite challenge and struggled to regulate its body temperature during the searing heat of Iraqi summers. Its dark skin absorbs heat; its thick skin reduces its ability to get rid of surplus heat. Compounding the thermal stress is the fact that the water buffalo is a milch animal. To keep pace with the milk demands of its calves and owners, it will continually feed and fuel metabolic heat production in its body. Humans and some other mammals sweat buckets to extricate themselves from such situations and cool their bodies. The water buffalo, however, has a small number of sweat glands per unit area, a handicap it shares with the hippo. It nonetheless found a remedy for its thermal conundrum in the natural wallows of the marshes, the center of the buffalo's world and most effective means to shelter it from exposure to intense solar radiation. A leisurely life in this aquatic environment allowed the water buffalo to triumph and become one of nature's giants in the entire Tigris-Euphrates basin.[41]

The marshes shielded the water buffalo from a harsh climate and offered ample supplies of fodder, in the form of torpedo grass, knotgrass, southern cattail, and, most important, common reed. Early modern buffalo grazing systems likely resembled modern ones that have been well documented in twentieth-century ethnographic research. Marsh dwellers foraged on their animals' behalf on a daily basis to secure most of the required fodder. The daily search for pasture typically began soon after dawn with the buffalos' departure from the islets of their owners, wading alone or with children through the waterways toward the nearest patches of land. Meanwhile, herding families set out in canoes, leaving behind their houses guarded by a watchdog, and spent the day cutting reeds and grass. Both parties of foragers—the buffalo herds and their owners—returned to their islets late in the afternoon, where the daily fodder harvest was spread for the buffalo to consume at night. In the winter and spring, when the water is high and the weather is cold, the buffalos remained confined to their dry platforms throughout the day, groaning and rasping until their owners returned and laid before them their daily rations of grass. Marsh dwellers used fire to further sculpt their landscape and make it more favorable to buffalo grazing. They burned large, coarse, and dead vegetation to stimulate young growth more palatable to the animal. Burning normally took place at the end of the year, so the young shoots could sprout during their natural growth cycle, beginning in January and continuing until June.[42]

Water buffalo husbandry allowed marsh dwellers to preserve their ecosystem and way of life. The constant exploitation of marsh vegetation as animal fodder kept the natural threats of eutrophication and overgrowth at bay.[43] A well-nourished water buffalo, moreover, could supply critical products for a sustainable human presence in the marshes. As far as marsh dwellers were concerned, the chief prize of tending a large mammal like the water buffalo was large amounts of dung, used as fuel to keep warm, cook, and ward off biting flies. Due to its consistency and ability to dry hard like cement, buffalo dung was also the primary mortaring material in the marshes, used in the building of houses and granaries, among other structures. Buffalo dung had medicinal uses as well, applied to the forehead for headaches and to cuts, wounds, and burns as a healing agent.[44]

Milk was the buffalo product second in importance for marsh dwellers. According to one estimate from the middle of the seventeenth century, the Iraqi female buffalo, presumably when in full lactation, could yield as much as twenty-two pints of milk daily.[45] Locals consumed buffalo milk fresh or sour. Using simple methods (whipping with a wooden spoon, churning by suspended animal skin, shaking in a hollowed-out gourd, heating over fire), the raw milk could be processed and converted into more durable and palatable dairy products, including curds, cream, cheese, and butter. Buffalo milk is very similar to cow milk in its chemical composition and physical properties and provides excellent supplements of protein, vitamins, mineral salts, and calories to the human diet.[46]

Regardless of the buffalo products herders could realize, what mattered to the Ottoman state was the beast itself, which it treated as a big cash cow. In some districts, each buffalo yielded the annual tax revenue collected from eight sheep.[47] Putting a price tag on the animal was the easy part; collecting the money was much more difficult. Dispersed among small islands and

Table 5.1. The Cemmasat

Year	1540	1544	1580
Households	—	2,221	3,026
Water Buffalo Population	—	17,881	19,134
Cow Population	—	1,544	2,187
Sheep Population	—	4,222	121,700
Tax Potential (akçe)	142,001	416,282	849,485

Sources: OA, TT 1028, 91; OA, TT 228, ff. 45r-46r; TKG.KK, TT 29, ff. 201v-215v.

separated by a trackless boggy ground, buffalo herders could easily escape the scrutiny of Ottoman officials. To enforce the buffalo levy more efficiently, the Ottoman administration followed a policy of aggregation similar to the one pursued in the grasslands. It aggregated one-fourth of marsh households into a state-sponsored herders' association called the Cemmasat, or the Water Buffalo Breeders.[48] Through a formal structure and routine bookkeeping that identified the names, locations, and leaders of Cemmasat affiliates, as well as their rights and obligations, the Ottoman state considerably improved its ability to intervene in their affairs, particularly the process of tax collection. The Cemmasat, as a result, emerged as a fiscal-administrative conglomerate, the size of Baghdad City's entire population and the second most important herders' association in the alluvium after the Ahşamat shepherds. Unlike the residents of Baghdad City, members of the Cemmasat did not share alleys or neighborhoods. Most of them perhaps never met and did not know each other. What brought them together in one Ottoman administrative unit was the black gold they owned—the water buffalo.

The marshes harbored the water buffalo, a sturdy aquatic athlete but with very special needs. In these watery oases it could moderate its body temperature and find plentiful pastures. Like the reindeer in the Arctic tundra and the yak in the Tibetan Plateau, the water buffalo carried some hitchhikers to an extreme ecosystem that only a few dared to penetrate alone. It helped tens of thousands of mixed farmers to make themselves at home on water, offering a constant supply of fuel and milk. The Ottoman state latched onto this resourceful beast as well to expand the reach of its extractive power into the heart of the marshes.

From the late fifteenth century, states worldwide reached new heights in size, complexity, and efficiency. With new powers they enhanced their capacity to disrupt the functioning of Earth's ecosystems, targeting wetlands in particular with an almost religious zeal. Ideas about improvement, health, and climate whipped up the global drainage frenzy. From the delta of the Yodo River in Japan, the marshes of Poitou in France, to Lake Texcoco in Mexico, early modern states initiated or supported the aggressive reclamation of wetlands to make room for more farms and settlements. This chapter has narrated a parallel, though less familiar, story—of a state that used its impressive bureaucratic machinery not to destroy, but to incorporate wetlands for its economic interests.[49]

On the flat alluvial plain, the Tigris and Euphrates scattered their liquid content to create wetlands, large and small, seasonal and permanent. Despite their many hazards, the marshes were popular destinations for settlement and work. Their settlers fulfilled their needs for food, housing, transport, and fuel without necessarily stepping outside their sheltered world in search of resources. Ottoman officials incentivized random, small-scale drainage efforts to make marshlands more suitable for growing wheat and barley. For the most part, however, they had to brave the elements and get their feet wet. They relied on time-honored systems of marsh exploitation, particularly rice cultivation and water buffalo raising, to endure one of the wettest and most remote ecosystems in the Ottoman domains.

PART III

The Rumblings of Nature

In the late seventeenth century, a deadly cocktail of natural and political disasters transformed the Ottoman Empire's relationship with the Tigris and Euphrates. At the heart of the turmoil was a major shift in the Euphrates' channel southwest of Baghdad. The chaos it unleashed undermined the institutions of the central state and the balance they had maintained in the countryside from the first years of Ottoman rule. The stage was set in the early eighteenth century for the development of a new system of river management based in the provinces, which came to exert a greater influence over the utilization of the Tigris-Euphrates waterways than the imperial capital of Istanbul. The localization of Ottoman authority in the drainage basin was a regional manifestation of an empire-wide trend. During the eighteenth century, according to historian Ali Yaycioglu, "the Ottoman polity experienced a turn from a vertical empire, in which the imperial elite sustained claims to power through a hierarchical system, to a horizontal and participatory empire, in which central and provincial actors combined to rule the empire together."[1] In the Tigris-Euphrates basin, the most consequential provincial actor was the Pashalik of Baghdad, which gradually dominated most policy decisions related to navigation and irrigation. The following two chapters detail the complicated dynamics behind this watershed moment in the history of the Tigris and Euphrates and how it transformed the Ottoman presence in the region.

6

Havoc

> The monstropolous beast had left his bed.... The sea was
> walking the earth with a heavy heel.
> —Zora Neale Hurston, *Their Eyes Were Watching God* (1937)

THE EUPHRATES RIVER mobilizes water and sediment into a dynamic and complex flow, powered by the energy of the sun and gravity of the earth. It delicately changes through space and time and adjusts to its variegated geographic setting. Spatially, runoff and sediment production predominate in the headwater regions of the Taurus Mountains before the river turns into a sediment sink and loses much of its water through dissipation and evaporation in the Iraqi alluvium. Temporally, the river displays markedly different identities in the summer and spring, with contrasting discharge rates and water-surface elevations.

Between 1687 and 1702, these timeless regional and seasonal changes were eclipsed by intense ecological disturbances that transformed the Euphrates' hydraulic architecture. A dramatic rupture in the river's flow occurred when a large segment of it, approximately a hundred miles in length, escaped its established channel and gushed into a new one. The abrupt relocation of the river, a process called avulsion by geologists and hydrologists, profoundly altered the ecology and politics of Iraq and imperiled the stability of the Ottoman Empire in the east, threatening traditional centers of power and permitting otherwise lesser tribes to enjoy temporary ascendance. Thousands of lives were lost during the intervening years, and numerous settlements were abandoned and left to ruin.

This chapter documents the metamorphosis of the Euphrates River in the late seventeenth century. Beginning in 1687, a prolonged meteorological anomaly and an ill-fated irrigation project divided the Euphrates in Iraq into two capricious branches. For a constellation of reasons, the Ottoman central and provincial administrations were incapable of resolving the environmental crisis until the river had completely abandoned its original bed in 1700. The channel pattern that emerged thereafter withstood the assault of an engineering expedition tasked with undoing the avulsion in 1701–1702, thus facilitating the fall of the Ottoman Rumahiyya fort southwest of Baghdad.

The environmental regime shift under analysis appears in the most comprehensive histories of Ottoman Iraq as a sudden, arbitrary rupture that took place in 1700, unconnected to any preceding crisis that had engulfed the region.[1] That dating is based on chronicles that recorded only the climax of a long-term process. The discussion herein, which combines analyses of tree-ring (dendrochronology) and speleothem (icicle-like cave rock) with untapped archival sources, argues that the Euphrates started to re-position its course at an earlier date, in part because of an exceptionally dry period in central Anatolia in 1687 and 1688. This earlier dating has major historiographical implications. It uncovers distinct causes that lay behind the river's tumultuous reconfiguration and connects the calamitous events that afflicted Iraq between 1687 and 1702, once thought to be random, into a coherent whole. As one scientist remarked, "Flowing water is not simply an unstructured chaos

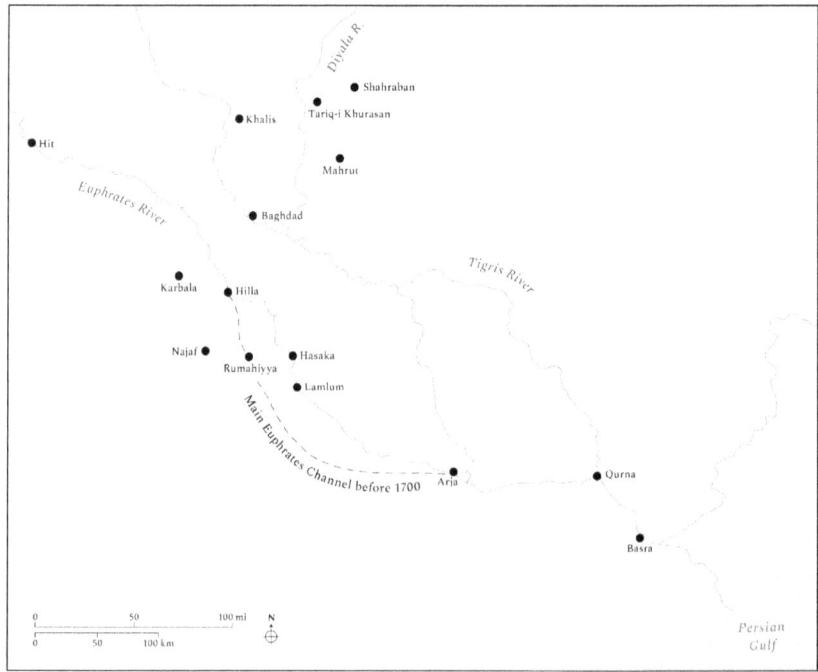

FIGURE 6.1. The Euphrates River's Avulsion, 1687–1702.

but contains persistent forms that can be recognized, recorded, analyzed—forms, moreover, that are of great beauty, of value to the artist as well as the scientist," not to mention the historian.[2]

The study of morphological change in the Tigris and Euphrates has been largely the domain of archaeologists and earth scientists.[3] Given that a significant portion of the literature is theoretical, this chapter aims to bring an empirical, holistic approach to the study of avulsion, concerned not only with causal factors but also with society's response and adaptation to it.[4] Archaeologists and scientists have deftly outlined the skeleton of avulsion in ancient Iraq, to which historians can add the sorely needed flesh and blood. The eyewitness accounts of the events offer a unique opportunity to move the analysis of channel evolution from the inanimate, all-seeing geographic information system (GIS) map to the living experience of anxious urban scribes, weary construction workers, and valiant tribal sheikhs situated in the middle of the changing alluvial environment.

Drought

The dread of famine gripped Baghdad on Tuesday, December 7, 1688. Starving families from Mosul and the Kurdish regions flocked to the city begging for aid. Epidemics infected the refugees, many of whom failed to find shelter in the increasingly overcrowded provincial capital despite the best efforts of its notables. Rumors and panic gave way to anarchy. A group of janissary soldiers approached and murdered an attendant at the shrine of Abu Hanifa for allegedly engaging in a monopoly that aggravated currency inflation. The prices of wheat, barley, meat, dates, and raisins skyrocketed in a year remembered by the people of Mosul as that of the Great Inflation.[5]

Famine, displacement, and inflated prices came in the wake of a dwindling flow in the Tigris and Euphrates. Change in the rivers' water levels occurred during a sharply defined and distinctive weather anomaly, registered in the pages of Ottoman documents, the rings of Anatolian trees, and the speleothems of Kurdish caves. As Ottoman historian Silahdar Mehmed Ağa (d. c. 1726) wrote, protracted wars and drought in the "Muslim lands" resulted in a great famine and inflation that became extremely intense after 1687.[6] Reports penned by Ottoman judges and officials in Diyarbakır and Akşehir, and as far west in Anatolia as Eskişehir and Seferihisar, conveyed to Istanbul the subjects' tribulations.[7] A dendroclimatic study corroborates Silahdar's account and the official reports, revealing that Sivas near the catchment area for the twin rivers had experienced drought in 1687 and 1688.[8]

A fifteen-inch stalagmite collected from the Gejkar Cave, within 200 miles north of Baghdad, offers additional confirmation of the region's environmental plight at the time. The thickness of its annual growth bands in 1687 and—in particular—1688 drops far below the preceding 600-year average, a sign of low cave drip rates and rainfall during these two years. In addition, the isotopic oxygen and carbon values of the same stalagmite experience a sharp positive shift between 1685 and 1691, which indicates increased evaporation and decreased plant cover and soil microbial activity—the trappings of a dry spell.[9]

Well documented in the written and natural records, this anomalous precipitation occurred during what paleoclimatologists call the Late Maunder Minimum (1675–1715), a period characterized by considerably low sunspot numbers, colder temperatures, and highly variable climatic conditions with a tendency toward extremes, especially in the continents of the Northern Hemisphere.[10] As precipitation declined in Anatolia, drought struck in Crimea, central Russia, and India; sea ice accumulated on the coast of Iceland; and hailstorms and gales smashed fields and vineyards in Switzerland. Firm connections between these extreme weather events are elusive, but they conformed to the prevailing, unstable climatic norm.[11]

Contemporary Baghdadi chroniclers establish the necessary correlation between the proxy records from Sivas and Gejkar and the Tigris-Euphrates flow in 1687–1688. "By the wisdom of the One who cannot be called to account for anything He does," Murtaza Nazmizade wrote, "dread and fear of inflation overcame the hearts of people when the dearth of quenching rainfall brought the Tigris and Euphrates to a halt."[12] Ahmad Ghurabzade added that the rivers flooded their banks.[13] These terse yet invaluable observations are crucial to comprehend the events that unfolded in the following months. Nazmizade's remark that the rivers came to a halt alludes to a crippling effect on their role as agents of transport. Rainfall shortage diminished stream power, the means by which rivers overcome friction, transport sediment, and perform the basic geomorphic work that creates their channels. A feebler stream intensified sediment accumulation within the channels, raised the riverbeds, and prompted a spillover, as Ghurabzade noted.[14]

The Euphrates is more typical of the hydraulically vulnerable rivers of the arid zone than the Tigris, and it was therefore at a greater risk under drought conditions. No significant tributaries join it between the Khabur River in Syria and Shatt al-Arab, a distance of some 750 miles. The feather-like catchment basin of the Tigris, by contrast, receives tributary contributions draining the Zagros Mountains from its eastern bank along most of its course, notably from the Greater Zab, Lesser Zab, and Diyala. Furthermore, the Euphrates' gentler slope makes it more sluggish than the Tigris, which has carved a deeper bed at a significantly lower altitude. The Euphrates is also longer than the Tigris; in fact, it is the longest river in West Asia. The mountains near Erzurum from which it emerges are more distant from lower Iraq than is the source of the Tigris. Mount Karaca, a shield volcano in southeast Anatolia that rises to 6,300 feet, deflects the Euphrates westward and further extends its length, forcing it to traverse several hundred miles through the parching heat of the Syrian desert. Therefore, whereas the Tigris, like the majority of river systems in humid environments, receives a push from tributaries that preserve the vigor of its flow through most of its course, the Euphrates—longer in channel, gentler in gradient, and without hydrologic support in Iraq—enters the alluvial plain as a much weaker river. Through seepage, evaporation, and irrigation, it suffers a considerable diminution in volume and power downstream—vulnerable to the resistance of the channel bed sediments to its movement—and loses definition in Iraq as the ratio of sediment to flow surges.[15]

As a result, precipitation decline in 1687–1688 disproportionately affected the Euphrates, weakened its already feeble stream, and aggravated its vulnerability to sedimentation and riverbed erosion. It also brought the river closer to what geologists term the threshold of critical power—the precarious point at which the power available for a stream to transport its sediment load equals the power needed to accomplish it.[16] In other words, prolonged drought further destabilized an inherently unstable river downstream and increased the risk of clogging, overbank spillage, and channel migration. In such perilous situations, rivers can adjust and find a way to provide, under the new controlling variables, the velocity needed to transport the load supplied from upstream. In this case, help came unexpectedly from a man named Sheikh Dhiyab.

Sheikh Dhiyab

Like twig-eating moose and dam-building beavers, humans perturb specific components of their environments, often inadvertently and with unintended consequences.[17] Around the turn of 1689, Sheikh Dhiyab attempted to direct irrigation waters down the levee slopes of the Euphrates north of Rumahiyya through a controlled levee break. Pressured by a hydraulically unbalanced stream since 1687, his project ultimately burst out of control and ushered in a new channel configuration.

Dhiyab was an ordinary sheikh engaged in an ordinary agrarian pursuit. Yet the unintended consequence of his action earned him ill repute in Istanbul and eternal condemnation in the form of one sentence in the official history of the Ottoman Empire compiled at the time. "One of the Arab sheikhs issued and channeled water by creating a fissure in the [Euphrates] River in order to irrigate his gardens," wrote the court historian Raşid Efendi (d. 1735) before recounting the ramifications.[18] The nameless sheikh's fateful action sheds light on his predicament and that of other farmers in Iraq.

A number of environmental complications stood in the way of Sheikh Dhiyab's basic aim to water his gardens. Canal excavation was essential under the alluvium's meager annual rainfall, less than the 200-millimeters minimum required for dry farming, but difficult on its low gradients. Conducting water from the Euphrates required a channel that sloped at a steeper gradient than the landscape to avoid sedimentation, but not so steep as to invite accelerated erosion. Sheikh Dhiyab was technically ill-equipped for the task and had to balance between the extremes of sedimentation and erosion to solve this engineering dilemma.

Fortunately, the Euphrates lent a hand. After passing Hit, it flowed, as it does still, several feet above plain level in an elevated bed of its own making. Instead of cutting deeply, it transformed into a sediment-sorting machine, depositing coarser and heavier materials adjacent to the banks and lighter ones into distant basins, the distance traveled for each sediment class being inversely related to grain size.[19]

The resultant river levees have historically bestowed numerous advantages on local inhabitants. The most compelling one, at least for Sheikh Dhiyab, was a lateral gradient steeper than the vexingly flat plain itself, along which he could direct irrigation waters down the levee slope toward his gardens. The Euphrates had endowed Sheikh Dhiyab with natural levees that he could break to lay the basis for a flexible and cost-effective irrigation module easily excavated, resistant to siltation, and manageable by the kin group to which he belonged.[20]

The scheme did not go as planned, however. All of their advantages notwithstanding, levee breaks, whether initiated by human or natural agency, represent points of weakness in the levee bank that could expand to the detriment of the river. The weak point provided by Sheikh Dhiyab became a node of avulsion, from which the Euphrates branched, producing an offshoot that joined the trunk channel (already under stress since 1687, due to diminished discharge) in conveying all of the water and sediment delivered to the river. Sheikh Dhiyab initiated the levee crevasse sometime after October 1688. By

September 1689, Ottoman officials were reporting that it was growing at an exponential rate, flooding villages on its way and depriving those on the main course of sufficient water. Thus, in this first, partial phase of avulsion, the Euphrates bifurcated below Hilla into two low-energy, erratic branches—the parent channel and the Dhiyab Canal—before reuniting above Arja. The onset of this new river regime brought an ecological quandary that strained the Ottoman provincial administration.[21]

Provincial Crises

Under stable conditions and careful management, governments at the time could turn a partial avulsion into an opportunity to expand irrigation networks, transportation routes, and settlement centers by maintaining sufficient and controlled flow in the parent and new channels. During the late seventeenth century, however, the Ottoman administration in Baghdad found itself entrapped in a vicious cycle difficult to break. Epidemiological, political, and financial crises impaired its ability to mount an effective response to the environmental shift, which amplified those crises.

The dwindling water supply of the Tigris and Euphrates after 1687 created a panic that propelled more and more people into Iraq's urban centers, thereby maximizing the virulence of the plague when it hit the region.[22] Raging in northern and western Persia as early as 1684, the plague surfaced in Baghdad in March 1690 to claim 100,000 lives, according to an eyewitness account. Several months later, it reached Basra, where it killed 500 daily, and hit Baghdad again in early 1691.[23] The damage and loss of life in Baghdad were of apocalyptic proportions. "Friends and loved ones eschewed each other, the dues of fellowship were ignored, and everyone was preoccupied with his own life and careless about the conditions of others," wrote the historian Nazmizade. In his eyes, the scene was reminiscent of the Quran's portrayal of doomsday: "The Day man flees from his own brother, his mother, his father, his wife, and his children" (80:35–37).[24]

"Plague wields a power that is disproportionate to the deaths that it causes," one scientist remarks.[25] In the late seventeenth century, the high mortality it had caused in Baghdad and Basra compounded the damage that the Euphrates had done, creating an opening for mobile pastoral groups to prey on the afflicted urban populations.[26] The extraordinary career of Sheikh Maniʿ, leader of the Muntafiq tribe, epitomized the rural-urban change in power dynamics. In 1691, he marched audaciously toward Basra with an armed force of 2,000 to 3,000 men. Standing in his way was an Ottoman contingent of no more than 500 men headed by Basra's governor Ahmad Pasha. Terrified and outnumbered, 400 of the Ottoman troops deserted; the remaining 100 stood loyally with the pasha and died at his side on the battlefield. Another humiliating defeat in 1693 ended the mission of Baghdad's governor to re-assert control over the Basra region. The threat of Sheikh Maniʿ and other emboldened tribal leaders hung over Baghdad until the end of the seventeenth century.[27]

After the Euphrates veered off course, plague broke out, and the countryside revolted, Baghdad plunged into a financial meltdown, its treasury "totally broken."[28] The budgets of 1689 and 1690 sounded the alarm bells. Losses totaled 68,310 and 95,542 guruş, respectively, more than triple the budget surplus that the provincial treasury recorded in 1670.[29] As late as 1702, officials at the Baghdad treasury were still dealing with a loss of 134,413 guruş that they attributed to the damage made by the relocation of the Euphrates.[30]

The myriad forces of disease, politics, and finance conspired to sustain the perilous, segmented pattern of the Euphrates, allowing it to evolve unmolested. Depopulation and financial distress handicapped the provincial administration's ability to mobilize sufficient labor and resources to control the shifting riverbed, creating a political vacuum around Baghdad that rural forces exploited. What was initially an engineering complication (levee failure) gradually became a major military threat that called for an armed force large enough to defeat an emboldened rural enemy.

Imperial Crises

The Imperial Council in Istanbul complained to Baghdad's governor in late 1694, "From all sides, the cursed infidels are at present daringly assaulting the Muslim lands."[31] The Ottoman Empire, mired in an epic struggle against armies of the Holy League (1683–1699), suffered crushing defeats and enormous fiscal strains.[32] The debilitating demands of an exacting and prolonged war severely hampered the attempts of Ottoman policymakers to devise a proper response to the Iraqi crisis. In June 1692, the Imperial Council ordered Baghdad's governor to dam the Dhiyab Canal and use state money to reconsolidate the Euphrates. Experts in water control from the Rumahiyya region who had witnessed the development of the Dhiyab Canal appeared before the governor in Baghdad to offer their assessment of the project's feasibility and potential costs. The provincial administration concluded that as long as the Dhiyab Canal did not completely drain the main river, it could focus on other urgent goals requiring less time, effort, and resources. Hence, provincial officials ignored Istanbul's order for the time being.[33]

Willingly or otherwise, the Imperial Council followed suit and shelved any major hydraulic project, focusing instead on uprisings in the countryside. Two daring tribal leaders, Sheikh Maniʿ in the south and Bebe Süleyman in Şehrizor to the north, had blatantly undermined what the authorities regarded as the very purpose of the Ottoman state—safeguarding the welfare of the subjects. Rural rebellion became even more urgent when it spiraled out of control around the Safavid border, threatening the peace with Persia established by the Zuhab Treaty back in 1639. Ottoman decision makers unequivocally affirmed their interest in maintaining friendly relations with the Safavids and avoiding provocations, at least as long as the main Ottoman army was embroiled on the western front.[34]

The embattled Imperial Council adopted various measures to contain the fallout caused by rural turmoil—among them partial tax exemptions and resettlement plans for the disadvantaged, diplomatic exchanges with Safavid authorities, and constant admonishments to provincial officials to conduct themselves with clemency and justice toward the subjects. Its strategy regarding Bebe Süleyman and Sheikh Maniʿ oscillated between the politics of appeasement and outright warfare. Although both rebels had the blood of Ottoman officials on their hands, Istanbul sent Sheikh Maniʿ a conciliatory letter in 1693, granted him additional landholdings, and appointed Bebe Süleyman as bey of the Bebe Kurds around 1695. In more acrimonious moments, the council called for Bebe Süleyman's head and supported two half-hearted military expeditions against Sheikh Maniʿ in 1695 and 1698, both of which ended in failure.[35]

Thus, the imperial administration, like Baghdad, was trapped in a profoundly difficult situation during the late seventeenth century. Lacking the capacity to mend the disorderly flow of the Euphrates, it decided to give priority to immediate political action over long-term hydraulic management. As a result, from its inception, the Euphrates' partial avulsion threatened to push the system toward another dire condition too costly to reverse. That grim prospect materialized by 1700, forcing Istanbul to change its political calculations.

Daltaban Mustafa Pasha

In August 1701, an Ottoman equerry arrived in Baghdad carrying an imperial edict to governor Daltaban Mustafa Pasha, for whom Sultan Mustafa II held high hopes. Four months earlier, while performing the Friday prayer at the Selimiye Mosque in Edirne, Mustafa II received news that Daltaban

Mustafa Pasha, described as "a vigorous and tyrannical Serbian, illiterate but thrusting," had just re-established Ottoman control in Iraq and re-conquered Qurna and Basra.[36] Welcomed with attention and ceremony and dressed in sable fur, the emissary opened and read the imperial edict, praising the pasha's loyal service and relaying to him his new task—to dam the Dhiyab Canal and restore the Euphrates to its former channel.[37]

The long-awaited order arrived at a time when the map of the Ottoman world was being redrawn. The signing of the Karlowitz Treaty in 1699 ended hostilities between the Ottoman Empire and the Holy League and diminished Ottoman influence in eastern and central Europe.[38] In the east, Mustafa II and Shah Sultan Husayn of the Safavid Empire resolved border violations stemming from the activities of Bebe Süleyman and Sheikh Maniʿ diplomatically, despite deep mistrust on both sides.[39] More important than the geopolitical scene was the emergence of a new channel configuration in the Euphrates. Ottoman authorities had not kept track of intensified sedimentation occurring since 1687. By 1700, when war in the west was over and the trunk channel had become completely "filled and shut" and left dry and sandy, chroniclers in Istanbul and Baghdad were moved to issue grim accounts of the situation. The sight of sand hills standing in the middle of the derelict conduit later mesmerized even the Ottoman engineers.[40] The Euphrates had entered its second, full avulsion phase, diverting its flow entirely to the Dhiyab Canal, which suddenly expanded and, in the words of Raşid Efendi, "came to possess the strength and vigor of a large river."[41]

The central authorities could not tolerate this development. The new Euphrates played havoc with the agrarian and commercial sectors of the economy important to the Ottomans while empowering rural outcasts and rebels. Farmers had their fields either flooded and submerged or completely stripped of their water supplies, depending on their location; merchants had the movement of their vessels and caravans along the rivers interrupted; mobile pastoralists, meanwhile, acquired an auspicious opportunity to seize power. The Euphrates' new course refueled the process of wetland formation, creating new marshes and engorging old ones. Soon, the waterlogged sites turned into places of refuge and escape for rural groups to reassert their autonomy and challenge state authority.[42]

The ending of the long war in the west and the settlement of border disputes in the east finally moved the rehabilitation of the Euphrates to become a top priority. The Ottoman emissary to Baghdad informed Daltaban Mustafa Pasha that Istanbul was prepared to offer all of the financial support and labor required to achieve the Herculean task.[43] Rural uprisings made the initiative all the more urgent. Given a lull in major military engagements, confidence was high that the empire could finally crush the despised pastoral enemies within its borders. "When the degree of ignorance displayed by the barefoot and naked wandering Arabs becomes evident," Nazmizade wrote, "nothing is easier than having their ears pulled."[44]

An Engineering Feat

The Ottoman state brought into play what historian Sam White has termed an "imperial ecology," tapping into vast reserves of natural resources and workers throughout the empire to remake the Euphrates.[45] The provisioning of wood, which took precedence, fell on the backs of the governors of Maraş and Malatya, where they cut 42,200 logs of different kinds and sizes and transported them via animals to various points along the Euphrates' shores. Rafts picked up the logs from each spot and met at the Birecik shipyard in the upper Euphrates. Lighter wood, 12,000 palm and mulberry trees, came from Hilla in Iraq. The rough fibers surrounding the bases of palm fronds produced thousands of ropes and baskets. Aleppo's tax collector was in charge of providing oakum, hawser, marline, and iron nails. When all of the resources collected from Maraş, Malatya, and Aleppo assembled in the Birecik shipyard—together with cauldrons for cooking pitch, carpentry tools, forges, spades, pickaxes, sacks, bags, cannons, explosive bombs, and mallets—Ottoman authorities loaded the supplies into boats and shipped them downstream to Iraq.[46]

The manpower assembled under the command of Baghdad's governor was no less impressive. Steersmen, rowers, caulkers, and blacksmiths from the areas around Birecik descended upon Iraq via the Euphrates with cargo. Infantrymen escorted the boats as they traveled. Governors and pashas from Kütahya, Diyarbakır, Şehrizor, Mosul, and Köysancak joined their forces with Baghdad's janissaries. From Istanbul came the deputy commander of the janissaries (*kul kethüdası*) with a regiment, gunners, and armorers. Artisans, including brass workers and carpenters, and a military band took part in the campaign. After four months of travel, preparation, and ceremony, this motley crew departed from its meeting point in Baghdad in early December 1701 and crossed to the western Euphrates bank through Hilla, where it unloaded provisions from the boats.[47]

After violently crushing a rebel force, the expedition was ready to embark on the grand project, which it sought to accomplish in two major steps—(1) dredging and re-digging the sediment-choked channel and (2) damming and closing the Dhiyab Canal that by 1700 had absorbed all the flow. As soldiers and workers pitched their tents, Daltaban Mustafa Pasha surveyed the defunct channel with the high-ranking officers, blacksmiths, and carpenters. The plan was to dig a channel approximately 2.5 miles long, 300 feet wide, and 50 feet deep. The followers of each military leader, as well as locals from Hilla, Hasaka, Karbala, and Najaf, had specific areas to dig. The project began on December 22, 1701.[48]

The imperial administration had drafted traders and craftsmen to accompany the expedition and supply it with provisions. Bakers, grocers, vegetable sellers, butchers, drapers, and silk manufacturers established a marketplace in the campsite and attracted customers with "beautiful melodies." The authorities neglected neither hygiene nor morale; they established several baths and coffee shops and recruited confectioners and perfume makers. Ships carrying food from Baghdad, Hasaka, and Hilla on a daily basis sustained the camp. "As famously said," Nazmizade wrote, "even bird milk was found."[49] Known by other historians as the military market (*ordu pazarı*), this organizational method for supporting large campaigns suited the circumstances that the Ottoman Empire had faced from the late sixteenth century—an increasingly hostile political environment, financial crises, and ecological pressures.[50]

About 4,000 workers were employed on the project, carrying soil on their backs in sacks and bags. Transported earth formed hills on the sides of the river. A military band played with gusto to lift everyone's spirits. In the afternoon of February 7, 1702, after about forty-eight days of digging, water started to burst into the channel under reconstruction. After consultations, leaders concluded, "Good is in what God has chosen," and removed the barrier between the Euphrates and the channel to allow the river to run its new course. In celebration of their accomplishment, officers and workers read Quranic chapters, made sacrificial offerings, and raised their hands in prayer. Meanwhile, 200 naked men braved the gushing waters with rafts to continue clearing the new channel. More than 20,000 people on both sides watched the spectacle with awe, crying, "God is great! There is no god but God!"[51]

Following the reclamation of the old riverbed, blacksmiths, carpenters, and weavers, with the assistance of twelve ships from the imperial fleet and a few thousand oxen, focused their efforts on damming the Dhiyab Canal to complete the diversion. Nature, however, refused to cooperate. The Euphrates became unruly in the spring, the season when melting snow brings high floods in the river, and swallowed the thousands of soil-filled mats and baskets that workers had rolled into the water to establish the damming structure. In late

March, the river inundated the Ottoman camp and carried away remaining timbers. Due to timber shortage, rising water levels, and the campsite's putrefaction, workers decided to abandon the project on March 30, 1702. After appointing guards to oversee the partially built dam, they met with Daltaban Mustafa Pasha to compose a letter to their superiors in Istanbul describing their ordeal and justifying their decision, and dispatched the remaining provisions back to Baghdad.[52]

The Fall of Rumahiyya

The new Euphrates emerged from the engineering expedition intact, dealing a death blow to the Ottoman fort of Rumahiyya. The fort emerged during the Ilkhanid period in Iraq (1258–1336) as a riverine transit hub and apparently a center for spear production.[53] Under Ottoman rule in the sixteenth century, it bustled with a dyeing workshop, a slaughterhouse, a market, a press (for making juice or oil), a spinning mill, a tannery, and even a gaming house.[54] The top Ottoman officer in the town sometimes ran Baghdad's affairs when its governor left on a military campaign.[55] Tax-collection figures indicate that the town received a major boost after the Ottoman conquest of Baghdad in 1534, flourishing throughout the sixteenth and seventeenth centuries despite periodic Ottoman-Safavid confrontations.[56]

When compared with other market towns in the Baghdad province around 1580, Rumahiyya appears to have been nothing special in terms of wealth and population. Its significance, however, extended beyond its fortified walls. As the medieval Jerusalemite geographer Muhammad al-Maqdisi (d. c. 1000) sagely observed, "You should know that an area does not become sublime by the number of its towns, but rather by the splendor of its rural villages. Do you not see the splendor of Nishapur and Bukhara despite the dearth of their towns?"[57] In its rural villages, Rumahiyya was the most profitable farming region in Baghdad, densely populated, cultivated, and grazed. The Rumahiyya fort functioned as the Ottoman administrative center for the far more prosperous rural settlements and tribal groups in its vicinity, organized in five districts—Khalid, Kabsha, Malik, Zubayd Gharbi, and Zubayd Sharqi. In total, their tax revenue potential was greater than any other farming region in Baghdad, followed by the plains adjoining the lower Diyala River, one of the primary tributaries of the Tigris. The chief granary of several bygone empires, the core farming region of the Diyala River in the sixteenth century consisted of the districts of Khalis, Mahrut, Tariq-i Khurasan, and Shahraban.

Table 6.1. Major Market Towns in the Baghdad Province, c. 1580

Town	Tax Potential (akçe)	Town	Taxpayers
Baghdad	4,522,125	Baghdad	3,309
Mandalijin	511,463	Hilla	1,498
Baladruzin	452,132	Mandalijin	1,335
Shahraban	442,902	Rumahiyya	890
Hilla	361,121	Shahraban	405
Rumahiyya	251,520	Baladruzin	364

Source: TKG.KK, TT 29.

The Rumahiyya sub-province, where ancient Borsippa and its hinterland once stood, was an ecological patchwork of arable lands and marshlands interspersed with semi-arid steppe and ephemeral channels.[58] The environmental conditions favored extensive over intensive systems of land use, evident in the poor yields in return for inputs of labor compared with the situation in Diyala, a more homogenous landscape with better productivity and more reliable irrigation agriculture. Highly vulnerable to a seasonal flow out of phase with the agricultural cycle and to periodic uncontrolled runoff under poor drainage conditions, most land tracts were more amenable to stockbreeding than crop cultivation. They supported a considerable sheep population, three times bigger than Diyala's, and offered natural wallows for water buffalo.

Animal raising was an efficient way to convert Rumahiyya's inedible and nontaxable coarse grasses, bulrush, reed shoots, and other vegetation into appetizing and taxable types of food and raw material. The water buffalo in particular is renowned for its capacity to digest low-quality roughage material otherwise not useful as livestock feed for protein synthesis and milk production.[59] Herders found uniquely hospitable conditions for buffalo husbandry in Rumahiyya's fragmented landscape. In return, Ottoman officials imposed a lucrative annual tax of twenty akçe on every buffalo in Rumahiyya, compared to one akçe on sheep and two akçe on the sale of buffalo hides used in various types of heavy leather manufacture. The sale of sheep garnered a tax of two akçe.[60]

Nevertheless, a region given to stockbreeding and semi-sedentary folk was fraught with environmental and political risks. Although a successful environmental adaptation and a contribution to biodiversity, managed mammal herbivory in Rumahiyya came at the expense of a loss of

Table 6.2. Rumahiyya and Diyala Regions Compared, c. 1580

Farming Region	Rumahiyya	Diyala
Tax Potential (akçe)	4,879,668	4,426,948
Taxpayers	12,135	4,492
Caprid Population	76,023	24,176
Water Buffalo Population	11,837	—

Note: Percentage figures have been rounded; all figures are approximate.
Source: TKG.KK, TT 29, 49v-110r, 304r-468r.

productivity and significantly lower yields in cereal crops. The farmers of Diyala at the northern end of the alluvium—where arable production, the highest in value per unit weight, ruled supreme—were able to avoid such drawbacks. More important, allowing or even encouraging mobile pastoral groups to occupy and pasture their flocks in Rumahiyya to maximize the land's tax revenues bestowed political advantages to herders in their highly contingent relationship with the ruling power. With their wealth on the hoof, they were "both inclined and able to resist or evade centralized government," particularly during times of political instability, when the chain of command inscribed in Ottoman cadasters and law codes eroded.[61] The following Anatolian poem finely captures pastoralists' penchant for independence and autarky: "Do not cultivate the vineyard; you'll be bound / Do not cultivate grains; you'll be ground / Pull the camel, herd the sheep / A day will come, you'll be crowned."[62]

An auspicious opportunity for Rumahiyya's pastoralists to crown a chief of their own came in 1700 when the original Euphrates branch fell into disuse and deprived the fort of its water supply. After the Baghdad administration suffered rapid financial losses in the Rumahiyya fort, tribal forces assumed control of the town in 1694.[63] Following the complete abandonment of the Euphrates trunk channel, Ottoman ledgers in 1702–1703 record for Rumahiyya, and its districts of Malik, Khalid, and Kabsha, an outstanding debt of 37,782 guruş, roughly double the budget surplus the Baghdad treasury recorded in 1670. After 1704, the fort dropped from the ledgers all together.[64] In 1765, a passerby noted that the Khazaʿil tribe was in control of the town's dwindling population, from which it collected tribute.[65] The town's decline became irreversible before the end of the eighteenth century, when the French consul in Baghdad described it as an "ancient city" that had "fallen into ruins."[66] By the nineteenth century, Rumahiyya had largely disappeared from the historical record.

———

The "fatal synergy between natural and human disasters," as historian Geoffrey Parker puts it, in late seventeenth-century Iraq is comparable to the Ottoman crisis in Anatolia nearly a century earlier, when the similar elements of drought, inflation, plague, rural rebellion, and military entanglement with the Habsburgs happened to coincide during the 1590s.[67] Nonetheless, this fatal synergy had a different tinge in Ottoman regions endowed with large river systems like Iraq and Egypt. Here, fluvial processes exerted enormous power in shaping and defining the experience of disaster.[68] The events between 1687 and 1702 in Iraq were no different. Dwindling flow in the Tigris and Euphrates created an atmosphere of crisis that crept over Baghdad and Basra, amplifying the impact of plague in the region. Change in the climatic, hydraulic, and sedimentary variables governing flow in the Euphrates triggered avulsion, abetting the military success of rural rebels in the countryside and contributing to the financial breakdown of the Baghdad treasury. Interaction between the Euphrates and Ottoman epidemiological, political, and financial crises was not a one-way process. The depopulation, rural unrest, and bankruptcy that changes in the Euphrates had fostered hampered effective hydraulic management and allowed the partial avulsion of the late 1680s to become full by 1700. When the dust had settled, literally (in the now-dry bed of the river), a new fluvial landscape was in place. Desert and river had changed places, Rumahiyya had crumbled, and the Ottoman Empire would need to start afresh in Iraq.

7

After the Flood

The governors of Baghdad certainly needed more power.
—Halet Efendi, *Hatt-ı Hümayun* (c. 1810)

WITHIN A FIFTEEN-YEAR period, feedback loops between natural and anthropogenic forces gave violent birth to a new hydraulic order in Iraq. This chapter outlines the gradual political changes that followed the environmental reconfiguration. Most abruptly, the Ottoman equilibrium in the countryside, based on a delicate balance between herders and cultivators, fell apart. In its place emerged a fragmented rural landscape divided into the private domains of ever more defiant pastoral groups. In response, Istanbul granted its provincial governors more prerogatives, which they used both to reassert Ottoman authority and to build their own power base in Iraq. This tension between central and local interests gave rise to an assertive military oligarchy referred to by European diplomats as the Pashalik of Baghdad. The Pashalik amassed its own private army and developed its own networks to secure grains, arms, and riverboats. By the end of the eighteenth century, this trend toward provincial autonomy localized the Ottoman management of the Tigris and Euphrates rivers, shifting its command post from Istanbul to the provinces, Baghdad in particular.

The Pashalik of Baghdad epitomized the rise of provincial dynasties throughout the Ottoman Empire during the eighteenth century. Until the 1980s, historians commonly viewed this empire-wide phenomenon as the hybrid offspring of nationalist awakening and Ottoman weakness.[1] Research since then has largely rejected this interpretation, emphasizing instead specific regional influences and imperial policies. In the Balkans and Anatolia, for example, military defeats with European powers (particularly Russia) and the spread of banditry in the countryside fed off each other, forcing the Ottoman state to rely on local magnates and their private armed retinues to maintain order and support Ottoman war efforts.[2] Other explanations emphasize the introduction of tax farms called *malikane*, lifetime tax-collection rights on most provincial revenue sources, including villages, markets, canals, mills, and public baths.[3] Conceived in 1695 and widespread throughout the Ottoman Empire by 1703, the malikane system aimed to eliminate the insecurity of short-term tax farms prevalent in the seventeenth century and to raise more cash for the Ottoman state. In return, investors in these tax contracts acquired administrative rights over their tax farm units, which they ruled with little interference from Istanbul.[4] Families that used the malikane system as a stepping stone to provincial power include the Caniklizades in Canik, the Çapanoğlus in Yozgat, the Azms in Damascus, and the Jalilis in Mosul.[5]

To these two popular explanations, this chapter adds that environmental change, at least in Iraq, contributed to the rise of "regional governance regimes" in the eighteenth-century Ottoman Empire.[6] The abrupt relocation of the Euphrates River between 1687 and 1702 fomented rural rebellion that could not be decisively quashed without a drastic overhaul of the Ottoman administration in the region. Istanbul initiated the reform process in 1704 by appointing a veteran statesman as governor of Baghdad and set the region on the path toward provincial autonomy for the rest of the century. Upon arrival, he would find the ruins of the old Ottoman order.

Whither the Royal Herds

The ecological crisis of the late seventeenth century produced a seismic shift in the Iraqi countryside. From the conquest of Baghdad in 1534, the Ottoman administration maintained some semblance of rural order by integrating a large segment of the mobile pastoral population into taxable, state-sponsored associations. Avulsion and its concomitant troubles, however, would inflict irreparable damage on this long-standing Ottoman arrangement.

During the seventeenth century, the Ottoman herders' associations continued to maintain their cohesion and strength. As late as 1670, revenues of the Baghdad province came from ninety-five towns, villages, businesses, and other revenue sources granted as tax farms (*mukata'a*). The most lucrative among them was the fixed sum that Ottoman Baghdad collected from the Ahşamat and Cemmasat associations, a staggering figure of 40,000 guruş, double the amount collected from the remaining ninety-four tax farms.[7] In the words of the Imperial Council, the Ahşamat and Cemmasat were

Baghdad's "largest and most profitable tax farm in terms of population size, resources, and [other] benefits."[8] Likewise, the two other shepherd groups in the region, the Qara Ulus and Qara'ul, were financially significant, appearing fifth (10,250 guruş) and seventh (7,000 guruş) on the list of Baghdad's revenue sources. Through these associations, the Ottoman administration welcomed a large segment of the mobile pastoral population into the fold and reduced the tensions between pastoralists and other Ottoman subjects, particularly those who tilled the soil in the countryside.

It was during the Euphrates' avulsion that the herders' associations started to unravel. Ottoman officials resorted to the wealth of the Ahşamat and Cemmasat in their attempt to steer Iraq through a political and financial breakdown. Increasingly subject to arbitrary exactions, most herders belonging to the Ahşamat and Cemmasat groups left their territories and sought refuge in different locations outside Iraq, aggravating the very financial problems that Ottoman officials were trying to solve.[9] Officially, the Ottoman administration did not abolish its herders' associations in Iraq until the introduction of the Tanzimat reforms in the early nineteenth century.[10] After 1700, however, the associations became shadows of their former selves.

State-sponsored herders' associations, the distinctive feature of the Ottoman political order in rural Iraq, came out of the ecological storm of the late seventeenth century weaker and poorer than ever before. In their place emerged a more divided rural landscape dominated by numerous tribal sheikhs with whom Ottoman authorities had to battle and negotiate individually on an ad hoc basis. No tribe was more creative in adapting to the new hydrography of Iraq than the Khaza'il. The story of their marshy kingdom typifies the intractable rural world that emerged in the eighteenth century.

Kings of the Middle Euphrates

"The flames of depravity and anarchy ignited, and a shepherd in every area called for independence and proclaimed sovereignty," wrote the Ottoman official Murtaza Nazmizade in the aftermath of the avulsion in 1700.[11] The time was ripe for the shepherds of Iraq, who could deploy their herding and hunting skills, finely honed for many decades with Ottoman acquiescence, to exploit the political advantages of a mobile pastoral life.

While many tribes fled the scene to evade the crushing exactions of Ottoman officials, others did not surrender easily. They held their ground and took advantage of the chaos to regroup into new and powerful confederations, better equipped to defend their political interests. Nazmizade highlights the lightning political success of three confederations during this tumultuous period. The first was a local confederation based in southern Iraq called the Muntafiq. By 1694, its eminent leader, Sheikh Mani', conquered Basra and extended his influence as far north as Samawa on the Euphrates.[12] The second was a confederation from northern Arabia called the Shammar, which exploited the upheaval to establish a permanent foothold in Iraq, initially between Falluja and Karbala along the Euphrates.[13] In 1695, the Ottoman governor captured around fifty Shammars and beheaded them in Baghdad following a battle with the tribe on the Euphrates.[14] In this bloody event, the Shammar confederation entered the chronicles of Iraq for the first time, becoming a major political player.[15] Joining the Muntafiq in the south and the Shammar in the north was the Khaza'il confederation, building its power base in the marshes of Hasaka and Lamlum along the middle Euphrates.

The Khaza'il confederation rose to eminence in the heart of the region ravaged by avulsion. As its channel moved eastward, away from the Rumahiyya fort and closer to Hasaka, the Euphrates flooded agricultural settlements on its way and created enormous marshes, the largest of which seemed to Nazmizade like a sea.[16] Tribesmen were quick to seize the opportunity. They refused to pay taxes and fortified their positions in the marshes. From the midst of an environmentally and politically fragmented landscape, a charismatic leader emerged, rocketing his tribe to fame across Iraq. Abdulrahman al-Suwaydi, a typically disdainful city dweller, reports that a man "of ugly face," "more debauched than a rat, and even more rotten," proclaimed his sovereign rule in the region between Hasaka and Najaf and spread his call for independence among the neighboring tribes.[17] The ugly, debauched, and rotten man was Sheikh Salman, son of Abbas and leader of the Khaza'il tribe. After transferring villagers to cultivate his newly conquered land, Sheikh Salman besieged the Hilla fortress, the last pillar of Ottoman rule in Iraq's middle Euphrates region. The move prompted a swift intervention by Ottoman commandos (*serdengeçti*) and janissary troops, successfully repelling the brazen incursion but failing to dislodge Sheikh Salman from the areas under his control, where he collected taxes and behaved like a sovereign.[18]

The rise of the Khaza'il in 1700 marked a dramatic change in their fortunes. Similar to other pastoral groups worldwide, the Khaza'il entered the historical record following their encounter with an agrarian, state-organized, and literate society. The Ottoman cadastral survey of Baghdad compiled around 1580 depicts them as a humble tribe in the Rumahiyya region, made up of 108 buffalo herders organized into seven groups. As a supplementary activity, they dabbled in sheep rearing. Even though they raised more sheep than buffalo, their sheep flocks (173 sheep in all) were significantly smaller than the average size maintained by the dedicated sheep breeders of Rumahiyya (about 521 sheep per group and village). They distinguished themselves primarily in buffalo rearing, maintaining a herd (113 head) almost double the average size of other buffalo breeders in the region (about 68 per group and village).[19]

Far from conferring prestige upon the Khaza'il, buffalo rearing more than likely made them the object of scorn. Iraqi pastoral and agricultural tribes generally considered buffalo herding to be beneath them, despite their recognition of its profitability.[20] The semi-aquatic buffalo were highly susceptible to infection from the helminth parasites that infest the wallows of the Euphrates as well as to sarcoptic mange in the dry season.[21] Anthropologist Shakir Mustafa Salim adds that the local disdain for buffalo breeding was traceable to the desert ethos of late antiquity, which prompted the Arab tribes encroaching on Iraq since the early first millennium to distinguish themselves from the local marsh population by occupation.[22] For whatever reasons, whether related to issues of health, hygiene, culture, or a combination of these and other factors, the Khaza'il enjoyed little prestige among their neighbors, far behind many of them in wealth and numbers.

The Safavid conquest of Iraq in 1623–1638 brought ephemeral prestige for the Khaza'il, who shared the Shi'i faith with the Persian power. Shah Abbas I appointed Sheikh Muhanna, son of Ali and leader at that time of the Khaza'il, as his vassal (khan) in the middle Euphrates. The tribe's political fortunes soon collapsed when the Ottoman Sultan Murad IV reconquered

Table 7.1. The Khaza'il Tribe, c. 1580

Group Name or Leader	Taxpayers	Water Buffalo Population	Sheep Population	Tax Potential (akçe)
Sheikh Khalifa	22	22	35	887
Sheikh Rumh	19	20	30	882
Sheikh Hasan	7	8	10	316
Al Abu 'Akkash	12	12	18	485
Sheikh Ghanim	14	15	25	589
Sheikh Ma'an	25	26	40	1,026
Askar Ra'is	9	10	15	393
Total	108	113	173	4,578

Source: TKG.KK, TT 29, ff. 332v-333v. See also OA, TT 1073, 18.

the region in late 1638. Three years later, an Ottoman platoon brutally crushed the Khazaʿil in their territory and presented Baghdad's governor with 600 severed heads as war trophies. Survivors of the massacre, including the Khazaʿil leader Sheikh Muhanna and his retinue, took refuge in the Safavid realm, where some remained while others returned to their homeland along the Euphrates in small groups during the following decades.[23]

Under the astute leadership of Sheikh Salman, the Khazaʿil seized the opportunity provided by the Euphrates' channel shift to radically transform their political standing in the region. A convicted tax felon and fugitive who broke out of a Baghdad prison around 1694, Sheikh Salman emerged in 1700 with a hefty military force of 10,000 men armed with muskets and spears. By 1701, he had brought most of Rumahiyya's districts under his sway.[24] The Khazaʿil, the once-stigmatized buffalo herders, had turned into "kings of the Middle Euphrates."[25]

The Khazaʿils' story demonstrates how the rural and fluvial realms of Iraq transformed in lockstep at the turn of the eighteenth century. On the ruins of state-sponsored herders' associations, assertive tribal confederations formed and challenged Ottoman hegemony in the region. Initially, the Muntafiq, Shammar, and Khazaʿil were the biggest rural sharks, but more tribes would join them later in the eighteenth century, particularly those escaping the onslaught of the fanatical Wahhabi movement in central Arabia.[26] Unprecedented political challenges, Istanbul quickly realized, required an unprecedented political response.

An Ottoman Restoration

After his appointment as governor of Baghdad in June 1704, Hasan Pasha entered a city worn out by violence and disasters. He did not stay long before setting out on his first expedition into the countryside. With the aid of 3,000 men, he stunned two Shammar tribes on the upper Tigris by climbing atop a nearby hill, from which he rained artillery shells onto the tribesmen's fortification.[27] After crushing them, the governor distributed a proclamation among the tribes of Iraq, arguing that those who refused to show loyalty and pay tribute to the Ottoman government were virtually betting their own lives on a lost cause. The governor concluded his proclamation with a severe warning cribbed from the Quran: "When We decide to destroy a town, We command those corrupted by wealth [to reform], but they [persist in their] disobedience; Our sentence is passed, and We destroy them utterly" (17:17).[28]

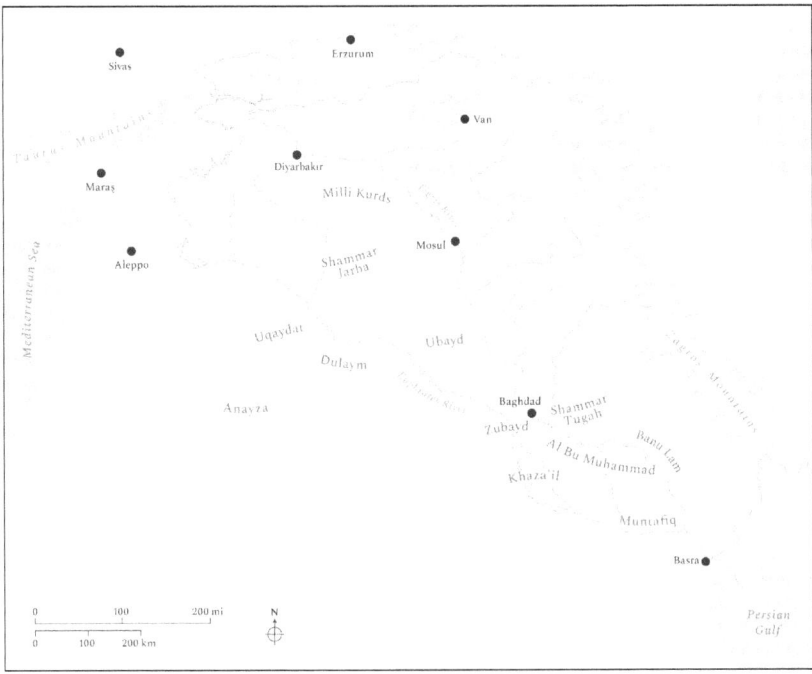

FIGURE 7.1. Dominant Tribal Confederation in the Tigris-Euphrates Basin, c. 1800. After Tom Nieuwenhuis, *Politics and Society in Early Modern Iraq: Mamluk Pashas, Tribal Shayks and Local Rule between 1802 and 1831* (The Hague: Martinus Nijhoff, 1982), Map V.

Hasan Pasha's shock and awe tactics announced the beginning of a new chapter in the history of Iraq. Born in the Greek town of Katerini to a Georgian officer, he rose through the echelons of Ottoman bureaucracy, holding the governorships of Konya, Aleppo, Urfa, and Diayrbakır before his appointment to Baghdad. His chief task was to rescue the Ottoman presence in Iraq after seventeen years of natural and political chaos that bred massive rural rebellion.[29] Istanbul emboldened the governorship of Baghdad to help the new appointee carry out his monumental mandate.

Hasan Pasha had at his disposal new privileges and opportunities possessed by none of his predecessors. The first was a lengthy term of office. Between 1638 and 1704, Istanbul installed thirty-eight governors in Baghdad, who occupied their office for less than two years on average. For the first time in history, Istanbul allowed Hasan Pasha to hold the governorship of Baghdad for twenty years (1704–1724), until his death, an extension that would become normalized throughout the eighteenth and early nineteenth centuries.[30]

Second, Istanbul extended to Baghdad the lifetime malikane tax farms, through which Hasan Pasha could control most offices and revenue sources in the province. He invested in these new tax grants, which proved instrumental in augmenting his political position, both within Baghdad and in relation to his superiors in Istanbul.[31]

Finally, Istanbul strengthened the hand of its governor in Baghdad by consolidating the imperial navy on the Tigris and Euphrates—the Shatt Fleet. Unlike the Ottoman fleets at Suez and on the Danube, this one lacked a permanent admiralty from its establishment.[32] Throughout the sixteenth and seventeenth centuries, the admiralty of the Shatt remained an ad hoc position, created temporarily to accomplish a particular mission, such as a major military campaign or a shipping order. In 1567, for instance, the imperial administration designated Canbulad Bey, the governor of Kilis and A'zaz, as admiral of the squadron joining an expedition against the marsh Arabs.[33] In 1577, Darende Mahmud was the admiral of boats transporting timber from Birecik to Basra and Lahsa.[34] A shift in policy occurred when Istanbul decided to deal with the aftermath of the Euphrates' avulsion at the turn of the eighteenth century. In 1700, Istanbul conferred the Shatt admiralty's office upon Aşçızade (Son of the Cook) Mehmed Pasha, who had been admiral of the Danube Fleet for the previous five years.[35] From his appointment onward, admiralty of the Shatt became an institutionalized position, called the Shatt Kapudanı and regularly staffed well into the nineteenth century.[36]

In the early eighteenth century, therefore, Istanbul created favorable conditions for the governor of Baghdad to cement his power and end the mayhem in Iraq once and for all. The incumbent could enjoy a long tenure of office and control most provincial posts and revenues. Backing his efforts to pacify a restive rural population was a consolidated naval bureaucracy operating year-round in Iraq. After assuming office in 1704, governor Hasan Pasha did not disappoint. He waged regular expeditions against rebellious tribes and brought about one of the most stable periods in the history of Ottoman Iraq.[37] The same opportunities that allowed Hasan Pasha to restore Ottoman authority, however, opened new doors for self-aggrandizement.

The Pashalik of Baghdad

Unlike previous governors who came and left within two years, Hasan Pasha struck deep roots in Baghdad during his long tenure in office. As he stabilized Iraq for Istanbul, he pursued his own political project through the purchase of offices and contracts. His powerful household would monopolize most positions within the Ottoman provincial administration, including the governor's office, until the early nineteenth century. In this great grab for provincial power, Hasan Pasha's household would take over the Admiralty of the Shatt and effectively gain control over the most strategic zone within the Tigris-Euphrates basin.

Educated in the Topkapı Palace in Istanbul, governor Hasan Pasha founded a provincial dynasty in Baghdad modeled on that of the Ottoman sultan and centered around his own person. His elaborate household included different chambers, schools, and even swimming pools, where recruits from a young age acquired a combination of martial and scribal skills. The goal was to establish a disciplined army and bureaucracy loyal to him and to outmaneuver the insubordinate regiments deployed by Istanbul. In the beginning, the sons of local families in Baghdad dominated the page corps in the household. Later, Hasan Pasha and his successors systematically recruited slave boys of Caucasian origin, mostly from the Tiflis area of Georgia. Purchased in the slave market, those Georgian boys rose to freedom during the training process and joined the ranks of the ruling elite as soldiers and bureaucrats, known as "mamluk" in Arabic and "köle" or "kölemen" in Turkish. Some of them would even marry into Hasan Pasha's family and occupy the governor's office.[38]

Members and clients of Hasan Pasha's household would fill the most important military and civil posts in the Pashalik's administration. Junior mamluks served in small districts and within the provincial palace. Veteran members, on the other hand, occupied the highest offices, including those of the lieutenant (*kethüda*) of Baghdad and the commissioner (*zabit*) of Hilla and Hasaka, the two richest agricultural areas on the middle Euphrates. The crème de la crème of the household seized the office of the governor. After Hasan Pasha's death in 1724, the governorship of Baghdad passed to his son Ahmad Pasha, then to the Georgian slave soldiers. Several times from the 1730s, Istanbul tried to interrupt this dynastic course by fielding its own candidates to the governorship, but to no avail. It had to acknowledge the governor backed by the provincial Georgian guard as a fait accompli.[39]

In addition to the governorship, lieutenancy, and other high offices in the provincial administration, the Pashalik over the course of the eighteenth century captured from Istanbul the Shatt Admiralty, the highest-ranking military officer operating on the Tigris and Euphrates rivers. This critical step gave Baghdad for the first time autonomous control over the navy and river transport in Iraq.[40] According to a British visitor in 1774, the Shatt Admiral was the "lord of the sea-coast in the part of the gulph of Persia dependent on the pasha of Bagdad, of the marine forces, and of the river from the bar to Bagdad on the Tigris, and Helah [Hilla] on the Euphrates. All ships that trade to and from Bussora [Basra] .... pay him a duty per ton on arrival and departure, as do all vessels that trade to and from Bagdad or Helah, or any intermediate place: even the fishing boats pay him so much per month."[41]

Effectively, the Pashalik of Baghdad came to resemble what one historian called a state within the Ottoman state.[42] Istanbul's hold on Iraq and the Tigris-Euphrates system loosened during the eighteenth century as the Pashalik amassed its own private army and navy and controlled the apparatus of the Ottoman provincial government, including taxation, finance, and policing. Iraq would become even less dependent on Istanbul when the Pashalik started to oversee its own provisioning networks.

The Northern March

In terms of natural resources, Iraq was inadequately endowed, at least compared to its northern neighbors in higher precipitation zones. The Pashalik of Baghdad could not live upon the country and hope to preserve its autonomy within the Ottoman Empire at the same time. Rather, it had to secure its access to resources from outside. The struggle for autonomy, thus, propelled an aggressive expansionist agenda.

The Pashalik turned its gaze northward. In the Kurdish region east of the Tigris, it appointed members of the mamluk household to the governorship of Şehrizor and installed its own Kurdish candidates to rule five critical districts: Qara Cholan, Zakho, Süleymaniye, Köysancak, and Amadiyya. In southeastern Anatolia, the Pashalik detached the districts of Mardin and Nusaybin from the Diyarbakır province through their purchase as lifetime tax farms. From here, the Pashalik subordinated the Milli Kurds, the dominant tribal group in the region, and sometimes granted the governorship of Urfa to their emir. Even Mosul, governed by the autonomous Jalili dynasty, fell within Baghdad's sphere of influence. Contenders for the Mosul governorship courted the Pashalik for its endorsement, supplying it with troops, gifts, and various goods in recognition of its dominance.[43]

Between its northern possessions and zones of influence, the Pashalik of Baghdad built and managed its own communication network. In the Kurdish east, it could control the Zagros pass, a vital artery between the Ottoman Empire and Persia. In the Anatolian west, it dominated the route between Mosul and Syria, using it as a leverage over the Jalili dynasty. A stony country more than 3,000 feet above sea level, Mardin proved particularly important,

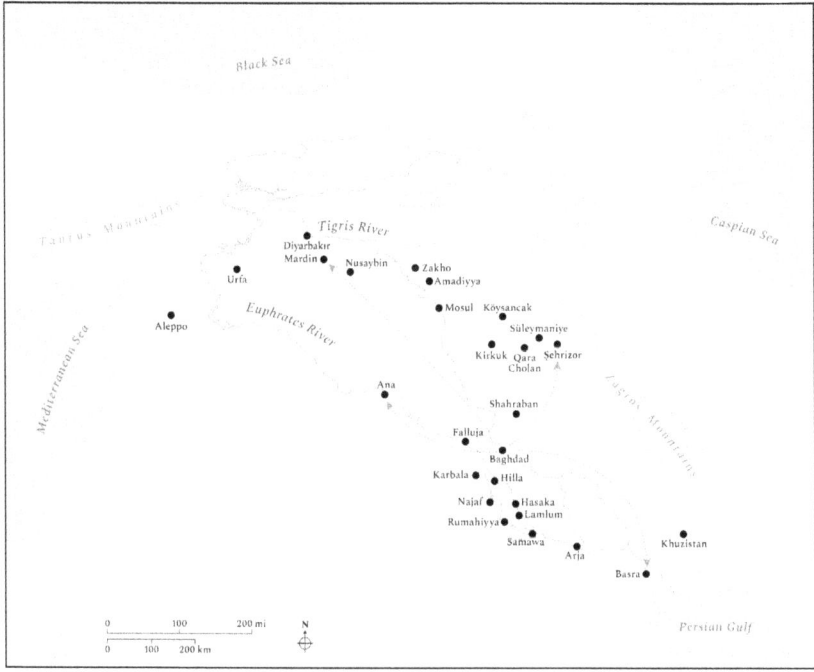

FIGURE 7.2. Expansion of the Pashalik of Baghdad.

forming Baghdad's watchtower in the north. "If I become the superintendent of Mardin," governor Ahmad Pasha (r. 1724–1734, 1736–1747) reportedly said, "I will be promptly aware of any imminent danger, and it will be difficult [for Istanbul] to dismiss me from my position in Baghdad."[44]

Through this northern network, Baghdad gained direct access to the resources that had been beyond its reach from the sixteenth century and thus reduced its dependence on Ottoman central authorities. Mardin and Nusaybin emerged as two of the most important granaries in the hands of the Pashalik, directly managed by its appointees.[45] Enclosed by the Pashalik's sphere of influence, the Jalili governors of Mosul, too, maintained constant supplies of grain whenever Baghdad needed it.[46] From the forested Kurdish mountains south of Süleymaniye, the Pashalik acquired valuable timber, principally plane trees (*çınar*). It sponsored local entrepreneurs to fell, clean, and dry the required timber. From the mountains, entrepreneurs would oversee the transportation of the logs to the banks of the Diyala River, where they could be floated downstream when the waters were high during the spring season. At the other end of the river near Baghdad, agents of the Pashalik on the watch would intercept and take the logs out. Aside from planes, Baghdad could obtain mulberry and nut trees from Kurdish gardens, poplar (*kavak*) from Amadiyya, and willow (*söğüt*) from the northern Euphrates.[47]

Northern expansion, in short, paid dividends. It created a buffer against Istanbul's attempts to depose the Pashalik from Baghdad and install its own candidates as governors. It placed critical nodes of communication under the control of the Pashalik, through which it could influence most trade flowing between Persia and the eastern Mediterranean. And last but not least, northern expansion balanced the natural resource deficit that plagued Baghdad without resorting to the assistance of Istanbul. Expansion into the south would introduce Baghdad to another world and another set of possibilities.

The Southern March

While pushing north to the foot of the Anatolian plateau, the Pashalik of Baghdad wrested control of the southern province of Basra. As early as 1710, Baghdad's governor Hasan Pasha appointed his son-in-law Mustafa Ağa to the Basra governorship.[48] A port city, Basra offered the Pashalik opportunities that inland provincial dynasties like the Jalilis lacked. Financially, the bustling commercial hub would generate the largest source of income for the Pashalik.[49] Commercially, it allowed the Pashalik to dominate the trade routes between the Persian Gulf and the eastern Mediterranean. Sometimes, the Pashalik would even force merchants traveling between Basra and Aleppo to make a detour to Baghdad along the way, so it could collect even more customs.[50] Most important, the port of Basra brought the Pashalik into close contact with maritime powers, on which it would rely to maintain the Shatt Fleet in virtual independence of the Imperial Naval Arsenal in Istanbul.

From Basra, the Pashalik resorted to the naval capabilities of three ascending maritime powers in the Persian Gulf. One was the Shi'i Arab tribe of Banu Ka'b, which built a power base in Khuzistan (the southern borderland region between Iraq and Persia) following the collapse of the Safavid Empire in 1722. At the zenith of their power in the 1760s, the Banu Ka'b expanded their control to the mouth of the Shatt al-Arab and established a powerful fleet of about eighty boats, larger than the Pashalik's Shatt Fleet at the time. During a conflict with the Pashalik in 1773, the Banu Ka'b's navy even destroyed much of the Shatt Fleet as well as the house of its admiral in Basra.[51] Nevertheless, as with all Arab tribes in the region, the Pashalik had a capricious relationship with the Banu Ka'b and solicited their naval support during its campaign against the Muntafiq tribe in 1769, during which the Banu Kaʿb provided a fleet of fourteen gallivats.[52]

The second maritime power to occasionally serve as an auxiliary for the Shatt Fleet was that of the Imam of Oman, Ahmad son of Saʿid. Two years after the death of Nadir Shah in 1747, Ahmad expelled a Persian occupation force and founded the Al Bu Saʿid dynasty, which has ever since ruled the country. During his reign, a considerable portion of the Persian Gulf trade shifted to Oman due to the political unrest in Basra and Persia and the abandonment of the British factory in Bandar Abbas. The commercial realignment made Oman an attractive destination to foreign merchants and allowed Omani merchants to build a fortune through the export of Mocha coffee and the transport of Batavian products to the Persian Gulf and western Indian Ocean. Around 1775, the fleet owned by Oman's Imam comprised four ships of forty guns each, twenty-five ketches and gallivats of eight to fourteen guns each, and a large yet unspecified number of dhows and trankies. During the Persian siege of Basra by the Zands in 1775, the Pashalik wrote to the Imam of Oman asking for his naval support to relieve the city. In response, he dispatched a fleet of some thirty ships under the command of his two sons. The fleet broke the blockade on Basra and offered limited relief after a long involvement with Persian forces that exhausted the fleet's provisions. Nevertheless, even the Ottoman Sultan Abdülhamid I appreciated Oman's naval aid to his vassals in Iraq and expressed his gratitude in a letter to its Imam.[53]

The third and most consequential maritime power that the Pashalik interacted with was the British East India Company. During intervals from the seventeenth century, the company maintained a factory and resident in the port of Basra, importing metal ware and woolens and bartering them for cash and Persian silk. Following its expansion into Basra, the Pashalik developed a special relationship with the company. In 1728, 1731, and 1759, it issued edicts outlining the (favorable) terms by which the company could continue to operate and trade in Iraq.[54] The relationship would become even stronger during the second half of the eighteenth century. First set ablaze by a French corsair and then involved in a bloody skirmish with the khan of Lar, the East India Company's factory in Bandar Abbas (southern Persia) was evacuated and moved to Basra in 1763, where its representative was promoted from resident to agent and strengthened by the grant of consular status.[55] The Pashalik immediately started to enlist the naval support of the company's agents in its conflicts with domestic and foreign enemies. In one campaign against the Banu Kaʿb in 1765, the company contributed a private vessel to the Pashalik's army and recruited two young Englishmen to command two vessels in the

Ottoman squadron. The two sides formalized this naval cooperation with the signing of a treaty in 1767, stipulating that a British fleet would protect the port of Basra and the Pashalik would pay for the services.[56] More joint naval operations would follow. During the Zand siege of Basra in 1775, for instance, British commanders took charge of two gallivats in the Pashalik's squadron, each carrying eight guns and eighty to a hundred men and hoisting British flags.[57]

Aside from hiring vessels out and providing mercenaries, the East India Company sold to the Pashalik brand new ships built by the Bombay Presidency. In 1773, its agent sold two ketches, each equipped with fourteen guns and able to carry 115 persons.[58] To operate both, the Pashalik hired an English naval officer from the company as a commander and recruited a crew of English and local sailors.[59] The Pashalik placed another order in 1780, after Baghdad's governor pleaded to Istanbul early in the year that he needed ten new vessels with six guns each to regain territories he had lost to Arab tribesmen. He said that the British "balyos," likely in reference to the company's resident in Basra William Digges La Touche, could oversee and complete the construction in Bombay within six months and asked Istanbul to make the payment to the British ambassador in Istanbul or to the consul in Aleppo.[60] Istanbul and Baghdad ended up purchasing six armed gallivats, which the company's agent delivered to Baghdad's governor in 1782 and 1783.[61] In the summer of 1785, the Pashalik acquired one more ship through the company and was still waiting for the arrival of five more.[62] After the Wahhabis sacked the holy city of Karbala, Baghdad's governor asked Istanbul in 1802 to order more gunboats from the Bombay Presidency to repel further Wahhabi encroachments.[63] The British resident in Baghdad, in a letter to the ambassador in Istanbul, supported the request, but there is no indication that the order was fulfilled.[64]

In addition to naval assistance, the East India Company became the Pashalik's primary arms supplier by the end of the eighteenth century.[65] It regularly armed Baghdad with mortars, shells, muskets, gunpowder, and other munitions of war and sent European bombardiers and gunners trained in the military service of the company in India to serve in the army of the Pashalik.[66] In 1799–1800, for instance, the governor of Baghdad Süleyman Pasha hired from India an English military instructor named John Raymond to train his army.[67] By 1810, an Ottoman commissioner noted that the Pashalik purchased most of its arms and ammunitions from British ships that docked at the port of Basra.[68]

The East India Company supported the Pashalik of Baghdad with its vessels and crews for different reasons. Enemies of the Pashalik often constituted a direct threat to the company's commercial interests in the area. Even when its interests were not directly at stake, agreements over pay and prizes for an intervention seemed attractive. The calculation radically changed when the Ottoman and Qajar dynasties were shaken by Napoleon's invasion of Egypt in 1798 and Russia's annexation of Georgia soon after, developments that Britain considered a threat to its position in India. Supporting the Ottomans and Qajars against their European rivals henceforward became an official British policy.

From the sixteenth century, the Ottoman navy on the Tigris and Euphrates owed its life to Istanbul, which assembled the capital and labor necessary for its maintenance. During the eighteenth century, the Pashalik of Baghdad severed this naval artery in the imperial body, allowing Iraq to drift further away from Istanbul. This audacious step could not have been possible without the takeover of Basra, where the Pashalik found alternative sources for naval support. Most important among them was the British East India Company, which provided the Shatt Fleet all the services that the Imperial Naval Arsenal in Istanbul once rendered. Aside from naval assistance, the company replaced Istanbul as Iraq's main supplier of military training and equipment. Dependence on the British arms industry became more entrenched during the nineteenth century, when iron replaced timber as the primary shipbuilding material and the steam engine replaced wind and human muscles as the primary means of propulsion.[69]

Abandoning Ship

Relentless expansion by Baghdad in the Tigris-Euphrates basin corresponded to the retreat of the Ottoman central state. The most striking sign of Istanbul's disengagement from the naval affairs of the region can be found in the Birecik shipyard. From the sixteenth century, Birecik emerged as the most vital naval installation and conduit for military provisions in the entire drainage basin. In the late seventeenth century, the Iraq crisis in the south and Ottoman war in the west impinged on Birecik's ability to command from out of town the timber and labor necessary for shipbuilding. As early as 1701, shipyard officials were airing their grievances, claiming that building a frigate in the imperial capital would be about 40 percent cheaper than building it in Birecik.[70] For a few decades, nevertheless, Birecik persisted. In 1733 and 1743, for instance, it built 300 and 124 vessels, respectively, in response to encroachments on

Iraq by the Persian conqueror Nadir Shah.[71] Amid the mounting political and financial pressures of the late eighteenth century, however, the central government finally relinquished this strategic shipyard, a move that reinforced Baghdad's control over the Tigris and Euphrates.

In May 1777, Istanbul and Baghdad were coordinating a military operation to rescue Basra from a brief Persian occupation. A chamberlain at the Ottoman court dispatched a letter to Baghdad's governor to vent his frustration over the lack of imperial support he found in the Birecik shipyard. None of the necessary material for ship construction was available. Timber from Maraş, Elbistan, Behisni, and Ayntab did not arrive, nor did the needed provisions from Aleppo and Raqqa. He had to purchase timber from the market at a high price, with which he constructed a mere twenty vessels. Had he had the necessary timber and provisions, he claimed, he could have constructed and loaded 150 vessels to join the war effort in Iraq. "By God I am bewildered," he grumbled, helpless to change a bitter reality.[72]

The chamberlain's letter is one of the last records of an Ottoman attempt to build river boats for the imperial Shatt Fleet on the Tigris and Euphrates. By the 1770s, Istanbul was no longer able to provision its eastern naval bases, nor was it able to meet the exigencies of its debilitating wars in eastern Europe without the goodwill of independent local grandees in the provinces. A few years later, the Ottoman central government ceased all shipbuilding operations in Birecik and Basra and devolved all naval duties to the Pashalik of Baghdad, which maintained a modest squadron in the waters of Iraq with the support of foreign maritime powers like the East India Company.

The breakdown in Ottoman military logistics, once the envy of Europe, was most visible during the Russo-Ottoman War of 1768–1774, a humiliating defeat that cost the Ottoman army tens of thousands of lives and its entire Mediterranean fleet in the bay of Çeşme. Many perished in the crossfire, but countless others died from famine and malnutrition caused by confusion and disorder in the supply system.[73] The Birecik shipyard was both far from Istanbul's main war theater and prohibitively expensive to maintain. The costs became more unaffordable and unwarranted than ever in the late eighteenth century, when the central government had to focus its resources and energies on fighting Russia, reforming its land and naval forces along Western lines, and relieving a strained central treasury—a crisis that prompted deliberations over raising a foreign loan for the first time in Ottoman history.[74] The change in priorities is best illustrated in Ottoman financial arrangements during the early nineteenth century, which funneled revenues from Baghdad's tax farms to support the Danube Fleet rather than that of the Tigris and Euphrates.[75]

Political and financial crises during the late eighteenth century crippled the Ottoman state's war machine and brought an end to its expensive naval industry in the Tigris-Euphrates basin. This predicament explains why Istanbul supported Baghdad's purchase of rivercraft from the East India Company, sometimes even chipping in funds. The abandonment of the Birecik shipyard sealed the position of the Pashalik as the dominant power in charge of the waters of Iraq.

At the turn of the eighteenth century, avulsion upset the political balance in Ottoman Iraq. State-sponsored herders' associations that once dominated the countryside suddenly broke up, giving way to powerful tribal confederations far less deferential to the Ottoman state. Istanbul responded to the challenge they posed by appointing a new governor to Baghdad in 1704 with greater discretion to save face and reassert Ottoman rule. The empowered governor fulfilled his duties with aplomb while establishing his own provincial dynasty. The Pashalik he founded tightened its grip on the Baghdad province and expanded its influence far and beyond. The end result of the Pashalik's expansionist agenda was the creation of what may be called greater Iraq, extending from Basra in the south to Mardin in the north, Ana in the west, and Şehrizor in the east. Within greater Iraq, Baghdad under the Pashalik emerged as a major metropolis that exercised the overwhelming influence over the region's waterways. It incorporated the Ottoman Shatt Fleet into its administrative structure and determined matters concerning navigation and irrigation.

As the Pashalik's influence waxed within the Tigris-Euphrates basin, that of the central government waned. The two related processes culminated in 1780, when Istanbul abandoned the Birecik shipyard altogether and combined the governorship of Baghdad, Basra, and Şehrizor into one position that it granted to the head of the Pashalik, Büyük Süleyman Pasha.[76] In its broad outlines, the formation of greater Iraq within the Ottoman Empire anticipated other regional adventures, such as that of Mehmed Ali in Egypt (r. 1805–1848), whose rule englobed the Sudan and the Morea, among other regions.[77]

Like Osman Pazvantoğlu of Vidin and Ali Pasha of Ioannina, the Pashalik of Baghdad cultivated its own relationships with other powerful households in neighboring provinces and even pursued its own foreign policy with European powers.[78] Through these newly forged connections, the Pashalik developed its own provisioning networks that would eclipse the imperial system of river transport once micromanaged from Istanbul. It acquired its grain supplies directly from its agents in Mardin and Nusaybin and allies in

Mosul. Arms and riverboats, from the sixteenth century delivered by Istanbul from the north, tended to mostly come from Basra in the south in coordination with the British East India Company. By the end of the eighteenth century, Istanbul relied entirely on the services of the Pashalik, rather than those of the Tigris and Euphrates, to bind Iraq to the imperial center.

Conclusion

The life in us is like the water in the river.
—Henry David Thoreau, *Walden* (1854)

FROM A DEEP history perspective, Ottoman rule in Iraq—the land of ancient Babylonia—was a political oddity. In its millennia-long history, Iraq was never ruled from Istanbul before the sixteenth century, unlike most Ottoman provinces in Anatolia, the Balkans, and the Mediterranean coastlands. Among the most distant imperial capitals to ever govern Babylonia for any considerable stretch of time were Persepolis and Antioch in the second half of the first millennium BC. But Achaemenid and Macedonian rule in Iraq pales into insignificance compared to what the Ottomans accomplished from the sixteenth century, ruling from a far more distant capital and for a far longer span.

This book has illuminated the role of Ottoman river management in making this political feat possible. In the long term, only a dynamic natural environment, through which fresh water and its constituents—sediment, chemicals, heat, and biota—are constantly on the move, could sustain Ottoman state-building and political survival in the sunbaked desert of Iraq. Most important, Istanbul relied on the newly unified twin rivers to rebalance the natural resource disparity along its eastern frontier. From the humid regions of the north to the arid regions of the south, it regularly organized the shipment of grains, arms, and riverboats to keep its most distant garrisons in the east afloat. In times of war, river transportation expanded the combat radius of Ottoman armies; in peace, it improved state policing and tax collection.

The Ottoman Empire used its river-based leverage to promote the most lucrative species in the alluvium. It did so by supporting and allying with the cultivators of four crops (rice, barley, wheat, and date palm) and the herders of four animals (sheep, goats, water buffalo, and cattle). For crop production, imperial support involved tax breaks for land reclamation projects and inputs of labor that assisted in the excavation and clearing of canals and in repairing dams and breaches in riverbanks. Moreover, it involved maintaining law and order and providing maximum security against pastoral raids. In the arena of animal rearing, the Ottoman administration modified the existing social structure of pastoral society and reorganized a considerable portion of it into herders' associations, the largest of which were the Ahşamat shepherds and the Cemmasat buffalo herders. The policy allowed imperial agents to monitor and regulate an otherwise intractable segment of the population and to capture its substantial wealth more efficiently. The Ottoman Empire's investment portfolio in the alluvium, diversified among different biomes, paid off. It generated substantial financial gains with minimal inputs of imperial labor and capital. Plus, social arrangements made with landholders, hydraulic engineers, herders' associations, and other rural groups expanded Ottoman authority from cities and towns into the countryside.[1]

As the Euphrates reconfigured its course in the late seventeenth century, it reconfigured the political landscape around it. The agitated river wreaked havoc on centuries-old rural arrangements, creating an opening for powerful tribal confederations to carve out their zones of influence in defiance of Ottoman authority. Ottoman rule would persist against all the odds, but it would have to come up with new solutions to deal with a new fluvial landscape and new challenges. Istanbul conferred greater prerogatives to the governors of Baghdad, which they used both to restore law and order and to enrich themselves. Gradually during the eighteenth century, Baghdad became an autonomous hereditary dynasty and self-sustaining center of power with loose ties to the imperial center. As Istanbul receded into the background, the Pashalik of Baghdad expanded its territorial influence, seized control of the Ottoman naval squadron in the Tigris-Euphrates basin, and developed its own channels to secure grains, arms, and riverboats once provisioned by Istanbul.

The Pashalik's challenge to the writ of Istanbul may give the impression of a separatist impulse, but both parties shared mutual interests and needed each other. For Istanbul, the territorial and military might of the Pashalik effectively represented imperial interests in a volatile region along the eastern frontier. For Baghdad, Istanbul was a source of legitimacy and support against rivals, particularly Persia. Far from a rebellion against the central state, the rise of the Pashalik during the eighteenth century localized the hegemony of the Ottoman state in the Tigris-Euphrates basin.[2]

Going beyond a study of the Tigris and Euphrates alone, *Rivers of the Sultan* has offered a new interpretation of the history of the Ottoman eastern provinces, focusing on the kinship between them, a relationship forged not by blood and marriage, but by water. Süleyman I's political unification of the twin rivers brought imperial agents into contact with different settlements, ecologies, and forms of life that congregated around the same river system for their survival and growth. From its dominant position over a unified drainage basin, Istanbul relied on the kinetic energy of air and water flows to reorganize elements of the ecosystem and establish a new order that best suited its imperial project in the east. As a result, farmers harvested the grain fields of Nusaybin and Mardin to feed the janissary corps of Baghdad in the south; loggers selectively chopped down the pine groves of Malatya and Maraş for the sake of Birecik and Basra, where galleys armed with guns could be assembled; mobile pastoralists in the vicinity of Aleppo sacrificed their sheep and goats to supply the raft makers of Diyarbakır and Mosul with watertight animal skins. It was these imperial efforts to rebalance resources and promote some over others that made Ottoman rule in Iraq possible.

Understanding the myriad opportunities and risks that the Tigris and Euphrates presented to early modern societies has significant historiographical implications. Among other things, it suggests that the Ottoman Empire could not persist as long as it did in an ecological vacuum. In West Asia, its political power was embedded in the circulation of water within the Tigris-Euphrates system, which gave coherence and force to its institutions. The centrality of the hydrologic cycle to Istanbul's authority abroad explains why the apparatus of Ottoman government came to a sudden halt following the interruption of flow within the Euphrates. The history of the twin rivers, in addition, shows clearly that evaluations of environmental growth and decline in West Asia cannot be based solely on the expansion and retraction of arable lands. Instead, scholarly analysis should take into account important services that the Tigris and Euphrates provided for communication, policing, political integration, and productive activities other than arable farming, such as animal rearing and wetland exploitation. From this perspective, the sixteenth century marks the dawn of a new period in the history of the Tigris and Euphrates, when Ottoman integration allowed the potential of both rivers to be realized more fully than was possible in the previous 500 years.

Only in 1831 could Istanbul finally abolish the Pashalik of Baghdad and begin the process of restoring its central authority throughout the

Tigris-Euphrates basin. The goal was not to resurrect the imperial model that Süleyman I imposed on conquest in the sixteenth century. Rather, the newly restructured imperial bureaucracy returned with a new vision of the world, deeply influenced by European ideas and institutions. Many Ottoman governors who served in Iraq had previously spent an extended period in Europe as foreign students and diplomats and saw themselves as the bearers of progress and enlightenment to their benighted subjects. They stressed the need to introduce modern technologies and establish public works to put Iraq on the path toward civilization. Under their leadership, field cannon cast in the foundries of Istanbul, pinewood galleys assembled in the Birecik shipyard, and janissary corps manning the Baghdad and Basra fortresses succumbed to the allure of English mortars and shells, Belgian iron steamers, and bombardiers and gunners trained in the European arts of war. The Tigris and Euphrates, in particular, became major sites for Ottoman modernizing efforts. To improve the rivers' potential for navigation and irrigation, imperial officials established steam navigation companies, cleared old canals and excavated new ones, renovated dockyards, and built dams. More broadly, the Ottoman vision of modernity sought to reshape the diverse ecology of the alluvial plain into a more homogenous one—drier, permanently settled, and intensively cultivated. Pastoral tribes that once formed the backbone of the imperial economy in the region were cast as squanderers of Ottoman wealth and as impediments to the state's attempt to join European powers within the ranks of civilized nations. Beset by mounting debt and costly wars, the late Ottoman Empire could not realize all its plans to remake the Tigris and Euphrates before its collapse.[3]

The nation-state system introduced by Britain and France in the aftermath of World War I marked a more dramatic transition in the history of the region. Scholars and politicians often criticize the post-Ottoman political landscape of West Asia for its incongruence with its ethnic and sectarian makeup. Geographically, the new map was no less arbitrary, dismembering the Tigris and Euphrates and dividing the pieces among the nascent nation-states of Turkey, Syria, and Iraq. Impoverished by the Great War or constrained by their European colonial masters, the new riparian neighbors first focused on establishing institutions and legal frameworks for river management. Their plans to exploit and reengineer their water resources steadily picked up steam after independence and the oil windfall of the post-1945 era. Notable among their earliest projects were the flood escapes of Tharthar and Habbaniyya in Iraq (1956), the Tabqa Dam in Syria (1973), and the Keban Dam in Turkey (1974). Since the 1980s, Turkey's Southeastern Anatolia Project (known by its Turkish acronym GAP) has been most ambitious in scope and most controversial. Once completed, the project will include a network of twenty-two dams, nineteen hydroelectric power plants, and extensive social and economic development projects. This aggressive alteration of the river system has unleashed numerous environmental and public health crises. Within the country, it has increased the incidence of waterborne parasitic diseases such as malaria and schistosomiasis; south of the border, it has caused major water shortages. The demographic makeup of southeastern Anatolia further complicates the challenges created by Turkey's development aspirations. GAP is based in a predominantly Kurdish region with long-standing grievances against Ankara. Among other things, dam construction and the expansion of irrigation have drowned cultural sites, displaced villagers, and undermined animal husbandry. Most recently, the Ilısu Dam on the Tigris has submerged the ancient town of Hasankeyf and is expected to uproot about 80,000 people.[4]

Overall, the artificial political divisions of the modern era have impinged on the natural physical unity of the river basin and stoked competition between Turkey, Syria, and Iraq, each seeking to appropriate scarce freshwater resources in a zero-sum game. Haphazard water storage facilities and irrigation projects sprouting up in each country have added unprecedented environmental stresses and taken their toll on the availability and quality of river water. The climate crisis, increasingly making our planet warmer and drier, has exacerbated the hydrologic problem, to which war-torn governments in the basin could not mount an effective response. The Tigris and Euphrates' flow downstream, as a result, has plummeted by at least 30 percent since the 1970s, a decrease expected to reach 50 percent by 2030 if current conditions remain unchanged. Their waters fast depleting, the twin rivers are losing their natural capacity to cleanse themselves of salt as well as industrial and urban waste. Water scarcity in the south is fomenting conflict among the general population and between states sharing the river basin, all competing for access to a diminishing resource. Among them, Iraqi farmers in the south are bearing the brunt of the ecological breakdown. They are migrating to cities in droves; others have been driven to commit suicide.[5]

To raise awareness about this catalogue of woes, an Iraqi nongovernmental organization (NGO) called Nature Iraq took in 2013 a small fleet of traditional watercraft on an expedition down the Tigris. The Tigris River Flotilla—as the project was called—sailed from the Turkish town of Hasankeyf in September (before its submergence) and reached the Iraqi marshes a month later, a distance of some 750 miles. In between, the journey proved far more arduous than the activists anticipated. "Though the original plan for the Tigris River Flotilla had been to row, paddle, and float the entire way," according to the *Christian Science Monitor*, "dams and other blockages on the Tigris, security concerns, and visa and permit hassles have periodically forced the group to load its boats onto trucks and skirt the most difficult portions of the river."[6] Fighting near the Turkish-Syrian border, for instance, pushed the boats overland. "The voyagers re-joined the river in Iraq," *The Economist* reported, "only to be held up at the dilapidated Mosul dam. Forbidden from crossing the lake, they again packed up and drove to Baghdad, the Iraqi capital, where they were met by a deluge of bureaucracy before being allowed to sail on."[7]

The Tigris River Flotilla laid bare the countless challenges to sustainable water management that have been created by the political fragmentation of the river basin since the twentieth century. It also serves as a perfect foil for the early modern experience of the Tigris and Euphrates, when countless voyagers could sail on their free-flowing waterways with the blessing of a single authority. The Ottoman state had transformed otherwise provincial rivals—from the Latin word *rīvālis*, meaning "person using the same stream as another"—into imperial partners who received orders from a central administration based in Istanbul.[8] Ottoman unification institutionalized cooperation among them, which before the modern age of water scarcity focused on improving the navigational uses of the twin rivers. Managed through a joint effort, waterflow fostered the political and economic cohesion of the river basin and its integration within the Ottoman Empire. Major causes of discord today, the Tigris and Euphrates could promote unity and harmony when managed cooperatively and holistically. They certainly did when they were the rivers of the Ottoman sultan.

APPENDIXES

Appendix A. Tables

Table A.1. Imperial Orders for Grain Shipments by Water

Year	Source	Destination	Shipment
1565	Diyarbakır	Baghdad	4,000 *mud*[a] wheat and barley
1566	Birecik	Baghdad and Basra	150,000 *kile*[b] barley, 50,000 *kile* wheat
1567	Diyarbakır, Nusaybin, and Hasankeyf	Baghdad	31,472 *kile* wheat, 39,929 *kile* barley
1579	Egypt and Cyprus	Baghdad	25,000 *erdeb*[c] wheat, 20,000 *erdeb* barley
1609	Birecik	Baghdad	Not reported
1639	Birecik, Ayntab, and Aleppo	Baghdad	35,480 *kile* wheat
1696	Mosul	Baghdad	20,000 *kile* wheat, 18,000 *kile* barley
1699	Birecik, Diyarbakır, Mosul, Nusaybin, Mardin, Raqqa, and Maraş	Baghdad	60,000 *kile* wheat, 11,412 *kile* barley
1700	Diyarbakır and Raqqa	Baghdad	90,000 *kile* wheat
1701	Birecik, Diyarbakır, Mosul, Nusaybin, Mardin, and Maraş	Baghdad	150,000 *kile* wheat, 10,000 *kile* barley
1702	Mosul	Basra	6,090 *qintar*[d] hardtack
1708	Diyarbakır	Basra	2,011 *qintar* hardtack
1726	Mardin and Diyarbakır	Baghdad	60,000 *kile* wheat and barley, 40,000 *kile* wheat, 50,000 *kile* barley
1727	Ayntab, Samsat, Süpürek	Mosul and Baghdad	113,000 *kile* wheat, 116,000 *kile* barley
1733	Diyarbakır, Mardin, Raqqa, and Mosul	Baghdad	100,000 *kile* wheat and barley
1742	Mosul and Mardin	Baghdad	68,819.5 *kile* wheat, 46,256.5 *kile* barley
1744	Aleppo	Baghdad	Not reported
1745	Raqqa	Baghdad	9,928 *kile* wheat

Table A.1. Continued

Year	Source	Destination	Shipment
1748	Mosul	Baghdad	5,280 *kile* wheat
1749–1750	Diyarbakır, Mosul, Mardin, and Nusaybin	Baghdad	200,000 *kile* wheat, 200,000 *kile* barley

Note: Most Ottoman units of measurement varied greatly depending on the locality, time period, and item being measured, a fact that complicates attempts to convert them to US customary units. The rough conversions provided with this table are based on rates most relevant to the Ottoman capital Istanbul and should therefore be treated with caution. The conversions are based on Halil İnalcık, "Weights and Measures," in *An Economic and Social History of the Ottoman Empire, 1300–1914*, vol. 1, *1300–1600*, ed. Halil İnalcık and Donald Quataert (New York: Cambridge University Press, 1994), xxxvii-xliii.

Sources: OA, MAD 2775, 694 (6 C 973/29 December 1565); OA, MAD 2775, 1391 (3 L 973/23 April 1566); OA, MD 5/1028 (28 B 973/19 February 1566); OA, MD 7/305 (7 R 975/10 October 1567); OA, MD 32/519 (27 ZA 986/25 January 1579); OA, MD 32/631 (17 M 987/16 March 1579); OA, MD 38/72 (1 S 987/30 March 1579); OA, MD 38/263 (15 RA 987/12 May 1579); OA, MD 38/308 (12 Z 987/30 January 1580); OA, MD 78/1787 (18 R 1018/21 July 1609); OA, MD 78/1985 (19 ZA 1018/13 February 1610); OA, MAD 46 (1049/1639–1640); OA, İE. DH 15/1395 (13 B 1107/17 February 1696); OA, MAD 9885, 180 (15 CA 1111/7 November 1699); OA, MAD 3134, 69–70 (2 S 1112/19 July 1700); OA, MAD 9885, 204–205 (12 N 1111/3 March 1700); OA, MAD 9885, 323–324 (22 ZA 1111/12 May 1700); OA, MAD 9885, 354 (3 Z 1111/23 May 1700); OA, MD 111/1802 (Evail L 1112/11–21 January 1701); OA, D.BŞM 7651/36 (2 RA 1113/7 August 1701); OA, MAD 7915, 400–402 (27 N 1113/24 February 1702); OA, MAD 7915, 333 (27 M 1120/19 April 1708); OA, C.ML 522/21342 (20 CA 1138/24 January 1726); OA, C.AS 786/33308 (8 S 1139/4 October 1726); OA, İE.AS 81/7334 (18 Ş 1139/10 April 1727); OA, C.HR 68/3397 (12 N 1145/27 February 1733); OA, MAD 9934, 310, 314 (17 B 1146/24 December 1733); OA, MAD 9934, 126–127 (9 N 1146/13 February 1734); OA, MAD 9934, 155 (20 L 1146/26 March 1734); OA, MAD 9934, 157–158 (9 B 1146/16 December 1733); OA, MAD 9934, 161–162 (24 Ş 1146/30 January 1734); OA, MAD 9934, 189 (29 L 1146/4 April 1734); OA, MAD 9934, 196 (12 L 1146/18 March 1734); OA, C.AS 977/42578 (Evail N 1146/5–14 February 1734); OA, C.AS 572/24061 (9 L 1155/7 December 1742); OA, MAD 9952, 84 (16 ZA 1157/21 December 1744); OA, D.BŞM.BGH 2/28 (Evail R 1158/23 May–1 June 1745); OA, D.BŞM.MSH 8/1 (1 ZA 1161/24 October 1748); OA, C.DH 31/1544 (15 C 1162/2 June 1749); OA, MAD 9948, 54 (25 N 1162/8 September 1749); OA, MAD 9968, 196 (22 CA 1163/29 April 1750), OA, MAD 9968, 215 (10 C 1163/10 May 1750); OA, MAD 9968, 232–233 (21 C 1163/21 May 1750); OA, MAD 9968, 264–265 (5 B 1163/10 June 1750); OA, MAD 9968, 266 (12 B 1163/17 June 1750); OA, MAD 9968, 267 (5 B 1163/10 June 1750).

[a]*mud* = 1,130 lbs

[b]*kile* = 55 lbs

[c]*erdeb* = 90-198 liters

[d]*qintar* = 125 lbs

Table A.2. Imperial Orders for Arms Shipments by Water

Year	Source	Destination	Shipment
1552	Diyarbakır	Baghdad	500 firearms
1565	Istanbul	Baghdad	1,000–1,500 firearms
1565	Birecik and Aleppo	Basra	450 *qintar* lead, 400 *qintar* iron
1565	Baghdad	Basra	20 *kolumburina* guns, 5 *darbzen* guns, 4 general guns
1566	Istanbul	Basra	50 *qintar* copper, 240 *shakaloz* guns
1568	Aleppo	Lahsa, via Basra	1,000 firearm stocks, 40 cart axles
1571–1572	Baghdad and Basra	Lahsa	300 *qintar* gunpowder, 4 *shahi darbzen* guns
1576	Tripoli	Basra	300 *qintar* sulfur
1700	Istanbul	Shatt al-Arab Squadron	60 *koğuş* guns, 60 *koğuş* gun stocks, 120 *yan* guns, 4 mortars 4 mortar stocks, 18 *saçma* guns, 230 cast-iron *saçma* guns, 24 *eynek* guns, 472 cast-iron *eynek* guns, 552 bombshells, 18,000 cannonballs
1700	Istanbul	Baghdad	450 *qintar* gunpowder
1703	Diyarbakır and Mosul	Basra	Unspecified number of galleon guns
1724	Istanbul	Baghdad	Unspecific quantities of gunpowder, lead, cannonball, and bombshell
1726	Istanbul	Qurna and Basra	2,400 *qintar* gunpowder

Table A.2. Continued

Year	Source	Destination	Shipment
1733	Istanbul	Mosul	50 *shahi* guns
1743	Istanbul	Diyarbakır	5 mortars
1750	Mosul	Baghdad	11 unfinished stocks

Note: Historians of Europe and Asia have struggled to offer precise definitions for the terminology premodern armies applied to their firearms and artillery pieces. Guns with the same type name varied greatly in caliber and barrel and could have more in common with those classified under an entirely different name. The myriad languages spoken and inconsistent usages throughout the Ottoman Empire add to the confusion. For a comprehensive attempt to outline the technical variations in each piece of ordnance mentioned in this table, see Gábor Ágoston, *Guns for the Sultan: Military Power and the Weapons Industry in the Ottoman Empire* (New York: Cambridge University Press, 2005), 61-95.

Sources: "Hükümname Mecmuası," Topkapı Sarayı Müzesi Kütüphanesi, Koğuşlar 888, fol. 241r (29 CA 959/23 May 1552); OA, MAD 2775, 328 (27 S 973/22 September 1565); OA, MAD 2775, 339 (29 S 973/24 September 1565); OA, MAD 2775, 619 (14 CA 973/7 December 1565); OA, MAD 2775, 631–632 (14 CA 973/7 December 1565); OA, MAD 2775, 1050 (12 Ş 973/4 March 1566); OA, MD 7/1349–1350, 468 (8 ZA 975/6 May 1568); OA, MD 12/1094, 575 (11 ZA 979/26 March 1572); OA, MD 16/20, 10–11 (10 CA 979/30 September 1571); OA, MD 27/523 (27 L 983/29 January 1576); OA, MAD 7915, 382 (3 ZA 1111/22 April 1700); OA, MAD 975, 16 (28 S 1112/14 August 1700); OA, MAD 7915, 417–418 (22 S 1115/7 July 1703); OA, C.AS 124/5554 (15 M 1137/3 October 1724); OA, AE.SAMD.III 51/5150 (27 C 1138/2 March 1726); OA, MAD 9934, 20 (1 B 1145/8 December 1733); OA, MAD 9947, 261 (13 C 1156/4 August 1743); OA, MAD 9968, 191 (16 CA 1163/23 April 1750).

Table A.3. Naval Buildup of the Shatt Fleet

Year	Order Type	Shipyard	Number of Vessels	Vessel Types	Target
1552	Construction	Birecik	300	*Sandal*[a]	Not reported
1560	Construction	Basra	5	Galliots	Portugal
1565–1567	Construction	Birecik	400	250 army boats, 150 cargo boats	Tribesmen
1571	Construction	Basra	5	Galliots	Portugal
1575	Construction	Basra	8	Galleys	Portugal
1629	Construction	Birecik	100	Cargo boats	Persia
1637–1638	Construction	Birecik	610	Not reported	Persia
1695–1696	Construction	Raqqa	10	Army boats	Tribesmen
1699	Construction	Raqqa	30	Army boats	Tribesmen
1699–1701	Construction	Birecik	98	60 frigates, 25 felucca and *sandal*, 5 *kayık*[b], 8 *üstüaçık*[c]	Tribesmen
1703	Construction	Basra	Not reported	Galleons	Tribesmen
1706	Renovation	Basra	Not reported	Galleon	Tribesmen
1706	Renovation	Baghdad	16	14 *işkampoye*[d], 2 frigates	Tribesmen
1709	Renovation	Basra	12	Frigates	Tribesmen
1709	Construction	Birecik	16	*Üstüaçık*	Tribesmen
1711	Renovation	Basra	11	*İşkampoye*	Tribesmen
1719–1720	Construction	Birecik	20–39	Not reported	Tribesmen

Table A.3. Continued

Year	Order Type	Shipyard	Number of Vessels	Vessel Types	Target
1728	Construction	Basra	20	Frigates	Tribesmen
1730	Construction	Basra	15	5 frigates, 10 işkampoye	Tribesmen

Sources: "Hükümname Mecmuası," Topkapı Sarayı Müzesi Kütüphanesi, Koğuşlar 888, fol. 344r (9 Ş 959/31 July 1552); "Hükümname Mecmuası," fol. 344v (10 Ş 959/1 August 1552); "Hükümname Mecmuası," fol. 345v (19 Ş 959/20 August 1552); Ali Yılmaz, "XVI. Yüzyılda Birecik Sancağı" (Ph.D. diss., İstanbul Üniversitesi, 1996), 168–169, 175; OA, MD 3/1185 (7 N 967/2 June 1560); OA, MD 3/1355 (10 ZA 967/2 August 1560); OA, MD 5/353 (10 RA 973/14 October 1565); OA, MD 5/821 (28 C 973/20 January 1566); OA, MAD 2775, 179 (30 M 973/27 August 1565); OA, MAD 2775, 324 (29 S 973/24 September 1565); OA, MD 10/421 (28 B 979/16 December 1571); OA, MD 16/301 (28 B 979/16 December 1571); OA, KK 79, 252 (11 R 979/2 September 1571); OA, MD 27/203 (9 Ş 983/13 November 1575); OA, MD 27/436 (9 Ş 983/11 January 1576); OA, MD 27/450 (14 Ş 983/16 January 1576); OA, MD 27/465 (15 Ş 983/17 January 1576); OA, MD 27/748 (6 Z 983/7 March 1576); Katib Çelebi, *Fezleke-i Katib Çelebi* (Istanbul: Ceride-i Havadis Matbaası, 1286/1869), 2:115–116; OA, MD 87/437 (1 B 1047/19 November 1637); OA, MD 87/464–465 (3 Ş 1047/21 December 1637); OA, MD 87/474 (18 N 1047/3 February 1638); OA, MD 106/1419 (Evasıt CA 1107/18–27 December 1695); OA, MD 108/848 (Evasıt Ş 1107/16–25 March 1696); OA, MD 108/1248 (Evasıt ZA 1107/22 June–1 July 1696); OA, MAD 4879, 118 (29 Z 1110/27 June 1699); OA, MAD 7915, 268, 272–275, 375–376, 392 (2 R 1111–1 N 1112/26 September 1699–18 February 1701); OA, D.BŞM.TRE 14598, 8 (16 R 1112/29 September 1700); OA, MAD 3134, 67 (12 M 1112/29 June 1700); OA, MAD 3134, 69 (11 S 1112/28 July 1700); OA, MAD 3134, 95 (13 S 1112/30 July 1700); OA, MAD 5433 (6 N 1112/14 February 1701); OA, D.BŞM 7654/47 (26 CA 1114/18 October 1702); OA, MAD 7915, 416–418 (22 S 1115/7 July 1703); OA, AE.SMST.II 2/171 (22 S 1115/7 July 1703); OA, MAD 7915, 326–328 (4 L 1117–18 C 1118/19 January–26 September 1706); OA, AE.SAMD.III 210/20291 (7 L 1117/22 January 1706); OA, AE.SAMD.III 210/20292 (10 L 1117/25 January 1706); OA, AE.SAMD.III 4/277 (21 CA 1118/30 August 1706); OA, D.BŞM.MSH 4/184 (5 B 1118/12 October 1706); OA, MAD 7915, 334–341 (2 Z 1120/12 February 1709); OA, SAMD.III 146/14315 (25 C 1118/4 October 1706); OA, MAD 7915, 338–341; OA, MAD 7915, 344 (24 S 1123/13 April 1711); OA, C.BH 50/2377 (5 Z 1120/15 February 1709); OA, MAD 7915, 358   361 (9 B 1131/29 May 1719); OA, MAD 7915, 359 (22 N 1132/28 July 1720); OA, MAD 7915, 346   347 (20 C 1140/1 February 1728); TSMA 890/32 (10 M 1142/5 August 1729); OA, MAD 7915, 345 (10 R 1143/22 October 1730); OA, MAD 7915, 348 (7 S 1143/21 August 1730); OA, MAD 7915, 350 (8 CA 1143/19 November 1730).

[a] *Sandal*: small rowboat that could be carried by ship

[b] *Kayık*: general Turkish term for boat

[c] *Üstüaçık*: oared vessel common on the Danube River

[d] *İşkampoye*: oared vessel common on the the Danube River

Appendix B. Staffing Religious Institutions

Appointments made in Amid (Diyarbakır)

OA, İE.EV 1/39 (23 M 954/16 March 1547); OA, İE.EV 6/628 (Evail Z 1071/28 July–7 August 1661); OA, İE.EV 15/1818 (1 R 1078/19 September 1667); OA, İE.EV 2/205 (22 N 1079/23 February 1669); OA, İE.EV 2/196 (23 N 1079/24 February 1669); OA, İE.EV 17/2079 (13 RA 1083/8 July 1672); OA, İE.EV 52/5781 (6 Ş 1106/21 March 1695); OA, AE.SMST.II 13/1225 (5 RA 1107/14 October 1695); OA, İE.EV 28/3215 (Evahir B 1109/3 February 1698); OA, İE.EV 30/3424 (Evail B 1110/3–12 January 1699); OA, İE.EV 32/3717 (1 N 1110/3 March 1699); OA, İE.EV 30/3504 (12 R 1114/5 September 1702); OA, AE.SAMD.III 126/12348 (10 R 1118/21 July 1706); OA, AE.SAMD.III 176/17090 (5 R 1119/5 July 1707); OA, AE.SAMD.III 186/18038 (13 N 1121/15 November 1709); OA, İE.EV 52/5734 (18 C 1124/22 July 1712); OA, AE.SAMD.III 161/15809 (22 B 1124/25 August 1712); OA, AE.SAMD.III 32/3048 (29 C 1124/2 August 1712); OA, AE.SAMD.III 126/12443 (5 RA 1125/1 April 1713); OA, İE.EV 53/5830 (9 RA 1125/5 April 1713); OA, İE.TCT 23/2462 (11 L 1125/31 October 1713); OA, İE.EV 39/4429 (6 N 1127/4 September 1715); OA, İE.EV 56/6192 (17 L 1132/22 August 1720); OA, İE.EV 38/4386 (1 M 1139/29 August 1726); OA, AE.SAMD.III 181/17643 (3 Ş 1139/26 March 1727); OA, C.EV 249/12464 (15 L 1141/14 May 1729), OA, C.ADL 85/5107 (13 R 1168/26 January 1755); OA, AE.SMST.III 130/10098 (28 B 1172/28 March 1759); OA, AE.SMST.III 131/10183 (19 N 1172/16 May 1759); OA, AE.SMST.III 189/14887 (21 CA 1175/18 December 1761); OA, AE.SABH.I 281/18900 (19 L 1199/24 August 1758).

Appointments made in Mosul

OA, MAD 2775, 391 (7 RA 973/2 October 1565); OA, AE.SAMD.I 1/33 (11 L 1014/19 February 1606); OA, İE.EV 17/2061 (Evasıt R 1083/5–14 August 1672); OA, İE.HAT 4/348 (Evahir C 1086/12–21 September 1675); OA, AE.SMMD.IV 6/619 (Evail Z 1088/24 January–2 February 1678); OA, AE.SMMD.IV 100/11727 (1058–1099/1648–1687); OA, AE.SAMD.II 22/2266 (1103/1691–1692); OA, AE.SAMD.II 21/2244 (15 Ş

1103/6 November 1691); OA, AE.SAMD.II 21/2252 (20 Z 1105/12 August 1694); OA, AE.SMST.II 16/1567 (Evahir M 1107/31 August–8 September 1695); OA, C.EV 63/3117 (1 C 1111/23 November 1699); OA, İE.EV 34/3897 (Evahir ZA 1111/11–20 May 1700); OA, AE.SAMD.III 118/11598 (Evahir RA 1115/4–13 August 1703); OA, İE.EV 46/5232 (5 ZA 1115/11 March 1704); OA, İE.EV 48/5341 (29 R 1117/19 August 1705); OA, AE.SAMD.III 106/10451 (3 Z 1117/18 March 1706); OA, AE.SAMD.III 106/10452 (3 Z 1117/18 March 1706); OA, AE.SAMD.III 106/10453 (3 Z 1117/18 March 1706); OA, AE.SAMD.III 28/2710 (16 R 1119/16 July 1707); OA, AE.SAMD.III 69/6921 (10 Z 1119/3 March 1708); OA, AE.SAMD.III 43/4218 (Evail B 1122/26 August–4 September 1710); OA, İE.EV 51/5617 (25 RA 1123/13 May 1711); OA, AE.SAMD.III 23/2146 (25 RA 1124/1 May 1712); OA, AE.SAMD.III 85/8567 (8 C 1124/12 July 1712); OA, AE.SAMD.III 179/17428 (27 L 1124/27 November 1712); OA, AE.SAMD.III 213/20609 (6 S 1124/14 March 1712); OA, AE.SAMD.III 165/16169 (6 RA 1127/12 March 1715); OA, AE.SAMD.III 165/16168 (10 RA 1127/16 March 1715); OA, AE.SAMD.III 159/15597 (27 Ş 1131/15 July 1719); OA, AE.SAMD.III 20/1864 (6 L 1131/22 August 1719); OA, AE.SAMD.III 95/9361 (23 R 1135/30 January 1723); OA, AE.SAMD.III 88/8829 (28 R 1135/4 February 1723); OA, İE.TCT 15/1717 (2 C 1135/10 March 1723); OA, AE.SAMD.III 176/17112 (3 Ş 1139/26 March 1727); OA, AE.SAMD.III 140/13613 (20 Ş 1140/31 March 1728); OA, AE.SAMD.III 58/5791 (17 Z 1140/25 July 1728); OA, İE.EV 61/6690 (25 N 1141/24 April 1729); OA, AE.SAMD.III 218/21002 (16 Z 1141/13 July 1729); OA, D.BŞM.MSH 8/19 (23 S 1159/16 March 1746); OA, AE.SOSM.III 86/6647 (4 RA 1168/18 December 1754); OA, D.BŞM 3383 (1169–1173/1755–1760); OA, AE.SOSM.III 69/5223 (22 C 1171/3 March 1758); OA, C.EV 122/6078 (9 Ş 1180/10 January 1767); OA, AE.SABH.I 230/15255 (7 M 1190/27 February 1776); OA, AE.SABH.I 325/22163 (5 R 1191/13 May 1777); OA, AE.SABH.I 325/22164 (5 R 1191/13 May 1777); OA, AE.SABH.I 325/22165 (5 R 1191/13 May 1777); OA, AE.SABH.I 92/6294 (9 L 1198/25 August 1784); OA, AE.SSLM.III 84/5094 (8 Ş 1216/14 December 1801); OA, C.MF 169/8420 (16 ZA 1222/14 January 1808).

Appointments made in Baghdad

"Hükümname Mecmuası," Topkapı Sarayı Müzesi Kütüphanesi, Koğuşlar 888, fol. 347v (22 Ş 959/13 August 1552); OA, MAD 2737, 198 (23 Z 1062/25 November 1652); OA, İE.EV 14/1607 (27 ZA 1067/6 September 1657); OA, AE.SMMD.IV 23/2579 (24 R 1087/6 July 1676); OA, İE.EV 16/1884 (Evasıt ZA 1087/15–24 January 1677); OA, İE.EV 8/964 (1 C 1088/31 July 1677); OA, AE.SMMD.IV 3/313 (8 C 1090/16 July 1679); OA, MAD 2915, 108 (24 B 1092/9 August 1681); OA, MAD 2926, 88 (9 CA 1093/17 May 1682); OA, MAD 2926, 88 (10 CA 1093/18 May 1682); OA, İE.EV 23/2731 (23 CA 1093/31 May 1682); OA, MAD 2926, 165 (13 L 1093/14 October 1682); OA, MAD 2926, 167 (25 L 1093/26 October 1682); OA, AE.SMMD.IV 2/177 (c. 1058–1099/c. 1648–1687); OA, D.BŞM.BGH 1/25 (23 N 1099/21 July 1688); OA, AE.SSÜL.II 23/2324 (19 CA 1102/18 February 1691); OA, AE.SAMD.II 21/2231 (c. 1102/c. 1690–1691);

OA, AE.SAMD.II 10/983 (19 M 1103/11 October 1691); OA, İE.EV 24/2883 (7 CA 1103/27 November 1691); OA, AE.SMST.II 83/8940 (13 N 1107/16 April 1696); OA, MAD 3134, 188–189 (6 CA 1112/19 October 1700); OA, AE.SMST.II 12/1115 (4 CA 1114/26 September 1702); OA, İE.EV 45/5064 (9 Ş 1114/28 December 1702); OA, AE.SMST.II 12/1148 (8 ZA 1114/26 March 1703); OA, AE.SMST.II 96/10361 (8 Z 1114); OA, İE.EV 46/5222 (1 C 1115/12 October 1703); OA, C.MF 1/42 (24 ZA 1118/27 February 1707); OA, AE.SAMD.III 194/18761 (14 M 1119/17 April 1707); OA, İE.AS 78/7027 (5 ZA 1119/28 January 1708); OA, AE.SAMD.III 131/12837 (20 ZA 1123/30 December 1711); OA, AE.SAMD.III 156/15350 (12 Ş 1131/30 June 1719); OA, İE.EV 59/6448 (25 L 1137/6 July 1725); OA, İE.EV 59/6446 (27 Z 1137/4 September 1725); OA, AE.SAMD.III 176/17175 (25 M 1138/2 October 1725); OA, D.BŞM.MSH 6/196 (5 M 1139/2 September 1726); OA, AE.SAMD.III 197/19016 (3 Ş 1139/26 March 1727); OA, AE.SAMD.III 198/19169 (16 RA 1141/20 October 1728); OA, D.BŞM.MSH 8/36 (5 N 1146/9 February 1734); OA, D.BŞM.MSH 8/41 (6 N 1146/10 February 1734); OA, D.BŞM.MSH 8/44 (28 M 1146/10 July 1733); OA, MAD 9970, 32 (29 M 1164/28 December 1750); OA, AE.SOSM.III 58/4192 (5 S 1171/19 October 1757); OA, D.BŞM.BGH 3/16 (11 M 1175/12 August 1761); OA, AE.SMST.III 174/13634 (11 L 1180/12 March 1767); OA, D.BŞM.BGH 3/52 (18 M 1182/4 June 1768); OA, C.EV 51/2522 (9 RA 1182/23 July 1768); OA, AE.SSLM.III 132/8022 (25 L 1206/15 June 1792); OA, C.EV 244/12198 (17 S 1234/16 December 1818).

Appointments made in Basra

OA, İE.EV 12/1467 (15 C 1080/9 November 1669); OA, AE.SMMD.IV 17/1929 (21 S 1081/9 July 1670); OA, İE.EV 21/2506 (Evasıt CA 1090/20–29 June 1679); OA, AE.SMMD.IV 95/11151 (19 CA 1090/28 June 1679); OA, İE.EV 23/2696 (21 C 1090/29 July 1679); OA, AE.SMMD.IV 96/11302 (19 N 1091/13 October 1680); OA, MAD 2926, 157 (18 N 1093/20 September 1682); OA, İE.EV 49/5448 (28 C 1119/29 June 1707).

Notes

ARCHIVAL CLASSIFICATIONS

OA: Osmanlı Arşivi

AE: Ali Emiri
    SABH.I: Abdülhamid I
    SAMD.I: Ahmed I
    SAMD.II: Ahmed II
    SAMD.III: Ahmed III
    SİBR: İbrahim
    SMHD.I: Mahmud I
    SMMD.IV: Mahmud II
    SMST.II: Mustafa II
    SMST.III: Mustafa III
    AE.SOSM.III: Osman III
    SSLM.III: Selim III
    SSÜL.II: Süleyman II

D.BŞM: Bab-ı Defteri Başmuhasebe Kalemi
    BGH: Bağdat Hazinesi
    BSH: Basra Hazinesi
    DBH: Diyarbakır Hazinesi
    MSH: Musul Hazinesi
    TRE: Tersane-i Amire Eminliği

C: Cevdet
    ADL: Adliye
    AS: Askeriye
    BH: Bahriye
    DH: Dahiliye

EV: Evkaf
HR: Hariciye
ML: Maliye
MF: Maarif
NF: Nafia
HAT: Hatt-ı Hümayun

İE: İbnülemin
AS: Askeriye
BH: Bahriye
DH: Dahiliye
EV: Evkaf
HAT: Hatt-ı Hümayun
ML: Maliye
SH: Sıhhiye
TCT: Tevcihat

KK: Kamil Kepeci

MAD: Maliyeden Müdevver Defterler

MD: Mühimme Defterleri

TT: Tapu Tahrir Defterleri

TKG.KK: Tapu ve Kadastro Genel Müdürlüğü, Kuyud-ı Kadime Arşivi

TT: Tapu Tahrir Defterleri

TSMA: Topkapı Sarayı Müzesi Arşivi

IOR: India Office Records

G: East India Company Factory Records
L/P&S: Political and Secret Department Records

TNA: The National Archives

FO: Foreign Office

MONTHS OF THE MUSLIM CALENDAR

M: Muharrem
S: Safer

RA: Rebiülevvel
R: Rebiülahir
CA: Cemaziyelevvel
C: Cemaziyelahir
R: Recep
Ş: Şaban
N: Ramazan
L: Şevval
ZA: Zilkade
Z: Zilhicce

INTRODUCTION

1. Matrakçı Nasuh, *Beyan-i Menazil-i Sefer-i Irakeyn-i Sultan Süleyman Han*, ed. Hüseyin G. Yurdaydın (Ankara: Türk Tarih Kurumu, 1976), fol. 4r. "Turan" in Ottoman literature is synonymous with Tartary in English, the putative homeland of the Turks in Central Asia.
2. Like all twins, though, the Tigris and Euphrates have their distinctive qualities that make each unlike the other. We will learn more about their differences as we get to know them better in the following chapters. For now, let us ponder the marvel of their likeness.
3. Perry A. Frey and George H. Reed, "The Ubiquity of Iron," *ACS Chemical Biology* 7, no. 9 (2012): 1474–1476; Jack Goody, *Metals, Culture and Capitalism: An Essay on the Origins of the Modern World* (New York: Cambridge University Press, 2012), 85–119; Richard A. Gabriel and Karen S. Metz, *From Sumer to Rome: The Military Capabilities of Ancient Armies* (New York: Greenwood Press, 1991), 19–46; Jane C. Waldbaum, "The Coming of Iron in the Eastern Mediterranean: Thirty Years of Archaeological and Technological Research," in *The Archaeometallurgy of the Asian Old World*, ed. Vincent C. Pigott (Philadelphia: University Museum, University of Pennsylvania, 1999), 27–57; Mario Liverani, *The Ancient Near East: History, Society and Economy*, trans. Soraia Tabatabai (New York: Routledge, 2014), 390–396; J. R. McNeill and William H. McNeill, *The Human Web: A Bird's-Eye View of World History* (New York: W. W. Norton, 2003), 55–60.
4. Wayne E. Lee, *Waging War: Conflict, Culture, and Innovation in World History* (New York: Oxford University Press, 2016), 215–292; Giancarlo Casale, "The Islamic Empires of the Early Modern World," in *The Cambridge World History*, vol. 6, *The Construction of a Global World, 1400–1800*, ed. Jerry H. Bentley, Sanjay Subrahmanyam, and Merry E. Wiesner-Hanks (Cambridge: Cambridge University Press, 2015), 337–339; Linda T. Darling, "Political Change and Political Discourse in the Early Modern Mediterranean World," *Journal of Interdisciplinary History* 38, no. 4 (2008): 505–531; Gábor Ágoston, *Guns for the Sultan: Military*

Power and the Weapons Industry in the Ottoman Empire (New York: Cambridge University Press, 2005); Andrew C. Hess, "The Ottoman Conquest of Egypt (1517) and the Beginning of the Sixteenth-Century World War," *International Journal of Middle East Studies* 4, no. 1 (1973): 58; William Hardy McNeill, *The Age of Gunpowder Empires, 1450–1800* (Washington, DC: American Historical Association, 1989); Marshall G. S. Hodgson, *The Venture of Islam: Conscience and History in a World Civilization*, vol. 3, *The Gunpowder Empires and Modern Times* (Chicago: University of Chicago Press, 1974), 16–27; McNeill and McNeill, *The Human Web*, 192–200.

5. Philip Ball, *Life's Matrix: A Biography of Water* (New York: Farrar, Straus and Giroux, 2001), 25.

6. Aşık Mehmed, *Menazırü'l-Avalim*, ed. Mahmut Ak (Ankara: Türk Tarih Kurumu, 2007), 2:274.

7. See, for example, Tahir Öğüt, *18–19 Yüzyıllarda Birecik Sancağında İktisadi ve Sosyal Yapı* (Ankara: Türk Tarih Kurumu, 2013), 243–330; Hala Fattah, *The Politics of Regional Trade in Iraq, Arabia, and the Gulf, 1745–1900* (Albany: State University of New York Press, 1997), 123–138; Salih Özbaran, "XVI. Yüzyılda Basra Körfezi Sahillerinde Osmanlılar Basra Beylerbeyliğinin Kuruluşu," *Tarih Dergisi* 25 (1971): 51–73; Thabit A. J. Abdullah, *Merchants, Mamluks, and Murder: The Political Economy of Trade in Eighteenth-Century Basra* (Albany: State University of New York Press, 2001), 74–77; S. H. Winter, "The Province of Raqqa under Ottoman Rule, 1535–1800: A Preliminary Study," *Journal of Near Eastern Studies* 68, no. 4 (2009): 253–268; Fasih Dinç, "Osmanlı Diyarbakır'ında Keleğin Yapımı ve Kullanımı," in *Osmanlıdan Günümüze Diyarbakır*, ed. İbrahim Özcoşar et al. (Istanbul: Ensar, 2018), 61–100; Ebubekir Ceylan, *The Ottoman Origins of Modern Iraq: Political Reform, Modernization and Development in the Nineteenth-Century Middle East* (London: I. B. Tauris, 2011), 189–201. One major exception is Cengiz Orhonlu and Turgut Işıksal, "Osmanlı Devrinde Nehir Nakliyatı Hakkında Araştırmalar: Dicle ve Fırat Nehirlerinde Nakliyat," *Tarih Dergisi* 13 (1962–1963): 79–102.

8. Nearly four decades ago, the late Robert Adams, former secretary of the Smithsonian and one of the greatest archaeologists of his generation, hinted at the potential rewards of writing a history of the Tigris and Euphrates, noting, "A watershed is only another topographic unit, of course, and need never have coincided with a historical and cultural unit of any significance. But in this case there are many leads to a recurrent unity within the Euphrates basin that deserve to be explored, even if that far-ranging task cannot be adequately undertaken here." Adams, *Heartland of Cities: Surveys of Ancient Settlement and Land Use on the Central Floodplain of the Euphrates* (Chicago: University of Chicago Press, 1981), 2. On the value of studying a river basin as an integrated unit, see Stanley A. Schumm, *The Fluvial System* (New York: John Wiley, 1977); Richard J. Chorley and Barbara A. Kennedy,

Physical Geography: A Systems Approach (London: Prentice-Hall International, 1971); Robin L. Vannote et al., "The River Continuum Concept," *Canadian Journal of Fisheries and Aquatic Sciences* 37, no. 1 (1980): 130–137. On the significance of building administrative frameworks around water resources, see Leila M. Harris and Samer Alatout, "Negotiating Hydro-Scales, Forging States: Comparison of the Upper Tigris/Euphrates and Jordan River Basins," *Political Geography* 29, no. 3 (2010): 148–156.

9. François Molle, "River-Basin Planning and Management: The Social Life of a Concept," *Geoforum* 40, no. 3 (2009): 484–494; Ludwik A. Teclaff, "Evolution of the River Basin Concept in National and International Water Law," *Natural Resources Journal* 36, no. 2 (1996): 359–391.

10. Ahmet T. Karamustafa, "Military, Administrative, and Scholarly Maps and Plans," in *The History of Cartography*, vol. 2, bk. 1, *Cartography in the Traditional Islamic and South Asian Societies*, ed. J. B. Harley and David Woodward (Chicago: University of Chicago Press, 1987), 222–223. According to one scholar, the map belongs to the renowned traveler Evliya Çelebi. See Zekeriya Kurşun, "Does the Qatar Map of the Tigris and Euphrates belong to Evliya Çelebi?" *Osmanlı Araştırmaları* 39 (2012): 1–15.

11. Thomas Dunne and Luna B. Leopold, *Water in Environmental Planning* (San Francisco: W. H. Freeman, 1978), 495; Molle, "River-Basin Planning and Management"; Ludwik A. Teclaff, *The River Basin in History and Law* (The Hague: Martinus Nijhoff, 1967), 7–14; J. David Allan and María M. Castillo, *Stream Ecology: Structure and Function of Running Waters*, 2nd ed. (Dordrecht: Springer, 2007), 2–4.

12. Kris Verhoeven, "Geomorphological Research in the Mesopotamian Flood Plain," in *Changing Watercourses in Babylonia: Towards a Reconstruction of the Ancient Environment in Lower Mesopotamia*, ed. Hermann Gasche and Michel Tanret (Ghent: University of Ghent and the Oriental Institute of the University of Chicago, 1998), 210; W. L. Powers, "Soil and Land-Use Capabilities in Iraq: A Preliminary Report," *Geographical Review* 44, no. 3 (1954): 373.

13. Evliya Çelebi, *Evliya Çelebi Seyahatnamesi: Topkapı Sarayı Bağdat 304 Yazmasının Transkripsiyonu-Dizini*, ed. Yücel Dağlı and Seyit Ali Kahraman (Istanbul: Yapı Kredi Yayınları, 2001), 4:251.

14. Jean-Baptiste Tavernier, *Les Six Voyages de Jean-Baptiste Tavernier* (Paris: G. Clouzier et C. Barbin, 1676), 1:217. See also Jean Baptiste Louis Jacques Rousseau, *Description du Pachalik de Bagdad* (Paris: Treuttel et Würtz, 1809), 57.

15. Hans H. Boesch, "El-'Iraq," *Economic Geography* 15, no. 4 (1939): 339. In the 1830s, a British geographer noted that the Euphrates town of Ana in northwestern Iraq marked the frontier between the olive tree and date palm. See William Ainsworth, *Researches in Assyria, Babylonia, and Chaldæa* (London: John W. Parker, 1838), 48–49.

16. Katib Çelebi (d. 1657) explicitly states that Arab Iraq comprises the Baghdad and Basra provinces. See Katib Çelebi, *Cihannüma* (Istanbul: Darü't-Tıbaati'l-Amire, 1145/1732–1733), 451.
17. In premodern geographical literature written in Arabic, Iraq normally referred to the alluvium and excluded the northern regions like Mosul. See Ali al-Wardi, *Dirasa fi Tabiʿat al-Mujtama al-Iraqi* (London: Dar al-Warraq, 2009), 166.
18. Henry D. Thoreau, *Journal*, vol. 4, *1851–1852*, ed. Leonard N. Neufeldt and Nancy Craig Simmons (Princeton, NJ: Princeton University Press, 1992), 4:412. I am grateful to Laura Dassow Walls for bringing this quote to my attention and for introducing me to Thoreau's life and work.
19. On the shortcomings of the hydrologic cycle as a concept, see Jamie Linton, "Is the Hydrologic Cycle Sustainable? A Historical-Geographical Critique of a Modern Concept," *Annals of the Association of American Geographers* 98, no. 3 (2008): 630–649.
20. Piet Buringh, *Soils and Soil Conditions in Iraq* (Baghdad: Ministry of Agriculture, 1960), 34–42; Trond H. Torsvik and L. Robin M. Cocks, *Earth History and Palaeogeography* (Cambridge: Cambridge University Press, 2017), 261–264. Geologists still debate the timing of the Arabia-Eurasia continental collision. A recent study suggests a minimum age of about 26 million years. See Renas I. Koshnaw, Daniel F. Stockli, and Fritz Schlunegger, "Timing of the Arabia-Eurasia Continental Collision—Evidence from Detrital Zicron U-Pb Geochronology of the Red Bed Series Strata of the Northwest Zagros Hinterland, Kurdistan Region of Iraq," *Geology* 47, no. 1 (2018): 47–50. An earlier study suggests that the collision was initiated 36 million years ago and accelerated 17.5 million years ago. See Paolo Ballato et al., "Arabia-Eurasia Continental Collision: Insights from Late Tertiary Foreland-Basin Evolution in the Alborz Mountains, Northern Iran," *Geological Society of America Bulletin* 123, no. 1/2 (2011): 106–131. I am grateful to Thamer Aldaajani for the references and for educating me about plate tectonics.
21. Guillermo Algaze, *Ancient Mesopotamia at the Dawn of Civilization: The Evolution of an Urban Landscape* (Chicago: University of Chicago Press, 2008), 49; T. J. Wilkinson, *Archaeological Landscapes of the Near East* (Tucson: University of Arizona Press, 2003), 89–91; Verhoeven, "Geomorphological Research in the Mesopotamian Flood Plain," 159–245; Adams, *Heartland of Cities*, 17–18.
22. Halil İnalcık, "Osman Ghazi's Siege of Nicaea and the Battle of Bapheus," in *The Ottoman Emirate (1300–1389)*, ed. Elizabeth Zachariadou (Rethymnon: Crete University Press, 1993), 77–99.
23. Quoted in Rudi Paul Lindner, *Explorations in Ottoman Prehistory* (Ann Arbor: University of Michigan Press, 2007), 20. Ottoman sources commonly referred to the nomads of Anatolia as *yürük*, from the Turkish verb *yürümek*, which means to walk or wander.
24. For assessments of Ottoman historiography today, see Alan Mikhail and Christine M. Philliou, "The Ottoman Empire and the Imperial Turn," *Comparative Studies*

in Society and History 54, no. 4 (2012): 721–745; Virginia H. Aksan, "What's Up in Ottoman Studies?" *Journal of the Ottoman and Turkish Studies Association* 1, no. 1/2 (2014): 3–21. On the latest developments in the environmental history of the Tigris-Euphrates basin, see Mark Altaweel et al., "New Insights on the Role of Environmental Dynamics Shaping Southern Mesopotamia: From the Pre-Ubaid to the Early Islamic Period," *Iraq* 81 (2019): 23–46. On the revolutionary changes in the discipline of history more generally, see J. R. McNeill, "Peak Document and the Future of History," *American Historical Review* 125, no. 1 (2020): 1–18.

25. Kenneth Burke, *The Philosophy of Literary Form: Studies in Symbolic Action* (Baton Rouge: Louisiana State University Press, 1941), 110–111. Quoted in Gerald Graff and Cathy Birkenstein, *"They Say/I Say": The Moves That Matter in Academic Writing*, 3rd ed. (New York: W. W. Norton, 2014), 13.

26. Halil İnalcık, "Ottoman Methods of Conquest," *Studia Islamica* 2 (1954): 103–129.

27. Gábor Ágoston, "A Flexible Empire: Authority and Its Limits on the Ottoman Frontiers," *International Journal of Turkish Studies* 9, no. 1/2 (2003): 15–31. For a similar view, see Daniel Goffman and Christopher Stroop, "Empire as Composite: The Ottoman Polity and the Typology of Dominion," in *Imperialisms: Historical and Literary Investigations, 1500–1900*, ed. Balachandra Rajan and Elizabeth Sauer (New York: Palgrave Macmillan, 2004), 129–145.

28. Karen Barkey, *Empire of Difference: The Ottomans in Comparative Perspective* (New York: Cambridge University Press, 2008), 28–66. See also Heath W. Lowry, *The Nature of the Early Ottoman State* (Albany: State University of New York Press, 2003).

29. Baki Tezcan, *The Second Ottoman Empire: Political and Social Transformation in the Early Modern World* (New York: Cambridge University Press, 2010).

30. Ali Yaycioglu, *Partners of the Empire: The Crisis of the Ottoman Order in the Age of Revolutions* (Stanford, CA: Stanford University Press, 2016).

31. On the rise (and limits) of the pragmatism lexicon in Ottoman studies since the late 1990s, see Murat Dağlı, "The Limits of Ottoman Pragmatism," *History and Theory* 52, no. 2 (2013): 194–213.

32. Stephanie Rost, "Water Management in Mesopotamia from the Sixth till the First Millennium B.C.," *Wiley Interdisciplinary Reviews: Water* 4, no. 5 (2017); Mark Altaweel, "Southern Mesopotamia: Water and the Rise of Urbanism," *Wiley Interdisciplinary Reviews: Water* 6, no. 4 (2019).

33. Jennifer R. Pournelle, "Marshland of Cities: Deltaic Landscapes and the Evolution of Early Mesopotamian Civilization" (PhD diss., University of California, San Diego, 2003); Pournelle, "KLM to CORONA: A Bird's-Eye View of Cultural Ecology and Early Mesopotamian Urbanization," in *Settlement and Society: Essays Dedicated to Robert McCormick Adams*, ed. Elizabeth C. Stone (Los Angeles: Cotsen Institute of Archaeology, 2007), 29–62; Pournelle and Guillermo Algaze, "Travels in Edin: Deltaic Resilience and Early Urbanism in Greater Mesopotamia," in *Preludes to Urbanism: The Late Chalcolithic of Mesopotamia*, ed. Augusta McMahon

and Harriet E. W. Crawford (Cambridge: McDonald Institute for Archaeological Research, 2014), 7–34; Angela Greco, "The Taming of the Wilderness: Marshes as an Economic Resource in 3rd Millennium BC Southern Mesopotamia," *Water History* 12, no. 1 (2020): 23–38; Algaze, *Ancient Mesopotamia*, 43–49; Wilkinson, *Archaeological Landscapes*, 87–88.

34. Magnus Widell et al., "Land Use of the Model Communities," in *Models of Mesopotamian Landscapes: How Small-Scale Processes Contributed to the Growth of Early Civilizations*, ed. T. J. Wilkinson, McGuire Gibson, and Widell (Oxford: Archaeopress, 2013), 56–80; T. J. Wilkinson, Louise Rayne, and Jaafar Jotheri, "Hydraulic Landscapes in Mesopotamia: The Role of Human Niche Construction," *Water History* 7, no. 4 (2015): 397–418; Jaafar Jotheri, "Recognition Criteria for Canals and Rivers in the Mesopotamian Floodplain," in *Water Societies and Technologies from the Past and* Present, ed. Yijie Zhuang and Mark Altaweel (London: UCL Press, 2018), 111–126.

35. Stephanie Rost, "Watercourse Management and Political Centralization in Third-Millennium B.C. Southern Mesopotamia: A Case Study of the Umma Province of the Ur III Period (2112–2004 B.C.)" (PhD diss., Stony Brook University, 2015).

36. The most extreme illustration of the imperial irrigation system was the Nahrawan complex east of the Tigris between Samarra and Kut. The total length of its two main canals, the Katul al-Kisrawi (upper course) and the Nahrawan proper (lower course), is estimated to have been more than 185 miles. Rost, "Water Management in Mesopotamia."

37. Peter Christensen, *The Decline of Iranshahr: Irrigation and Environments in the History of the Middle East, 500 B.C. to A.D. 1500* (Copenhagen: Museum Tusculanum Press, 1993), 67–72; Robert McC. Adams, "Intensified Large-Scale Irrigation as an Aspect of Imperial Policy: Strategies of Statecraft on the Late Sasanian Mesopotamian Plain," in *Agricultural Strategies*, ed. Joyce Marcus and Charles Stanish (Los Angeles: Cotsen Institute of Archaeology, 2006), 17–37; Wilkinson, *Archaeological Landscapes*, 92–97.

38. Peter Christensen, *The Decline of Iranshahr*, 85–104; Muhammad Rashid al-Feel, *The Historical Geography of Iraq between the Mongolian and Ottoman Conquests, 1258–1534* (Najaf: al-Adab Press, 1965), 3–53; David Waines, "The Third Century Internal Crisis of the Abbasids," *Journal of the Economic and Social History of the Orient* 20, no. 3 (1977): 284–285; Michele Campopiano, "Cooperation and Private Enterprise in Water Management in Iraq: Continuity and Chance between the Sasanian and Early Islamic Periods (Sixth to Tenth Centuries)," *Environment and History* 23, no. 3 (2017): 385–407; Marshall Hodgson, *The Venture of Islam*, vol. 1, *The Classical Age of Islam* (Chicago: University of Chicago Press, 1974), 483–485; Robert McC. Adams, *Land behind Baghdad: A History of Settlement on the Diyala Plains* (Chicago: University of Chicago Press, 1965), 84–111; E. Ashtor, *A Social and Economic History of the Near East in the Middle Ages* (Berkeley: University of California Press, 1976), 168–177.

39. For major exceptions, see Orhonlu and Işıksal, "Osmanlı Devrinde Nehir Nakliyatı Hakkında Araştırmalar," 79–102; Rhoads Murphey, "The Ottoman Centuries in Iraq: Legacy or Aftermath? A Survey Study of Mesopotamian Hydrology and Ottoman Irrigation Projects," *Journal of Turkish Studies* 11 (1987): 17–29.
40. Wilkinson, *Archaeological Landscapes*, 97.
41. Ahmed Susa, *Tarikh Hadarat Wadi al-Rafidayn* (Baghdad: Wizarat al-Ray, 1986), 2:254. For similar views, see Christensen, *The Decline of Iranshahr*; Robert McC. Adams, "Historic Patterns of Mesopotamian Irrigation Agriculture," in *Irrigation's Impact on Society*, ed. Adams, Theodore E. Downing, and McGuire Gibson (Tucson: University of Arizona Press, 1974), 1–5.
42. Hellmut Ritter, "Autographs in Turkish Libraries," *Oriens* 6, no. 1 (1953): 65.
43. Osmanlı Arşivi Daire Başkanlığı, *Başbakanlık Osmanlı Arşivi Rehberi* (Istanbul: Seçil Ofset, 2017), 3.
44. *Başbakanlık Osmanlı Arşivi Rehberi*, 7–8, 22–29.
45. *Başbakanlık Osmanlı Arşivi Rehberi*, 81–83. For the origins of the cadastral surveys, see İnalcık, "Ottoman Methods of Conquest," 109–112.
46. Metin M. Coşgel, "Ottoman Tax Registers (*Tahrir Defterleri*)," *Historical Methods* 37, no. 2 (2004): 87–100; Ömer Lutfi Barkan, "Research on the Ottoman Fiscal Surveys," in *Studies in the Economic History of the Middle East: From the Rise of Islam to the Present Day*, ed. M. A. Cook (London: Oxford University Press, 1970), 163–171; Osman Gümüşçü, "The Ottoman *Tahrir Defters* as a Source for Historical Geography," *Belleten* 72, no. 265 (2008): 911–941; J. Káldy-Nagy, "The Administration of the *Sanjaq* Registrations in Hungary," *Acta Orientalia Academiae Scientiarum Hungaricae* 21, no. 2 (1968): 181–223; Rhoads Murphey, "Ottoman Census Methods in the Mid-Sixteenth Century: Three Case Histories," *Studia Islamica* 71 (1990): 115–126; Margaret L. Venzke, "The Ottoman Tahrir Defterleri and Agricultural Producitivty," *Osmanlı Araştırmaları* 17 (1997): 1–51; Bistra Cvetkova, "Early Ottoman Tahrir Defters as a Source on the History of Bulgaria and the Balkans," *Archivum Ottomanicum* 8 (1983): 133–213; M. Mehdi İlhan, "The Process of Ottoman Cadastral Surveys during the Second Half of the Sixteenth Century: A Study Based on Documents from Muhimme Defters," *A. D. Xenopol* 24, no. 1 (1987): 17–25; Heath W. Lowry, *Studies in Defterology: Ottoman Society in the Fifteenth and Sixteenth Centuries* (Istanbul: Isis Press, 1992); Amy Singer, "*Tapu Tahrir Defterleri* and *Kadı Sicilleri*: A Happy Marriage of Sources," *Tarih* 1 (1990): 95–125; Colin Heywood, "Between Historical Myth and 'Mythohistory': The Limits of Ottoman History," *Byzantine and Modern Greek Studies* 12 (1988): 322–345; Molly Greene, "An Islamic Experiment? Ottoman Land Policy on Crete," *Mediterranean Historical Review* 11, no. 1 (1996): 60–78; Bernard Lewis, "The Ottoman Archives as a Source for the History of the Arab Lands," *Journal of the Royal Asiatic Society* 83, no. 3–4 (1951): 139–155; Lewis, "Studies in the Ottoman Archives—I," *Bulletin of the School of Oriental and African Studies* 16, no. 3 (1954): 469–501.

47. On the sources' limitations, see Lowry, *Studies in Defterology*, 3–18; Gümüşçü, "The Ottoman *Tahrir Defters*," 931–934; Singer, "*Tapu Tahrir Defterleri*," 101–102. For a rebuttal of the criticism, see Coşgel, "Ottoman Tax Registers."
48. Sam White, *The Climate of Rebellion in the Early Modern Ottoman Empire* (New York: Cambridge University Press, 2011), 64; Alan Mikhail, *Nature and Empire in Ottoman Egypt: An Environmental History* (New York: Cambridge University Press, 2011), 38, 296. The modernization of agriculture from the late nineteenth century is documented in great detail in Syria by Elizabeth Rachel Williams, "Cultivating Empires: Environment, Expertise, and Scientific Agriculture in Late Ottoman and French Mandate Syria" (PhD diss., Georgetown University, 2015).
49. On the use of analogy as a tool to interpret the past, see Ian Hodder, *The Present Past: An Introduction to Anthropology for Archaeologists* (Barnsley, UK: Pen & Sword Archaeology, 2012). On the use of analogy in the archaeological literature of the ancient Near East, see Patty Jo Watson, "The Theory and Practice of Ethnoarchaeology with Special Reference to the Near East," *Paléorient* 6 (1980): 55–64; Marc Verhoeven, "Ethnoarchaeology, Analogy, and Ancient Society," in *Archaeologies of the Middle East: Critical Perspectives*, ed. Susan Pollock and Reinhard Bernbeck (Malden, MA: Blackwell, 2005), 251–270. I am grateful to Stephanie Rost for directing me to some of these references and for educating me about the field of ethnoarchaeology.
50. For a couple of ethnoarchaeological studies based in Iraq, see Stephanie Rost and Abdulamir Hamdani, "Traditional Dam Construction in Modern Iraq: A Possible Analogy for Ancient Mesopotamian Irrigation Practices," *Iraq* 73 (2011): 201–220; E. L. Ochsenschlager, "Village Weavers: Ethnoarchaeology at al-Hiba," *Bulletin on Sumerian Agriculture* 7 (1993): 43–62.
51. For a critique of this tendency in the study of the civilizations of ancient Iraq, see Roger Matthews, *The Archaeology of Mesopotamia: Theories and Approaches* (New York: Routledge, 2003), 123–126.
52. For a recent assessment, see Mona Damluji et al., "Roundtable: Perspectives on Researching Iraq Today," *Arab Studies Journal* 23, no. 1 (2015): 236–265.

PART I

1. Chester G. Starr, *The Roman Imperial Navy, 31 B.C.–A.D. 325*, 3rd ed. (Chicago: Ares, 1993), 124–166; Brian Campbell, *Rivers and the Power of Ancient Rome* (Chapel Hill: University of North Carolina Press, 2012), 160–199.
2. Ewan W. Anderson, *The Middle East: Geography and Geopolitics* (London: Routledge, 2000), 48.
3. Richard White, *The Organic Machine* (New York: Hill and Wang, 1995).
4. Anderson, *The Middle East*, 74.
5. On the Safavid Empire's limited success with gunpowder technology, see Rudi Matthee, "Unwalled Cities and Restless Nomads: Firearms and Artillery in Safavid Iran," in *Safavid Persia: The History and Politics of an Islamic Society*, ed. Charles Melville (London: I. B. Tauris, 1996), 389–416.

6. Baghdad and Basra briefly fell to Persian forces in 1623–1638 and 1776–1779, respectively.

CHAPTER 1

1. OA, MD 6/549 (27 CA 972/31 December 1564).
2. For the role of the Jazira desert in restricting cultural, political, and military contact between Anatolia and Iraq, see J. N. Postgate, *Early Mesopotamia: Society and Economy at the Dawn of History* (New York: Routledge, 1994), 18.
3. John Cartwright, "Observations of Master John Cartwright in his Voyage from Aleppo to Hispaan, and backe againe: published by himselfe, and here contracted," in *Hakluytus Posthumus: Or Purchas His Pilgrimes: Contayning a History of the World in Sea Voyages and Lande Travells by Englishmen and Others*, ed. Samuel Purchas (Glasgow: James MacLehose and Sons, 1905), 8:483.
4. Klára Hegyi, "The Ottoman Networks of Fortresses in Hungary," in *Ottomans, Hungarians, and Habsburgs in Central Europe: The Military Confines in the Era of Ottoman Conquest*, ed. Pál Fodor and Géza Dávid (Leiden: Brill, 2000), 163–193; Gábor Ágoston, "The Costs of the Ottoman Fortress-System in Hungary in the Sixteenth and Seventeenth Centuries," in Fodor and Dávid, *Ottomans, Hungarians, and Habsburgs*, 195–228; Ágoston, "Where Environmental and Frontier Studies Meet: Rivers, Forests, Marshes and Forts along the Ottoman-Hapsburg Frontier in Hungary," in *The Frontiers of the Ottoman World*, ed. A. C. S. Peacock (Oxford: Oxford University Press, 2009), 58–60; Mark L. Stein, *Guarding the Frontier: Ottoman Border Forts and Garrisons in Europe* (New York: I. B. Tauris, 2007).
5. A. C. S. Peacock, "Introduction: The Ottoman Empire and Its Frontiers," in Peacock, *The Frontiers of the Ottoman World*, 1–27; Suavi Aydın and Oktay Özel, "Power Relations between State and Tribe in Ottoman Eastern Anatolia," *Bulgarian Historical Review* 34, no. 3–4 (2006): 51–67; Tom Sinclair, "The Ottoman Arrangements for the Tribal Principalities of the Lake Van Region of the Sixteenth Century," *International Journal of Turkish Studies* 9, no. 1–2 (2003): 119–143; Hakan Özoğlu, "State-Tribe Relations: Kurdish Tribalism in the 16th- and 17th-Century Ottoman Empire," *British Journal of Middle Eastern Studies* 23, no. 1 (1996): 5–27; Rhoads Murphey, "The Resumption of Ottoman-Safavid Border Conflict, 1603–1638: Effects of Border Destabilization on the Evolution of State-Tribe Relations," in *Shifts and Drifts in Nomad-Sedentary Relations*, ed. Stefan Leder and Bernhard Streck (Wiesbaden: L. Reichert, 2005), 308–323; Alfred J. Rieber, *The Struggle for the Eurasian Borderlands: From the Rise of Early Modern Empires to the End of the First World War* (New York: Cambridge University Press, 2014), 28–29.
6. My basin-wide perspective of Ottoman fortresses is inspired by Craig E. Colten, "Fluid Geographies: Urbanizing River Basins," in *Urban Rivers: Remaking Rivers, Cities, and Space in Europe and North America*, ed. Stéphane Castonguay and Matthew Evenden (Pittsburgh: University of Pittsburgh Press, 2012), 201–218.

7. Halil İnalcık, "The Question of the Emergence of the Ottoman State," *International Journal of Turkish Studies* 2 (1980): 71–79; Cemal Kafadar, *Between Two Worlds: The Construction of the Ottoman State* (Berkeley: University of California Press, 1995), 88–89.
8. The following account is a summary of the events. Readers interested in a more detailed discussion can consult Donald Edgar Pitcher, *An Historical Geography of the Ottoman Empire from Earliest Times to the End of the Sixteenth Century* (Leiden: E. J. Brill, 1972); Ebru Boyar, "Ottoman Expansion in the East," in *The Cambridge History of Turkey*, vol. 2, *The Ottoman Empire as a World Power, 1453–1603*, ed. Suraiya N. Faroqhi and Kate Fleet (Cambridge: Cambridge University Press, 2006), 74–140.
9. Nejat Göyünç, "Diyarbekir Beylerbeyiliği'nin İlk İdari Taksimatı," *Tarih Dergisi* 23 (1969): 23–34; Ahmet Gündüz, *Osmanlı İdaresinde Musul (1523–1639)* (Elazığ: Fırat Üniversitesi Basımevi, 2003), 32–43; Ali Yılmaz, "XVI. Yüzyılda Birecik Sancağı" (PhD diss., İstanbul Üniversitesi, 1996), 40–45; S. H. Winter, "The Province of Raqqa Under Ottoman Rule, 1535–1800: A Preliminary Study," *Journal of Near Eastern Studies* 68, no. 4 (2009): 253–268.
10. Ömer Lutfi Barkan devised the most popular formula for calculating the total population from the number of taxpayers recorded in Ottoman cadasters, proposing a multiplier of five for each household (*hane*) and the addition of 10 percent to account for tax-exempt bureaucrats and religious men. See Ömer Lutfi Barkan, "Essai sur les données statistiques des registres de recensement dans l'Empire ottoman aux XVe et XVIe siècles," *Journal of the Economic and Social History of the Orient* 1, no. 1 (1957): 9–36. For alternative views, see Osman Gümüşçü, "The Ottoman *Tahrir Defters* as a Source for Historical Geography," *Belleten* 72, no. 265 (2008): 921–922; Bekir Kemal Ataman, "Ottoman Demographic History (14th–17th Centuries): Some Considerations," *Journal of the Economic and Social History of the Orient* 35, no. 2 (1992): 187–198; Leila Erder, "The Measurement of Preindustrial Population Changes: The Ottoman Empire from the 15th to the 17th Centuries," *Middle Eastern Studies* 11, no. 3 (1975): 284–301; Nejat Göyünç, "'Hane' Deyimi Hakkında," *Tarih Dergisi* 32 (1979): 331–348; Huri İslamoğlu-İnan, *State and Peasant in the Ottoman Empire: Agrarian Power Relations and Regional Economic Development in Ottoman Anatolia during the Sixteenth Century* (Leiden: Brill, 1994), 27–29; Linda T. Darling, *Revenue-Raising and Legitimacy: Tax Collection and Finance Administration in the Ottoman Empire, 1560–1660* (Leiden: E. J. Brill, 1996), 100–108; Oktay Özel, *The Collapse of Rural Order in Ottoman Anatolia: Amasya, 1576–1643* (Leiden: Brill, 2016), 120.
11. Gülru Necipoğlu, *The Age of Sinan: Architectural Culture in the Ottoman Empire* (Princeton, NJ: Princeton University Press, 2005), 355–362; Bruce Masters, "Aleppo: The Ottoman Empire's Caravan City," in *The Ottoman City between East and West: Aleppo, Izmir, and Istanbul*, ed. Edhem Eldem, Daniel Goffman, and Bruce Masters (Cambridge: Cambridge University Press, 1999), 28–29;

Joshua M. White, "Shifting Winds: Piracy, Diplomacy, and Trade in the Ottoman Mediterranean, 1624–1626," in *Well-Connected Domains: Towards an Entangled Ottoman History*, ed. Pascal W. Firges et al. (Leiden: Brill, 2014), 37–41.

12. The topic is examined in detail by Donna Landry, *Noble Brutes: How Eastern Horses Transformed English Culture* (Baltimore: Johns Hopkins University Press, 2009).

13. For an eighteenth-century description of the three routes between Aleppo and the river basin, see William Beawes, "Remarks and Occurrences in a Journey from *Aleppo* to *Bassora*, by the Way of the Desert," in *The Desert Route to India: Being the Journals of Four Travellers by the Great Desert Caravan Route between Aleppo and Basra, 1745–1751*, ed. Douglas Carruthers (London: Hakluyt Society, 1929), 5–9.

14. *Encyclopaedia Iranica*, s.v. "Tigris River" (Daniel T. Potts), accessed November 2, 2017, http://www.iranicaonline.org/articles/tigris-river. Early modern travelers noted the Persian etymology of the Tigris. See, for example, Vincenzo Maria Murchio, *Il viaggio all' Indie Orientali* (Rome: Filippomaria Mancini, 1672), 88.

15. Emrullah Güney, "Dicle Irmağında Kelek Taşımacılığı," *Coğrafya Araştırmaları* 2, no. 2 (1990): 323–328; Kenan Ziya Taş, "Osmanlı'nın Son Döneminde Fırat ve Dicle Nehirlerinde Kelek ile Ulaşım," in *Osmanlı Devleti'nde Nehirler ve Göller*, ed. Şakir Batmaz and Özen Tok (Kayseri: Not Yayınları, 2015), 1:413–428. The kelek fascinated Ottoman and European authors alike. For some of their early descriptions of the raft, see Katib Çelebi, *Cihannüma* (Istanbul: Darü't-Tıbaati'l-Amire, 1145/1732), 468–469; Evliya Çelebi, *Evliya Çelebi Seyahatnamesi: Topkapı Sarayı Bağdat 304 Yazmasının Transkripsiyonu-Dizini*, ed. Yücel Dağlı and Seyit Ali Kahraman (Istanbul: Yapı Kredi Yayınları, 2001), 4:54; M. John Eldred, "The Voyage of M. John Eldred to Tripolis in Syria by Sea, and from thence by Land and River to Babylon, and Balsara, Anno 1583," in *The Principal Navigations, Voyages, Traffiques & Discoveries of the English Nations*, ed. Richard Hakluyt (London: J. M. Dent, 1907), 3:321–328; Jean-Baptiste Tavernier, *Les Six Voyages de Jean-Baptiste Tavernier* (Paris: Gervais Clouzier, 1781), 184; Domenico Sestini, *Voyage de Constantinople à Bassora, en 1781, par le Tigre et l'Euphrate: Et Retour a Constantinople, en 1782, par le Désert et Alexandrie* (Paris: Chez Dupuis, 1798), 153. Early modern descriptions of the kelek varied little from that provided by Herodotus in the fifth century BC and underline the powerful continuities in human-river relations in the region. See Herodotus, *The Landmark Herodotus: The Histories*, ed. Robert B. Strassler, trans. Andrea L. Purvis (New York: Anchor Books, 2009), 105.

16. Jean de Thévenot, *The Travels of Monsieur de Thévenot into the Levant in Three Parts*, trans. Archibald Lovell (London: Printed by H. Clark, H. Faithorne, J. Adamson, C. Skegnes, and T. Newborough, 1687), 2:54.

17. For an elaboration on this point, see Robert M. Adams, "Designed Flexibility in a Sewn Boat of the Western Indian Ocean," in *Sewn Plank Boats: Archaeological and Ethnographic Papers Based on Those Presented to a Conference at Greenwich in November, 1984*, ed. Sean McGrail and Eric Kentley (Oxford: BAR, 1985), 289–302.

18. On this point, see Pietro della Valle, *Viaggi di Pietro della valle il Pellegrino* (Rome: n.p., 1650), 1:693–694.
19. In one decree, Istanbul explicitly ordered officials in Diyarbakır to recycle animal skin used in keleks for the construction of new ones. See OA, C.AS 71/3355 (n.d.).
20. OA, MD 87/448 (29 B 1047/16 December 1637); OA, C.HR 68/3397 (12 N 1145/27 February 1733); OA, MAD 9934, 173–174, 177–178 (9 L 1146/15 March 1734); OA, AE.SMHD.I 85/5750 (Evasıt M 1147/13–23 June 1734); OA, HAT 7/223 (11 CA 1159/1 June 1746); OA, D.BŞM.MSH 8/1 (1 ZA 1161/24 October 1748); OA, MAD 9948, 45–46 (17 CA 1162/4 May 1749); OA, MAD 9948, 54 (25 N 1162/8 September 1749); OA, MAD 9968, 93 (10 ZA 1163/11 October 1750); OA, MAD 9968, 389 (29 L 1163/30 September 1750); OA, MAD 9970, 7 (3 M 1164/2 December 1750); OA, MAD 9970, 29 (21 M 1164/20 December 1750).
21. OA, MD 87/448 (29 B 1047/16 December 1637).
22. OA, AE.SİBR 5/560 (c. 1050/1640–1641); Evliya Çelebi, *Evliya Çelebi Seyahatnamesi*, 4:54.
23. TSMA 1046/44 (Evasıt B 1049/17–26 November 1638). For other orders of animal skins delivered to the kelekçiyam, see OA, MAD 3134, 170–171 (21 R 1112/4 October 1700); OA, İE.ML 107/10135 (14 M 1138/21 September 1725); OA, C.ML 522/21342 (20 CA 1138/24 January 1726); OA, C.AS 954/41442 (24 CA 1139/17 January 1727); OA, MAD 9934, 309 (28 B 1146/4 January 1734); OA, MAD 9934, 173–174, 177–178 (9 L 1146/15 March 1734); OA, AE.SMHD.I 85/5750 (Evasıt M 1147/13–22 June 1734); OA, C.ML 221/9191 (21 ZA 1147/6 April 1735).
24. TSMA 1046/44 (Evasıt B 1049/17–26 November 1638).
25. Özlem Başarır cites evidence of a formal kelekçiyan guild in Diyarbakır that dates to 1741. See Başarır, "Diyarbekir Voyvodası Mustafa Ağa'nın Terekesi Üzerine Bazı Düşünceler," *Bilig* 65 (2013): 35. On the kelekçiyan guild in Mosul, see OA, D.BŞM.MSH 8/1 (1 ZA 1161/24 October 1748).
26. Suraiya Faroqhi, *Artisans of Empire: Crafts and Craftspeople Under the Ottomans* (New York: I. B. Tauris, 2009).
27. The internal organization of the raft makers' guild is better documented in the eighteenth century and was likely analogous to practices followed in the previous century. See OA, D.BŞM.MSH 8/33 (Evail RA 1166/6–15 January 1753); OA, C.ML 753/30679 (2 M 1196/18 December 1781); OA, C.NF 38/1891 (2 M 1196/18 December 1781).
28. Location in relation to the Tigris and Euphrates is one of the reasons the Abbasid dynasty established Baghdad in this particular spot. See Hugh Kennedy, "The Feeding of the Five Hundred Thousand: Cities and Agriculture in Early Islamic Mesopotamia," *Iraq* 73 (2011): 189.
29. Sir Robert Ker Porter, *Travels in Georgia, Persia, Armenia, Ancient Babylonia, & c. & c. during the Years 1817, 1818, 1819, and 1820* (London: Longman, Hurst, Rees, Orme, and Brown, 1822), 2:265. Along the same lines, a British commercial consul described Baghdad in 1774 as "the grand mart for the produce of India and Persia,

Constantinople, Aleppo, and Damascus" and "the grand oriental depository." Abraham Parsons, *Travels in Asia and Africa* (London: Longman, Hurst, Rees, and Orme, 1808), 127.

30. Matrakçı Nasuh, *Beyan-ı Menazil-i Sefer-i Irakeyn Sultan Süleyman Han* (Ankara: Türk Tarih Kurumu, 2014), fol. 61r; "Dastan-i Sultan Süleyman," Topkapı Sarayı Müzesi Kütüphanesi, Revan Köşkü 1286, ff. 238v–239r; OA, MD 12/146 (12 L 978/9 March 1571); OA, MD 12/152 (12 L 978/9 March 1571).

31. For a historical background on the bridges of Baghdad, see Guy Le Strange, *Baghdad during the Abbasid Caliphate from Contemporary Arabic and Persian Sources* (Oxford: Oxford University Press, 1900), 177–186.

32. Dr. Leonhart Rauwolff, "Travels into the Eastern Countries," in *A Collection of Curious Travels and Voyages*, ed. John Ray (London: Royal Society, 1693), 1:180. Two European travelers in the late eighteenth century estimated the channel width to be between 600 and 800 feet. See Carsten Niebuhr, *Reisebeschreibung nach Arabien und andern umliegenden Ländern* (Copenhagen: Gedruckt bey N. Möller, 1778), 2:298–299; Parsons, *Travels in Asia and Africa*, 118–119.

33. TKG.KK, TT 29, fol. 4r. Toll rates frequently changed between the sixteenth and eighteenth centuries but were always calculated based on cargo items and their mode of transportation. See Parsons, *Travels in Asia and Africa*, 119.

34. Evliya Çelebi, *Evliya Çelebi Seyahatnamesi*, 4:255–256.

35. John Newberie, "Two Voyages of Master John Newberie, One into the Holy Land; The Other to Balsara, Ormus, Persia, and Backe Thorow Turkie," in *Hakluytus Posthumus*, 8:454.

36. The reasons behind the rise of Ridwaniyya at the expense of Falluja are unclear. By 1703, a document notes that Falluja was on the brink of collapse. OA, MAD 2510, 152 (20 ZA 1114/7 April 1703).

37. Population figure is based on a taxpaying population of about 5,300. TKG.KK, TT 29, ff. 218r–265r; Parsons, *Travels in Asia and Africa*, 120.

38. OA, TT 282, 15, 21, 25; TKG.KK, TT 30, ff. 183r, 203r; Eldred, "The Voyage of M. John Eldred," 3:326.

39. On this point, see Janet L. Abu Lughod, *Before European Hegemony: The World System A.D. 1250–1350* (New York: Oxford University Press, 1989), 185–211; E. Ashtor, *A Social and Economic History of the Near East in the Middle Ages* (Berkeley: University of California Press, 1976), 263–279; Bernard Lewis, "The Mongols, the Turks and the Muslim Polity," *Transactions of the Royal Historical Society* 18 (1968): 54–55; Nükhet Varlık, *Plague and Empire in the Early Modern Mediterranean World: The Ottoman Experience, 1347–1600* (New York: Cambridge University Press, 2015), 177–178.

40. OA, MD 7/1312 (2 ZA 975/30 April 1568).

41. OA, MAD 9948, 45–46 (17 CA 1162/4 May 1749). For similar observations, see OA, MAD 9970, 7 (3 M 1164/2 December 1750); OA, MAD 9970, 97 (19 ZA 1164/9 October 1751).

42. OA, MD 19/293 (13 S 980/24 June 1572). See also Dina Rizk Khoury, "The Introduction of Commercial Agriculture in the Province of Mosul and Its Effects on the Peasantry, 1750–1850," in *Landholding and Commercial Agriculture in the Middle East*, ed. Çağlar Keyder and Faruk Tabak (Albany: State University of New York Press, 1991), 157–158.
43. "Correspondence, mainly of Sir Harford Jones, with 1st and 2nd Viscounts Melville: 1785–1820," British Library, Add. 41767, fol. 82v.
44. Pietro della Valle, *The Travels of Sig. Pietro della Valle, A Noble Roman, into East-India and Arabia Deserta, In Which, the Several Countries, Together with the Customs, Manners, Traffique, and Rites both Religious and Civil, of Those Oriental Princes and Nations, Are Faithfully Described*, trans. G. Havers (London: Printed by J. Macock, 1665), 243.
45. Marshall Hodgson, *The Venture of Islam: Conscience and History in a World Civilization*, vol. 1, *The Classical Age of Islam* (Chicago: University of Chicago Press, 1974), 485; Kennedy, "The Feeding of the Five Hundred Thousand"; Michele Campopiano, "State, Land Tax and Agriculture in Iraq from the Arab Conquest to the Crisis of the Abbasid Caliphate (Seventh–Tenth Centuries)," *Studia Islamica* 3 (2012): 5–50; Campopiano, "Cooperation and Private Enterprise in Water Management in Iraq: Continuity and Change between the Sasanian and Early Islamic Periods (Sixth to Tenth Centuries)," *Environment and History* 23, no. 3 (2017): 385–407; David Waines, "The Third Century Internal Crisis of the Abbasids," *Journal of the Economic and Social History of the Orient* 20, no. 3 (1977): 282–306; Peter Christensen, *The Decline of Iranshahr: Irrigation and Environments in the History of the Middle East, 500 B.C. to A.D. 1500* (Copenhagen: Museum Tusculanum Press, 1993), 85–104; Robert McC. Adams, *Land behind Baghdad: A History of Settlement on the Diyala Plains* (Chicago: University of Chicago Press, 1965), 84–111; Ashtor, *A Social and Economic History of the Near East*, 168–177; Guillermo Algaze, *Ancient Mesopotamia at the Dawn of Civilization: The Evolution of an Urban Landscape* (Chicago: University of Chicago Press, 2008), 74–77; Michael Roaf, *Cultural Atlas of Mesopotamia and the Ancient Near East* (New York: Facts On File, 1996), 35.
46. See Tables A.1 and A.2 in the appendix; Cengiz Orhonlu and Turgut Işıksal, "Osmanlı Devrinde Nehir Nakliyatı Hakkinda Araştırmalar: Dicle ve Fırat Nehirlerinde Nakliyat," *Tarih Dergisi* 13 (1962–1963): 79–102; Nejat Göyünç, "Dicle ve Fırat Nehirlerinde Nakliyat," *Belleten* 65, no. 243 (2001): 655–660; Ali Yılmaz, "16 ve 17 Yüzyıllarda Fırat'ta Nehir Nakliyatı," in *Osmanlı Devleti'nde Nehirler ve Göller*, 1:591–611. The raw metals used in the Ottoman arms industry were primarily iron, copper, bronze, and lead. The definitive study on the Ottoman arms industry is Gábor Ágoston, *Guns for the Sultan: Military Power and the Weapons Industry in the Ottoman Empire* (New York: Cambridge University Press, 2005).

47. Rossitsa Gradeva, "War and Peace along the Danube: Vidin at the End of the Seventeenth Century," *Oriente Moderno* 81, no. 1 (2001): 151.
48. OA, MD 6/1418 (18 Z 972/16 July 1565); OA, D.BŞM.MSH 1/88 (12 RA 1071/15 November 1660); OA, D.BŞM.MSH 1/190 (12 RA 1071/15 November 1660); OA, D.BŞM.MSH 2/103 (1661[?]); OA, D.BŞM.MSH 2/149 (22 R 1105/20 December 1693); OA, D.BŞM.MSH 2/191 (27 R 1105/20 December 1693); OA, MAD 3242 (2 CA 1116/2 September 1704); OA, İE.AS 54/4873 (7 ZA 1118/7 February 1707). On the militarization of the eastern frontier, see Rhoads Murphey, "The Functioning of the Ottoman Army under Murad IV (1623–1639/1032/1049): Key to the Understanding of the Relationship between Center and Periphery in Seventeenth-Century Turkey" (PhD diss., University of Chicago, 1979), 171–208.
49. Evliya Çelebi, *Evliya Çelebi Seyahatnamesi*, 4:249–252; Niebuhr, *Reisebeschreibung nach Arabien*, 2:293–295; Tavernier, *Les Six Voyages*, 1:208–209; Parsons, *Travels in Asia and Africa*, 124–125. By Evliya Çelebi's count, Baghdad's wall had 24,000 crenels.
50. Evliya Çelebi, *Evliya Çelebi Seyahatnamesi*, 4:251–252.
51. Niebuhr, *Reisebeschreibung nach Arabien*, 2:295; Jean Baptiste Louis Jacques Rousseau, *Description du Pachalik de Bagdad* (Paris: Treuttel et Würtz, 1809), 3.
52. Faisal H. Husain, "The Tigris-Euphrates Basin under Early Modern Ottoman Rule, c. 1534–1830" (PhD diss., Georgetown University, 2018), 64–65.
53. Evliya Çelebi, *Evliya Çelebi Seyahatnamesi*, 4:249.
54. Evliya Çelebi, *Evliya Çelebi Seyahatnamesi*, 4:252; Selim Güngörürler, "Diplomacy and Political Relations between the Ottoman Empire and Safavid Iran, 1639–1722" (PhD diss., Georgetown University, 2016), 113–114. I am grateful to Selim Güngörürler for sharing his dissertation with me.
55. Our knowledge of the local units (*yerlü neferat*) is patchy. We know that in 1733, for instance, they numbered 4,051 men. OA, MAD 9934, 86 (15 L 1146/21 March 1734).
56. Husain, "The Tigris-Euphrates Basin," 66; Ágoston, *Guns for the Sultan*, 148–149, 180; Turgut Işıksal, "Gunpowder in Ottoman Documents of the Last Half of the 16th Century," *International Journal of Turkish Studies* 2, no. 2 (1981–1982): 81–91.
57. Evliya Çelebi, *Evliya Çelebi Seyahatnamesi*, 4:249.
58. M. Mehdi İlhan, "XVI. Yüzyılın İlk Yarısında Diyarbakır Şehrinin Nüfusu ve Vakıfları: 1518 ve 1540 Tarihli Tapu Tahrir Defterlerinden Notlar," *Tarih Araştırmaları Dergisi* 16, no. 27 (1994): 52–56; Gündüz, *Osmanlı İdaresinde Musul*, 105–139; Erdinç Gülcü, "Osmanlı İdaresinde Bağdat (1534–1623)" (PhD diss., Fırat Üniversitesi, 1999), 170–210.
59. For an examination of several endowment projects established in Diyarbakır, Baghdad, and Basra during the second half of the sixteenth century, see Necipoğlu, *The Age of Sinan*, 462–470.
60. The Ottoman Archive alone holds thousands of documents related to staffing religious foundations in Diyarbakır, Mosul, Baghdad, and Basra. I have closely

examined a sample of 135 documents dating between the sixteenth and eighteenth centuries and listed them in the appendix.

61. See, for example, OA, AE.SMST.II 13/1225 (5 RA 1107/14 October 1695); OA, AE.SAMD.III 118/11598 (Evahir RA 1115/4–13 August 1703). Other rights to official commissions and land titles had to be renewed following the enthronement of a new sultan. The logic behind the custom stems from the Ottoman theory that the entire realm belonged to the sultan, whose will and absolute authority validated all rights. See Halil İnalcık, "Ottoman Methods of Conquest," *Studia Islamica* 2 (1954): 112–113.
62. A document dated to 1682 details how officials in the Imperial Treasury retrieved information related to the post of the prayer leader in the Khulafa Mosque Complex in Baghdad to resolve a conflict between two contenders for the post. See OA, MAD 2926, 165 (13 L 1093/14 October 1682).
63. Ayfer Karakaya-Stump, *The Kizilbash-Alevis in Ottoman Anatolia: Sufism, Politics and Community* (Edinburgh: Edinburgh University Press, 2019), 256–319; Derin Terzioğlu, "How to Conceptualize Ottoman Sunnitization: A Historiographical Discussion," *Turcica* 44 (2012–2013): 301–338; C. H. Imber, "The Persecution of the Ottoman Shiʿites according to the Mühimme Defterleri, 1565–1585," *Der Islam* 56 (1979): 245–273. On the Ottoman Empire's systematic effort to create an imperial religious hierarchy of its own, see Guy Burak, *The Formation of Islamic Law: The Hanafi School in the Early Modern Ottoman Empire* (New York: Cambridge University Press, 2015).
64. Lewis Mumford, *The City in History: Its Origins, Its Transformations, and Its Prospects* (New York: Harcourt Brace Jovanovich, 1961), 31.
65. Robert Stewart Castlereagh, *Correspondence, Despatches, and Other Papers of Viscount Castlereagh*, ed. Charles William Vane (London: William Shoberl, 1851), 5:186.

CHAPTER 2

1. The following narrative is based on the vivid contemporary account of Feridun Bey, *Nüzhet-i esrarü'l-ahyar der-ahbar-ı sefer-i Sigetvar: Sultan Süleyman'ın son seferi*, ed. H. Ahmet Arslantürk, Günhan Börekçi, and Abdülkadir Özcan (Istanbul: Zeytinburnu Belediyesi, 2012), 383–410. Katib Çelebi provides a more condensed version of the events a few decades later in his *Tuhfetü'l-Kibar fi Esfari'l-Bihar* (Istanbul: Matbaa-i Bahriye, 1329/1911), 83–85. For secondary examinations, see Abdurrahman Sağırlı, "Cezayir-i Irak-ı Arab veya Şattü'l-Arab'ın Fethi—Ulyanoğlu Seferi—1565-1571," *Tarih Dergisi* 41 (2005): 43–94; Nicolas Vatin, "Un territoire 'bien gardé' du sultan? Les Ottomans dans leur vilayet de Basra, 1565-1568," in *The Ottoman Middle East: Studies in Honor of Amnon Cohen*, ed. Eyal Ginio and Elie Podeh (Leiden: Brill, 2014), 63–91; İ. Metin Kunt, "An Ottoman Imperial Campaign: Suppressing the Marsh Arabs,

Central Power and Peripheral Rebellion in the 1560s," *Journal of Ottoman Studies* 43 (2014): 1–18.
2. Feridun Bey, *Nüzhet*, 401.
3. Feridun Bey, *Nüzhet*, 404.
4. Matrakçı Nasuh, "Tarih-i Al-i Osman," Österreichische Nationalbibliothek, Vienna, Cod. Mixt. 339 Han, fol. 241v. This manuscript used to be attributed to Grand Vizier Rüstem Paşa, but Hüseyin Yurdaydın convincingly argued that Matrakçı Nasuh is the actual author. See Yurdaydın, "Matrakçı Nasuh'un Hayatı ve Eserleri ile İlgili Yeni Bilgiler," *Belleten* 29, no. 114 (1965): 349–354.
5. Jan Glete, *Navies and Nations: Warships, Navies, and State Building in Europe and America, 1500–1860* (Stockholm: Almqvist & Wiksell International, 1993), 2:5–21.
6. Michael Roberts, *Essays in Swedish History* (Minneapolis: University of Minnesota Press, 1967), 195–225; Geoffrey Parker, *The Military Revolution: Military Innovation and the Rise of the West, 1500–1800*, 2nd ed. (New York: Cambridge University Press, 1996), 82–114; Glete, *Navies and Nations*; John F. Guilmartin Jr., *Galleons and Galleys* (London: Cassel, 2002); Carlo M. Cipolla, *Guns, Sails and Empires: Technological Innovation and the Early Phases of European Expansion, 1400–1700* (New York: Pantheon Books, 1965); Gábor Ágoston, *Guns for the Sultan: Military Power and the Weapons Industry in the Ottoman Empire* (New York: Cambridge University Press, 2005), 48–56; Kenneth Chase, *Firearms: A Global History to 1700* (New York: Cambridge University Press, 2003), 56–82. Since Michael Roberts introduced the idea of a military revolution in a lecture delivered in 1955, historians have hotly debated its meaning, extent, and chronology. Others have rejected it altogether. On this debate, see Clifford J. Rogers, ed., *The Military Revolution Debate: Readings on the Military Transformation of Early Modern Europe* (Boulder, CO: Westview, 1995). All disagreements aside, most proponents of the Military Revolution agree that the fusion of improved artillery and sailing ship technologies in the fifteenth century marked a turning point in the conduct of war, paving the way for more dramatic and widespread developments in the sixteenth and seventeenth centuries. For a concise chronology of the Military Revolution at sea, see John F. Guilmartin Jr., "The Military Revolution in Warfare at Sea during the Early Modern Era: Technological Origins, Operational Outcomes and Strategic Consequences," *Journal of Maritime Research* 13, no. 2 (2011): 129–137.
7. On "steamboat imperialism," see Daniel R. Headrick, *Power over Peoples: Technology, Environments, and Western Imperialism, 1400 to the Present* (Princeton, NJ: Princeton University Press, 2010), 177–225.
8. Katib Çelebi, *Cihannüma* (Istanbul: Darü't-Tıbaati'l-Amire, 1145/1732), 457.
9. Matrakçı Nasuh, "Tarih-i Al-i Osman," ff. 240v-243r; Lokman bin Hüseyin, "Mücmel-ul-Tumar," British Library, Or. 1135, ff. 67v–69r.
10. Dom Manuel de Lima's letter is translated into English in full in Salih Özbaran, "The Ottoman Turks and the Portuguese in the Persian Gulf, 1534–1581," *Journal of Asian History* 6, no. 1 (1972): 71–79, quotation on 73.

11. OA, TT 276, 17–19; OA, TT 501, 19–20; OA, TT 496, 29–30.
12. "Hükümname Mecmuası," Topkapı Sarayı Müzesi Kütüphanesi, Koğuşlar 888, fol. 344r (9 Ş 959/31 July 1552); "Hükümname Mecmuası," fol. 344v (10 Ş 959/1 August 1552); "Hükümname Mecmuası," fol. 345v (19 Ş 959/20 August 1552); Ali Yılmaz, "XVI. Yüzyılda Birecik Sancağı" (PhD diss., İstanbul Üniversitesi, 1996), 168–169.
13. Tahir Öğüt, *18–19 Yüzyıllarda Birecik Sancağında İktisadi ve Sosyal Yapı* (Ankara: Türk Tarih Kurumu, 2013), 1–20; Yılmaz, "Birecik Sancağı," 9–16; İdris Bostan, "Birecik," *Türkiye Diyanet Vakfı Ansiklopedisi* 6 (1992): 187–189.
14. William C. Brice, *South-West Asia* (London: University of London Press, 1966), 229–230.
15. Özbaran, "The Ottoman Turks and the Portuguese," 73–74.
16. Pedro Teixeira, *The Travels of Pedro Teixeira*, trans. William F. Sinclair (London: Hakluyt Society, 1902), 28.
17. OA, MD 3/1446 (25 ZA 967/17 August 1560).
18. OA, MD 3/463 (28 M 967/30 October 1559); OA, MD 3/751 (5 CA 967/2 February 1560). The initial order was for twenty galliots but was later reduced to eight, then to five. See OA, MD 3/834 (8 C 967/6 March 1560); OA, MD 3/1185 (7 N 967/2 June 1560).
19. OA, MD 3/764 (13 CA 967/10 February 1560); OA, MD 3/834 (8 C 967/6 March 1560); OA, MD 3/849 (12 C 967/10 March 1560).
20. OA, MD 3/1355 (10 Z 967/2 August 1560); OA, MD 3/1446 (25 Z 967/17 August 1560).
21. Ágoston, *Guns for the Sultan*, 50.
22. *Encyclopaedia of Islam*, 3rd ed. (Leiden: Brill Online, 2016), s.v. "Imperial Arsenal" (İdris Bostan); Colin Imber, "The Navy of Süleyman the Magnificent," *Archivum Ottomanicum* 6 (1980): 211–282.
23. The relationship between warfare and nature's energy is elaborated in Micah S. Muscolino, *The Ecology of War in China: Henan Province, the Yellow River, and Beyond, 1938–1950* (New York: Cambridge University Press, 2015), 1–20.
24. Population figures are derived from Sanjay Subrahmanyam, *The Portuguese Empire in Asia, 1500–1700*, 2nd ed. (Malden, MA: Wiley-Blackwell, 2012), 38; Ömer Lutfi Barkan, "Essai sur les données statistiques des registres de recensement dans l'Empire ottoman aux XVe et XVIe siècles," *Journal of the Economic and Social History of the Orient* 1, no. 1 (1957): 20–23. For a recent review of Ottoman demographic trends during the sixteenth century, see Sam White, *The Climate of Rebellion in the Early Modern Ottoman Empire* (New York: Cambridge University Press, 2011), 52–59.
25. Palmira Brummett, *Ottoman Seapower and Levantine Diplomacy in the Age of Discovery* (Albany: State University of New York Press, 1994), 69, 75, 111–121, 171–174; Giancarlo Casale, *The Ottoman Age of Exploration* (New York: Oxford University Press, 2010), 25–29; Salih Özbaran, *The Ottoman Response to European Expansion: Studies on Ottoman-Portuguese Relations in the Indian Ocean and Ottoman Administration in the Arab Lands during the Sixteenth Century*

(Istanbul: Isis Press, 1994), 89–97; Özbaran, "Ottoman Naval Policy in the South," in *Süleyman the Magnificent and His Age: The Ottoman Empire in the Early Modern World*, ed. Metin Kunt and Christine Woodhead (New York: Longman, 1995), 55–70.

26. For renovation orders, see OA, MAD 2775, 618 (14 CA 973/7 December 1565); OA, MD 10/421 (28 B 979/16 December 1571). For the construction of five new galliots, see Table A.3 in the appendix.

27. OA, MD 22/631 (15 CA 981/12 September 1573); OA, MD 22/632 (15 CA 981/12 September 1573); OA, MD 22/633 (15 CA 981/12 September 1573); OA, MD 22/636 (15 CA 981/12 September 1573); OA, MD 22/638 (15 CA 981/12 September 1573); OA, MD 22/639 (15 CA 981/12 September 1573).

28. OA, MD 27/203 (9 Ş 983/13 November 1575); OA, MD 27/436 (9 Ş 983/11 January 1576); OA, MD 27/450 (14 Ş 983/16 January 1576); OA, MD 27/465 (15 Ş 983/17 January 1576); OA, MD 27/748 (6 Z 983/7 March 1576). For the context of the Ottoman-Portuguese conflict, see Salih Özbaran, *Ottoman Expansion toward the Indian Ocean in the 16th Century* (Istanbul: Istanbul Bilgi University Press, 2009), 95–101; M. Mehdi İlhan, "The Katif District (*Liva*) during the First Few Years of Ottoman Rule: A Study of the 1551 Ottoman Cadastral Survey," *Belleten* 51, no. 200 (1987): 781–798; João Teles e Cunha, "The Portuguese Presence in the Persian Gulf," in *The Persian Gulf in History*, ed. Lawrence G. Potter (New York: Palgrave Macmillan, 2009), 207–234; Rudi Matthee, "The Portuguese Presence in the Persian Gulf: An Overview," in *Imperial Crossroads: The Great Powers and the Persian Gulf*, ed. Jeffrey R. Macris and Saul Kelly (Annapolis, MD: Naval Institute Press, 2012), 3–11; Casale, *The Ottoman Age of Exploration*, 95–102.

29. I am grateful to Giancarlo Casale for answering my questions about Ottoman relations with European powers in the Persian Gulf in an email message dated 4 January 2017. See also Fariba Zarinebaf, *Mediterranean Encounters: Trade and Pluralism in Early Modern Galata* (Oakland: University of California Press, 2018), 115–124; M. N. Pearson, *The Portuguese in India* (New York: Cambridge University Press, 1987), 37–39; Willem Floor, *The Persian Gulf: A Political and Economic History of Five Port Cities, 1500–1730* (Washington, DC: Mage, 2006), 139–187, 479–597; Floor, "Dutch Relations with the Persian Gulf," in *The Persian Gulf*, 235–259; Cunha, "The Portuguese Presence in the Persian Gulf," 207–234; Matthee, "The Portuguese Presence in the Persian Gulf," 3–11.

30. Rudi P. Matthee, "The Safavid Economy as Part of the World Economy," in *Iran and the World in the Safavid Age*, ed. Willem Floor and Edmund Herzig (New York: I. B. Tauris, 2012), 33.

31. Rudi Matthee, "Unwalled Cities and Restless Nomads: Firearms and Artillery in Safavid Iran," *Safavid Persia: The History and Politics of an Islamic Society*, ed. Charles Melville (New York: I. B. Tauris, 1996), 389–416. In the middle of the eighteenth century, Nadir Shah possessed a fleet acquired from the British and Dutch, but it fell into disuse soon after his death in 1747. See Michael Axworthy,

"Nader Shah and Persian Naval Expansion in the Persian Gulf, 1700–1747," *Journal of the Royal Asiatic Society* 21, no. 1 (2011): 31–39; Willem Floor, "The Iranian Navy in the Persian Gulf during the Eighteenth Century," *Iranian Studies* 20, no. 1 (1987): 31–53.

32. Katib Çelebi, *Fezleke-i Katib Çelebi* (Istanbul: Ceride-i Havadis Matbaası, 1286/1869), 2:80–81; Mustafa Naima, *Tarih-i Naima* (Istanbul: Darü't-Tıbaati'l-Amire, 1147/1734-1735), 1:428.

33. See Table A.3 in the appendix.

34. See Table A.3 in the appendix; TSMA 643/72 (1047/1637-1638); TSMA 301/14 (13 ZA 1048/18 March 1639); TSMA 1067/55 (13 Z 1048/18 March 1639).

35. Jean-Baptiste Tavernier, *Les Six Voyages de Jean Baptiste Tavernier* (Paris: G. Clouzier et C. Barbin, 1676), 1:138. See also Peçevi İbrahim, *Tarih-i Peçevi* (Istanbul: Matbaa-i Amire, 1283/1866-1867), 2:445.

36. The period between 1639 and 1722 in Ottoman-Safavid relations is documented by Selim Güngörürler, "Diplomacy and Political Relations between the Ottoman Empire and Safavid Iran, 1639–1722" (PhD diss., Georgetown University, 2016).

37. OA, MD 5/353 (19 RA 973/14 October 1565).

38. On the advantages tribal populations had in their competition with the state, see Ira M. Lapidus, "Tribes and State Formation in Islamic History," in *Tribes and State Formation in the Middle East*, ed. Philip S. Khoury and Joseph Kostiner (Berkeley: University of California Press, 1990), 42.

39. Teixeira, *The Travels of Pedro Teixeira*, 28.

40. See Table A.3 in the appendix.

41. Lionel Casson, *Ships and Seamanship in the Ancient World* (Princeton, NJ: Princeton University Press, 1971), 22–29; P. R. S. Moorey, *Ancient Mesopotamian Materials and Industries: The Archaeological Evidence* (Oxford: Clarendon Press, 1994), 10–12; D. T. Potts, *Mesopotamian Civilization: The Material Foundations* (London: Athlone Press, 1997), 122–137; Georges Contenau, *Everyday Life in Babylon and Assyria*, trans. K. R. and A. R. Maxwell-Hyslop (New York: St. Martin's Press, 1954), 44–47.

42. Faisal H. Husain, "The Tigris-Euphrates Basin under Early Modern Ottoman Rule, c. 1534–1830" (PhD diss., Georgetown University, 2018), 97.

43. M. John Eldred, "The Voyage of M. John Eldred to Tripolis in Syria by Sea, and from thence by Land and River to Babylon, and Balsara, Anno 1583," in *The Principal Navigations, Voyages, Traffiques & Discoveries of the English Nations*, ed. Richard Hakluyt (London: J. M. Dent, 1907), 3:326.

44. Casson, *Ships and Seamanship*, 30–76.

45. Svat Soucek, "Certain Types of Ships in Ottoman-Turkish Terminology," *Turcica* 7 (1975): 234–235.

46. OA, MAD 2775, 618 (14 CA 973/7 December 1565). On the *kolumburina* and *darbzen*, see Ágoston, *Guns for the Sultan*, 81–85.

47. İdris Bostan, *Kürekli ve Yelkenli Osmanlı Gemileri* (Istanbul: Bilge, 2005), 224–228.

48. Floor, *The Persian Gulf*, 173; Özbaran, "The Ottoman Turks and the Portuguese," 56; Özbaran, *Ottoman Expansion*, 192.
49. OA, MAD 2775, 618 (14 CA 973/7 December 1565).
50. For accounts on the use of boat cannon in conflicts with Iraq's tribesmen, see Matrakçı Nasuh, "Tarih-i Al-i Osman," fol. 240v–243r; Lokman bin Hüseyin, "Mücmel-ul-Tumar," ff. 67v–69r; OA, MD 46/49 (27 B 992/4 August 1584); OA, MD 69/75 (22 R 1000/6 February 1592).
51. Jonathan Grant, "Rethinking the Ottoman 'Decline': Military Technology Diffusion in the Ottoman Empire, Fifteenth to Eighteenth Centuries," *Journal of World History* 10, no. 1 (1999): 179–201; Bostan, *Osmanlı Gemileri*, 103–147, 278–291.
52. Bostan, *Osmanlı Gemileri*, 278–291; Tuncay Zorlu, *Innovation and Empire in Turkey: Sultan Selim III and the Modernisation of the Ottoman Navy* (New York: Tauris Academic Studies, 2008), 1–13; Emir Yener, "Ottoman Seapower and Naval Technology during Catherine II's Turkish Wars, 1768–1792," *International Naval Journal* 9, no. 1 (2016): 4–15.
53. See Table A.3 in the appendix.
54. Bostan, *Osmanlı Gemileri*, 228–233.
55. See Table A.3 in the appendix.
56. Husain, "The Tigris-Euphrates Basin," 97.
57. OA, C.BH 90/4305 (18 CA 1112/31 October 1700); OA, MAD 7915, 268 (11 B 1111/2 January 1700).
58. See, for example, OA, MAD 975, 15 (28 Safar 1112/14 August 1700).
59. OA, MAD 7915, 350–351 (17 B 1143/26 January 1731).
60. OA, C.BH 90/4305 (18 CA 1112/31 October 1700). On *saçma*, see Ágoston, *Guns for the Sultan*, 86–87.
61. Bostan, *Osmanlı Gemileri*, 236–242.
62. See Table A.3 in the appendix; Ayşe Pul, "Osmanlı Tuna Donanmasının Üstüaçık Gemileri," *Tarih Okulu Dergisi* 18 (2014): 285–317; İdris Bostan, *Osmanlı Bahriye Teşkilatı: XVII. Yüzyılda Tersane-i Amire* (Ankara: Türk Tarih Kurumu, 1992), 89, 246–247; Husain, "The Tigris-Euphrates Basin," 97. Ottoman scribes highlighted the Danubian origins of the *üstüaçık* by occasionally referring to it as the "açık of the Danube" (*açık Tuna*). See OA, MAD 5433, 2 (6 N 1112/14 February 1701); Rossitsa Gradeva, *War and Peace in Rumeli: 15th to Beginning of 19th Century* (Istanbul: Isis Press, 2008), 92–94.
63. OA, MAD 75, ff. 16v–22v; OA, TT 184, 7–13; OA, TT 276, 10–19; OA, TT 501, 10–19; OA, TT 496, 20–29.
64. OA, TT 282, 42–72; TKG.KK, TT 30, ff. 15v–32v.
65. OA, MD 19/293 (13 S 980/24 June 1572).
66. OA, MD 27/511 (26 L 983/28 January 1576).
67. OA, MAD 7915, 350–352 (17 B 1143/26 January 1731). In this context, Qizilbash could also mean Shi'is in general.

68. Murat Çizakça, "The Ottoman Empire: Recent Research on Shipping and Shipbuilding in the Sixteenth to Nineteenth Centuries," in *Maritime History at the Crossroads: A Critical Review of Recent Historiography*, ed. Frank Broeze (St. John's, Canada: International Maritime Economic History Association, 1995), 221. For an alternative perspective, see Giancarlo Casale, "The Ethnic Composition of Ottoman Ship Crews and the 'Rumi Challenge' to Portuguese Identity," *Medieval Encounters* 13 (2007): 122–144.
69. OA, MD 27/511 (26 L 983/28 January 1576).
70. OA, MAD 5433 (6 N 1112/14 February 1701).
71. OA, MAD 2775, 642 (6 CA 973/29 November 1565).
72. Husain, "The Tigris-Euphrates Basin," 104.
73. OA, MD 3/751 (5 CA 967/2 February 1560). The location for the construction project later moved from Birecik to Basra. See OA, MD 3/849 (12 C 967/10 March 1560).
74. OA, D.BŞM.TRE 14598, 2 (16 R 1112/26 September 1700); OA, MAD 9885, 177–178 (15 R 1111/10 October 1699); OA, C.BH 149/7106 (15 R 1111/10 October 1699).
75. OA, C.BH 149/7106 (16 R 1111/11 October 1699); OA, MAD 7915, 276 (15 R 1111/9 October 1699); OA, MAD 7915, 278 (1111/1699–1700); OA, MAD 7915, 369 (28 B 1111/19 January 1700); OA, MAD 7915, 375 (10 Ş 1111/31 January 1700); OA, MAD 7915, 348–349 (7 S 1143/21 August 1730).
76. Chester G. Starr, *The Roman Imperial Navy, 31 B.C.–A.D. 325*, 3rd ed. (Chicago: Ares Publishers, 1993), 55–61.
77. Husain, "The Tigris-Euphrates Basin," 108.
78. OA, MAD 7915, 278 (1111/1699-1700).
79. OA, MAD 7915, 369 (28 B 1111/19 January 1700). The absence of wage discrimination based on religion is noted in the Imperial Arsenal in Istanbul as well. See Çizakça, "The Ottoman Empire," 219–220.
80. Sophia Laiou, "The Levends of the Sea in the Second Half of the 16th Century: Some Considerations," *Archivum Ottomanicum* 23 (2005–2006): 233–247.
81. Husain, "The Tigris-Euphrates Basin," 109.
82. OA, AE.SAMD.III 91/9069 (Evail S 1121/12-21 April 1709).
83. OA, MAD 7915, 380 (1111/1699/1700); OA, MAD 7915, 400–401 (27 N 1113/24 February 1702); D.BŞM.TRE 3/8 (8 L 1111/29 March 1700); OA, MAD 9885, 295–296 (28 L 1111/18 April 1700); OA, MAD 3134, 95 (13 S 1112/30 July 1700). On the use of hardtack in Ottoman Egypt, see Alan Mikhail, *Nature and Empire in Ottoman Egypt: An Environmental History* (New York: Cambridge University Press, 2011), 101–103.
84. OA, MD 87/465 (1 B 1047/19 November 1637); OA, MD 87/479 (18 N 1047/18 February 1638); OA, MAD 7915, 273 (2 R 1111/26 September 1699); OA, MAD 7915, 398 (9 B 1113/9 December 1701); OA, MAD 9885, 306 (10 ZA 1111/30 April 1700); OA, C.BH 149/7106 (2 B 1111/24 December 1699); OA, MAD 7915, 350–351 (17 B 1143/26 January 1731).

85. OA, MD 5/825 (28 C 973/20 January 1566); OA, MD 5/826 (28 C 973/20 January 1566); OA, MD 22/636 (15 CA 981/12 September 1573); OA, MD 24/228 (14 Z 981/6 April 1574); OA, MD 27/805 (3 Z 983/4 March 1576); OA, MD 53/556 (993/1558).
86. Fariba Zarinebaf, *Crime and Punishment in Istanbul, 1700–1800* (Berkeley: University of California Press, 2011), 164–168; Bostan, *Osmanlı Bahriye Teşkilatı*, 213–220; Imber, "The Navy of Süleyman the Magnificent," 268.
87. OA, D.BŞM.TRE 14598, 7 (16 R 1112/26 September 1700). See also Öğüt, *Birecik*, 268; Cengiz Orhonlu and Turgut Işıksal, "Osmanlı Devrinde Nehir Nakliyatı Hakkında Araştırmalar: Dicle ve Fırat Nehirlerinde Nakliyat," *Tarih Dergisi* 13 (1962–1963): 80–83.
88. J. R. McNeill, "Woods and Warfare in World History," *Environmental History* 9, no. 3 (2004): 395–399.
89. Bostan, *Osmanlı Bahriye Teşkilatı*, 102–118.
90. OA, MD 12/413 (10 ZA 978/5 April 1571); OA, MD 12/431 (18 ZA 978/13 April 1571); OA, C.BH 104/5049 (25 S 1160/8 March 1747).
91. OA, C.BH 193/9048 (1110/1699-1700); OA, MAD 7915, 372 (21 B 1111/12 January 1700).
92. Laura Mason, *Pine* (London: Reaktion Books, 2013), 102, 111; Joachim Radkau, *Wood: A History*, trans. Patrick Camiller (Cambridge: Polity Press, 2012), 39. On the utility of pine in early modern Korea, see John S. Lee, "Postwar Pines: The Military and the Expansion of State Forests in Post-Imjin Korea, 1598–1684," *Journal of Asian Studies* 77, no. 2 (2018): 319–332.
93. OA, C.BH 149/7106 (2 R 1111/27 September 1699); OA, MAD 9885, 130 (4 R 1111/29 September 1699); OA, MAD 3134, 67 (12 M 1112/29 June 1700); OA, MAD 5433, 19 (10 N 1112/18 February 1701); OA, MAD 7915, 334 (2 Z 1120/12 February 1709); OA, C.BH 222/10313 (26 RA 1156/20 May 1743).
94. OA, MD 12/413 (10 ZA 978/5 April 1571); OA, MD 12/431 (18 ZA 978/13 April 1571).
95. OA, MAD 7915, 273 (15 R 1111/9 October 1699); OA, C.BH 193/9048 (1110/1699-1700); OA, MAD 3134, 67 (12 M 1112/29 June 1700); OA, MAD 7915, 334 (2 Z 1120/12 February 1709); OA, AE.SMHD.I 192/14977 (Evahir S 1160/4-13 March 1747); Salih Özbaran, "XVI. Yüzyılda Basra Körfezi Sahillerinde Osmanlılar Basra Beylerbeyliğinin Kuruluşu," *Tarih Dergisi* 25 (1971): 60.
96. OA, MAD 2775, 619 (14 CA 973/7 December 1565); OA, MAD 7915, 334 (2 Z 1120/12 February 1709).
97. OA, C.BH 193/9048 (1110/1699-1700); Öğüt, *Birecik*, 254–261.
98. Moorey, *Ancient Mesopotamian*, 352; Khidr al-Duri, "Society and Economy of Iraq under the Seljuqs (1055–1160 A.D.)" (PhD diss., University of Pennsylvania, 1970), 322; George Hourani, *Arab Seafaring in the Indian Ocean in Ancient and Early Medieval Times* (Princeton, NJ: Princeton University Press, 1995), 90–91. Other Indian Ocean ports imported teak from India as well. For the use of teak in the Suez shipyard, see Mikhail, *Nature and Empire*, 121–122.

99. Charles Henry Snow, *The Principal Species of Wood: Their Characteristic Properties*, 2nd ed. (New York: John Wiley, 1910), 17–18, 121–122; Andrew Murray, *Ship-Building in Iron and Wood* (Edinburgh: Adam and Charles Black, 1863), 72.
100. Manuel Godinho, *Relação do novo caminho que fez por terra e mar: vindo da India para Portugal, no anno de 1663* (Lisbon: Sociedade propagadora dos conhecimentos uteis, 1842), 117; John Carmichael, "A Journey from Aleppo to Basra in 1751," in *The Desert Route to India: Being the Journals of Four Travellers by the Great Desert Caravan Route between Aleppo and Basra, 1745–1751*, ed. Douglas Carruthers (London: Hakluyt Society, 1929), 178; Domenico Sestini, *Voyage de Constantinople à Bassora en 1781 par le Tigre et l'Euphrate, et retour à Contantinople en 1782, par le desert et Alexandrie* (Paris: Chez Dupuis, 1798), 197.
101. Dejanirah Potache, "The Commercial Relations between Basrah and Goa in the Sixteenth Century," *Studia* 48 (1989): 158; Floor, *The Persian Gulf*, 173–174.
102. See Table A.3 in the appendix; OA, MD 10/421 (28 B 979/16 December 1571); OA, MAD 7915, 350 (4 Ş 1143/11 February 1731).
103. OA, MAD 7915, 328 (4 L 1117/19 January 1706); McNeill, "Woods and Warfare," 389–390.
104. OA, MD 27/748 (6 Z 983/7 March 1576); OA, MD 27/752 (6 Z 983/7 March 1576); OA, MAD 9885, 131 (7 R 1111/1 October 1699); OA, MAD 9885, 204 (12 N 1111/3 March 1700); OA, MAD 9885, 240 (4 L 1111/25 March 1700); OA, MAD 9885, 262 (16 L 1111/6 April 1700); MAD 7915, 376 (9 B 1111/31 December 1699); MAD 7915, 392 (10 N 1112/18 February 1701); OA, İE.BH 10/914 (12 N 1111/3 March 1700).
105. OA, D.BŞM.TRE 14598, 2 (16 R 1112/26 September 1700).
106. OA, D.BŞM.TRE 14598, 1 (16 R 1112/26 September 1700); OA, MAD 5433, 4 (6 N 1112/14 February 1701); OA, MAD 5433, 18 (6 N 1112/14 February 1701); OA, C.AS 922/39875 (Evahir Z 1146/25 May-4 June 1734).
107. OA, D.BŞM.TRE 14598, 1 (16 R 1112/26 September 1700); OA, C.AS 922/39875 (Evahir Z 1146/25 May-4 June 1734).
108. OA, MD 3/751 (5 CA 967/2 February 1560).
109. J. Horton Ryley, *Ralph Fitch, England's Pioneer to India and Burma; His Companions and Contemporaries, with His Remarkable Narrative Told in His Own Words* (London: T. F. Unwin, 1899), 52–53.
110. Vincenzo Maria Murchio, *Il viaggio all' Indie Orientali* (Rome: Filippomaria Mancini, 1672), 77. I am grateful to Maria Sole Costanzo for the translation.
111. John Newberie, "Two Voyages of Master John Newberie, One into the Holy Land; The Other to Balsara, Ormus, Persia, and Backe Thorow Turkie," in *Hakluytus Posthumus; or, Purchas His Pilgrimes: Contayning a History of the World in Sea Voyages and Lande Travells by Englishmen and Others*, ed. Samuel Purchas (Glasgow: James MacLehose and Sons, 1905), 8:453.
112. "Hükümname Mecmuası," ff. 340v-341r (10 Ş 959/1 August 1552); OA, MD 27/436 (9 L 983/11 January 1576); OA, MD 27/465 (15 L 983/17 January 1576); OA,

MAD 7915, 275 (12 R 1111/6 October 1699); OA, MAD 7915, 344 (24 S 1123/ 13 April 1711); OA, C.AS 922/39875 (Evahir Z 1146/25 May-4 June 1734); OA, C.BH 66/3118 (20 CA 1160/30 May 1747); OA, C.BH 83/3978 (6 Ş 1156/25 September 1743).

113. Tahir Öğüt implies that the history of the Shatt Fleet reflects environmental change over time, claiming that deforestation in the mountains of Maraş contributed to a decline in Ottoman shipbuilding in Birecik during the nineteenth century. Öğüt, *Birecik*, 329–330. I personally have not found evidence substantiating this claim during my research.

114. On boats as the largest machines in pre-industrial times, see Keith Muckelroy, *Maritime Archaeology* (New York: Cambridge University Press, 1978), 3.

PART II

1. Aridity is a defining feature of the fluvial landscape of Iraq; it left an imprint on many aspects of human organization and land use and will therefore be a recurring theme throughout this book. My approach echoes environmental histories of the American West rooted in the region's arid setting. See, in particular, Donald Worster, *Rivers of Empire: Water, Aridity, and the Growth of the American West* (New York: Oxford University Press, 1992).

2. Dale Stahl, "The Two Rivers: Water, Development and Politics in the Tigris-Euphrates Basin, 1920–1975" (PhD diss., Columbia University, 2014); Peter Beaumont, "Restructuring of Water Usage in the Tigris-Euphrates Basin: The Impact of Modern Water Management Policies," in *Transformations of Middle Eastern Natural Environments: Legacies and Lessons*, ed. Jeff Albert, Magnus Bernhardsson, and Roger Kenna (New Haven, CT: Yale School of Forestry and Environmental Studies, 1998), 168–186.

3. By redrawing the boundaries of Ottoman Iraq along ecological rather than provincial lines, I draw inspiration from generations of environmental historians who have transcended the political divisions of space to anchor their stories to their environmental setting, thereby offering new questions and answers about the past. On this trend in the North American context, see Dan Flores, "Place: An Argument for Bioregional History," *Environmental History Review* 18, no. 4 (1994): 1–18.

4. John V. Murra, "'El Archipiélago Vertical' Revisited," in *Andean Ecology and Civilization: An Interdisciplinary Perspective on Andean Ecological Complementarity: Papers from Wenner-Gren Foundation for Anthropological Research Symposium*, ed. Shozo Masuda, Izumi Shimada, and Craig Morris (Tokyo: University of Tokyo Press, 1985), 3–13; Murra, "The Limits and Limitations of the 'Vertical Archipelago' in the Andes," in *Andean Ecology and Civilization*, 15–20.

CHAPTER 3

1. Arguably, only breathable air is more critical to human life than water.
2. Karen Barkey, *Empire of Difference: The Ottomans in Comparative Perspective* (New York: Cambridge University Press, 2008); Gábor Ágoston, "A Flexible Empire: Authority and Its Limits on the Ottoman Frontiers," *International Journal of Turkish Studies* 9, no. 1/2 (2003): 15–31; Nicholas Doumanis, "Durable Empire: State Virtuosity and Social Accommodation in the Ottoman Mediterranean," *Historical Journal* 49, no. 3 (2006): 953–966; Daniel Goffman and Christopher Stroop, "Empire as Composite: The Ottoman Polity and the Typology of Dominion," in *Imperialisms: Historical and Literary Investigations, 1500–1900*, ed. Balachandra Rajan and Elizabeth Sauer (New York: Palgrave Macmillan, 2004), 129–145. For a critique of the pragmatism paradigm in Ottoman historiography, see Murat Dağlı, "The Limits of Ottoman Pragmatism," *History and Theory* 52, no. 2 (2013): 194–213.
3. The questions echo the protracted controversy caused by the sinologist Karl Wittfogel after the publication of his book *Oriental Despotism: A Comparative Study of Total Power* (New Haven, CT: Yale University Press, 1957). Wittfogel's thesis of hydraulic societies is now widely discredited. For an influential critique, see Robert C. Hunt, "Size and the Structure of Authority in Canal Irrigation Systems," *Journal of Anthropological Research* 44, no. 4 (1988): 335–355. On the shortcomings of Wittfogel's theory in the context of the Tigris-Euphrates basin, see Stephanie Rost, "Water Management in Mesopotamia from the Sixth till the First Millennium B.C.," *Wiley Interdisciplinary Reviews: Water* 4, no. 5 (2017).
4. Alan Mikhail, *Nature and Empire in Ottoman Egypt: An Environmental History* (New York: Cambridge University Press, 2011), 39.
5. Mehmet Genç, "Osmanlı İktisadi Dünya Görüşünün İlkeleri," *Sosyoloji Dergisi* 3, no. 1 (1989): 175–185. Ottoman traditionalism is comparable to American conservatism in the colonial era. See John Demos, *Circles and Lines: The Shape of Life in Early America* (Cambridge, MA: Harvard University Press, 2004), 22–23.
6. Gülru Necipoğlu, *The Age of Sinan: Architectural Culture in the Ottoman Empire* (Princeton, NJ: Princeton University Press, 2005), 190.
7. TKG.KK, TT 29, ff. 6r–v; Ahmed Akgündüz, *Osmanlı Kanunnameleri ve Hukuki Tahlilleri*, bk. 5, *Kanuni Devri Kanunnameleri: Eyalet Kanunnameleri* (Istanbul: FEY Vakfı, 1992), 159. On precedent as a guiding principle in Ottoman irrigation policy in Egypt, see Mikhail, *Nature and Empire in Ottoman Egypt*, 52–58.
8. Evliya Çelebi, *Evliya Çelebi Seyahatnamesi: Topkapı Sarayı Bağdat 304 Yazmasının Transkripsiyonu-Dizini*, ed. Yücel Dağlı and Seyit Ali Kahraman (Istanbul: Yapı Kredi Yayınları, 2001), 4:258.
9. Likewise, officials and supporters of the Rhône River Authority (CNR) in modern France frequently invoked historical precedent to justify their projects. See Sara B. Pritchard, *Confluence: The Nature of Technology and the Remaking of the Rhône* (Cambridge, MA: Harvard University Press, 2011), 63–66.

10. Erdinç Gülcü, "Osmanlı İdaresinde Bağdat (1534–1623)" (PhD diss., Fırat Üniversitesi, 1999), 138–141; OA, TT 1028, 5; TKG.KK, TT 29, ff. 6r–v, 51r, 78r, 87r, 121r; Akgündüz, *Kanuni Devri Kanunnameleri*, 159.
11. Akgündüz, *Kanuni Devri Kanunnameleri*, 158.
12. Linda T. Darling, *A History of Social Justice and Political Power in the Middle East: The Circle of Justice from Mesopotamia to Globalization* (New York: Routledge, 2012), 2.
13. Darling, *A History of Social Justice*, 127–154.
14. Akgündüz, *Kanuni Devri Kanunnameleri*, 158.
15. OA, TT 1028, 11; Akgündüz, *Kanuni Devri Kanunnameleri*, 160.
16. OA, MD 22/9 (17 M 981/19 May 1573); OA, MD 22/278 (10 S 981/10 June 1573).
17. OA, MD 7/375 (10 CA 975/11 November 1567).
18. OA, TT 282, 79–80, 152, 238, 278; "Hükümname Mecmuası," Topkapı Sarayı Müzesi Kütüphanesi, Koğuşlar 888, fol. 56v (11 S 959/6 February 1552); "Hükümname Mecmuası," fol. 346v (15 Ş 959/6 August 1552); OA, MD 6/276 (20 RA 972/26 October 1564); OA, MAD 2775, 1272 (12 N 973/3 April 1566); OA, MD 22/9 (17 M 981/19 May 1573); OA, MD 22/253 (4 RA 981/4 July 1573); OA, MD 22/278 (10 S 981/10 June 1573); OA, MD 30/628 (20 S 985/9 May 1577); TKG.KK, TT 29, ff. 9r, 74v–75r, 96v, 98v–99r, 109r, 233v, 236v–237r, 264v, 288r–v, 301v, 302v, 331v, 350r, 435v, 440r; TKG.KK, TT 30, fol. 134v.
19. Michael Cook, "The Long-Term Geopolitics of the Pre-Modern Middle East," *Journal of the Royal Asiatic Society* 26, no. 1–2 (2016): 33–41; Robert McC. Adams, "Intensified Large-Scale Irrigation as an Aspect of Imperial Policy: Strategies of Statecraft on the Late Sasanian Mesopotamian Plain," in *Agricultural Strategies*, ed. Joyce Marcus and Charles Stanish (Los Angeles: Cotsen Institute of Archaeology, 2006), 17–37; James Howard-Johnston, "The Two Great Powers in Late Antiquity: A Comparison," in *The Byzantine and Early Islamic Near East*, vol. 3, *States, Resources, and Armies*, ed. Averil Cameron (Princeton, NJ: Darwin Press, 1992), 157–226.
20. M. G. Ionides, *The Régime of the Rivers, Euphrates and Tigris* (London: E. & F. N. Spon, 1937), 1–5.
21. David A. Lytle and N. LeRoy Poff, "Adaptation to Natural Flow Regimes," *Trends in Ecology and Evolution* 19, no. 2 (2004): 94–100.
22. Raoul C. Mitchell, "Instability of the Mesopotamian Plains," *Bulletin de la Société de Géographie d'Égypte* 31 (1958): 129.
23. TKG.KK, TT 29, fol. 314v.
24. George Manwaring, *The Three Brothers; or, The Travels and Adventures of Sir Anthony, Sir Robert, and Sir Thomas Sherley, in Persia, Russia, Turkey, Spain, etc.* (London: Printed for Hurst, Robinson, & Co., 1825), 46–47. See also Pedro Teixeira, *The Travels of Pedro Teixeira*, trans. William F. Sinclair (London: Hakluyt Society, 1902), 56.

25. Margareta Tengberg, "Fruit-Growing," in *A Companion to the Archaeology of the Ancient Near East*, ed. D. T. Potts (Malden, MA: Wiley-Blackwell, 2012), 1:194; Hilda Simon, *The Date Palm: Bread of the Desert* (New York: Dodd, Mead, 1978), 76.
26. Simon, *The Date Palm*, 73; Paul B. Popenoe, *Date Growing in the Old World and the New* (Altadena, CA: West India Gardens, 1913), 36.
27. Evliya Çelebi, *Evliya Çelebi Seyahatnamesi*, 4:260. By the early twentieth century, Iraq had 132 varieties of date palms. See V. H. W. Dowson, *Dates and Date Cultivation of the Iraq*, pt. III, *The Varieties of Date Palms of the Shatt al 'Arab* (Cambridge: W. Heffer and Sons, 1923), 17.
28. Tengberg, "Fruit-Growing," 193; Simon, *The Date Palm*, 45; Domenico Sestini, *Voyage de Constantinople à Bassora, en 1781, par le Tigre et l'Euphrate: Et Retour a Constantinople, en 1782, par le Désert et Alexandrie* (Paris: Chez Dupuis, 1798), 204–205.
29. Evliya Çelebi, *Evliya Çelebi Seyahatnamesi*, 4:260.
30. M. P. Charles, "Onions, Cucumbers and the Date Palm: An Introduction to the Cultivation of Alliaceae, Cucurbitaceae and Fruit Trees in Modern Iraq," *Bulletin on Sumerian Agriculture* 3 (1987): 2; Simon, *The Date Palm*, 53.
31. In a similar vein, Ellen Stroud highlights the necessity of viewing trees and forests in the northeastern United States not simply as natural landscapes but also as artifacts of the city. See Stroud, *Nature Next Door: Cities and Trees in the American Northeast* (Seattle: University of Washington Press, 2012).
32. J. N. Postgate, "Notes on Fruit in the Cuneiform Sources," *Bulletin on Sumerian Agriculture* 3 (1987): 115–144; D. T. Potts, *Mesopotamian Civilization: The Material Foundations* (London: Athlone Press, 1997), 69–70; V. H. W. Dowson, *Dates and Date Cultivation of the Iraq*, pt. 1, *The Cultivation of the Date Palm on the Shat Al 'Arab* (Cambridge: W. Heffer and Sons, 1921), 58.
33. John Jackson, *Journey from India, Towards England, in the Year 1797* (London: Printed for T. Cadell, Jun. and W. Davies, Strand, 1799), 22–23.
34. J. N. Postgate, *Early Mesopotamia: Society and Economy at the Dawn of History* (New York: Routledge, 1994), 318n282.
35. TKG.KK, TT 29, fol. 29v; OA, TT 534, 10.
36. TKG.KK, TT 29, ff. 29v, 450v-455r; OA, TT 386, 40, 47, 57–59, 61–62, 65–66, 83–85, 92, 95–102, 157, 167–168, 177–180, 239; OA, TT 582, ff. 27r, 29r, 33r-40v, 97v-98r, 126v.
37. OA, TT 1028, 11.
38. Katib Çelebi, *Cihannüma* (Istanbul: Darü't-Tıbaati'l-Amire, 1145/1732), 452; Dowson, *The Cultivation of the Date Palm*, 20; Nawal Nasrallah, *Dates: A Global History* (London: Reaktion Books, 2011), 95.
39. OA, TT 282; TKG.KK, TT 29.
40. Stephanie Rost and Abdulamir Hamdani, "Traditional Dam Construction in Modern Iraq: A Possible Analogy for Ancient Mesopotamian Irrigation Practices,"

Iraq 73 (2011): 201–220; Peter Christensen, *The Decline of Iranshahr: Irrigation and Environments in the History of the Middle East, 500 B.C. to A.D. 1500* (Copenhagen: Museum Tusculanum Press, 1993), 56; T. J. Wilkinson, *Archaeological Landscapes of the Near East* (Tucson: University of Arizona Press, 2003), 47–51; J. Wilkinson and Carrie Hritz, "Physical Geography, Environmental Change and the Role of Water," in *Models of Mesopotamian Landscapes: How Small-Scale Processes Contributed to the Growth of Early Civilizations,* ed. T. J. Wilkinson, McGuire Gibson, and Magnus Widell (Oxford: Archaeopress, 2013), 23.

41. T. J. Wilkinson, Louise Rayne, and Jaafar Jotheri, "Hydraulic Landscapes in Mesopotamia: The Role of Human Niche Construction," *Water History* 7, no. 4 (2015): 400; Wilkinson, "Hydraulic Landscapes and Irrigation Systems of Sumer," in *The Sumerian World,* ed. Harriet E. W. Crawford (New York: Routledge, 2013), 42–45.

42. OA, TT 282, 79–80, 152, 238, 278; "Hükümname Mecmuası," fol. 56v (11 S 959/6 February 1552); "Hükümname Mecmuası," fol. 346v (15 Ş 959/6 August 1552); OA, MD 6/276 (20 RA 972/26 October 1564); OA, MAD 2775, 1272 (12 N 973/3 April 1566); OA, MD 22/9 (17 M 981/19 May 1573); OA, MD 22/253 (4 RA 981/4 July 1573); OA, MD 22/278 (10 S 981/10 June 1573); OA, MD 30/628 (20 S 985/9 May 1577); TKG.KK, TT 29, ff. 9r, 74v-75v, 96v, 98v-99r, 109r, 233v, 236v-237r, 264v, 288r-v, 301v, 302v, 331v, 350r, 435v, 440r; TKG.KK, TT 30, fol. 134v; OA, MAD 2926, 180 (15 Z 1093/15 December 1682).

43. OA, MD 33/364 (9 ZA 985/18 January 1578); OA, MD 40/436 (17 Ş 987/8 October 1579); OA, AE.SMMD.IV 9/900 (24 R 1075/13 November 1664).

44. The Ottoman administration initially adopted a fourth large canal, the Isa River, built on a horizontal topographical decline from the Euphrates north of Falluja to the Tigris near Baghdad. After several rehabilitation efforts, however, the administration realized that dealing with the chronic instabilities of this large canal was overly cumbersome, auctioning it off to the highest bidder in the 1570s. See OA, MD 3/1355 (10 ZA 967/2 August 1560); TSMA 813/44 (17 L 967/11 July 1560); "Hükümname Mecmuası," fol. 346v (15 Ş 959/6 August 1552); OA, MD 12/588 (17 ZA 978/12 April 1571); OA, MD 22/9 (17 M 981/19 May 1573); TKG.KK, TT 29, fol. 48v.

45. IOR/L/P&S/9/98, James B. Fraser, "On the Present Condition of the Pachalik of Baghdad," (12 November 1834): 108; William Willcocks, *The Restoration of the Ancient Irrigation Works on the Tigris, or, The Re-Creation of Chaldea* (Cairo: National Printing Department, 1903), 13. For partial archaeological reconstructions of the Dujayl Canal, see Alastair Northedge, T. J. Wilkinson, and Robin Falkner, "Survey and Excavations at Samarra 1989," *Iraq* 50 (1990): 121–147; Tony James Wilkinson and Louise Rayne, "Hydraulic Landscapes and Imperial Power in the Near East," *Water History* 2, no. 2 (2010): 115–144. For a description of the Dujayl based on textual sources, see Guy Le Strange, *The Lands of the Eastern Caliphate: Mesopotamia, Persia, and Central Asia from the Moslem Conquest to the*

Time of Timur (Cambridge: Cambridge University Press, 1905), 51, 65; Christensen, *The Decline of Iranshahr,* 55.

46. OA, TT 1028, 54; TKG.KK, TT 228, ff. 3r-4v; TKG.KK, TT 29, ff. 119v-140v.
47. Muhammad Husayn al-Uqayli, *Tarikh an-Najaf al-Ashraf,* ed. Abdulrazzaq Hirzuldin (Qum: Dalil-i Ma, 1427/2006-2007), 1:275–314; Ja'far al-Mahbuba, *Madhi an-Najaf wa Hadhiruha,* 2nd ed. (Beirut: Dar al-Adwa, 1986), 1:183–208.
48. Abdulhusayn Al Tu'ma, *Bughyat an-Nubala fi Tarikh Karbala* (Baghdad: Matba'at al-Irshad, 1966), 97–100.
49. OA, İE.EV 1/29 (15 Rebiüahir 943/1 October 1536); OA, MD 7/2316 (1 CA 976/22 October 1568); OA, MD 7/2331 (1 CA 976/22 October 1568); OA, MD 33/364 (9 ZA 985/18 January 1578); OA, AE.SMMD.IV 9/900 (24 R 1075/13 November 1664); OA, İE.ML 5/353 (Evasıt C 1075/29 December 1664–7 January 1665); OA, İE.ML 56/5309 (14 L 1082/13 February 1672); OA, MAD 2933, 13–14 (8 M 1097/5 December 1685); OA, AE.SAMD.III 4/277 (21 CA 1118/30 August 1706); OA, MAD 10151, 162 (25 C 1118/3 October 1706); OA, AE.SAMD.III 3/221 (12 Ş 1143/19 February 1731).
50. OA, MAD 4117, 57–58 (6 M 1071/11 September 1660). For a secondary treatment of Ottoman sediment clearance on the Dujayl, see Rhoads Murphey, "The Ottoman Centuries in Iraq: Legacy or Aftermath? A Survey Study of Mesopotamian Hydrology and Ottoman Irrigation Projects," *Journal of Turkish Studies* 11 (1987): 23.
51. OA, MD 26/434 (9 R 982/29 July 1574); OA, MD 40/665 (15 Ş 987/7 October 1579).
52. OA, MD 7/2316 (1 CA 976/22 October 1568); OA, MD 7/2331 (1 CA 976/22 October 1568); OA, MD 33/434 (12 ZA 986/10 January 1579); OA, MD 26/389 (24 R 982/13 August 1574); OA, MD 27/[order number illegible] (28 ZA 983/28 February 1576).
53. OA, TT 282; TKG.KK, TT 29.
54. Ágoston, "A Flexible Empire"; Barkey, *Empire of Difference,* 83–93.
55. On the role of salinization in the Tigris-Euphrates alluvium past and present, Thorkild Jacobsen and Robert M. Adams, "Salt and Silt in Ancient Mesopotamian Agriculture," *Science* 128, no. 3334 (1958): 1251–1258; Mark Altaweel, "Simulating the Effects of Salinization on Irrigation Agriculture in Southern Mesopotamia," in *Models of Mesopotamian Landscapes,* 219–238; M. P. Charles, "Irrigation in Lowland Mesopotamia," *Bulletin on Sumerian Agriculture* 4 (1988): 1–39; Piet Buringh, *Soils and Soil Conditions in Iraq* (Baghdad: Ministry of Agriculture, 1960), 83–114. For a useful global review of salinization, see William D. Williams, "Salinization of Rivers and Streams: An Important Environmental Hazard," *Ambio* 16, no. 4 (1987): 180–185.
56. Robert McC. Adams, "Historic Patterns of Mesopotamian Irrigation Agriculture," in *Irrigation's Impact on Society,* ed. Adams, Theodore E. Downing, and McGuire Gibson (Tucson: University of Arizona Press, 1974), 2.

CHAPTER 4

1. Tina Hesman Saey, "The Road to Tameness," *Science News* 191, no. 13 (2017): 20–27; Emily Louise Hammer and Benjamin S. Arbuckle, "10,000 Years of Pastoralism in Anatolia: A Review of Evidence for Variability in Pastoral Lifeways," *Nomadic Peoples* 21, no. 2 (2017): 214–267; Susan H. Lees and Daniel G. Bates, "The Origins of Specialized Nomadic Pastoralism: A Systemic Model," *American Antiquity* 39, no. 2 (1974): 187–193.
2. Philip Carl Salzman, "Pastoral Nomads: Some General Observations Based on Research in Iran," *Journal of Anthropological Research* 58, no. 2 (2002): 249.
3. On the challenges that tribalism, mobility, and location pose to state authority, see James C. Scott, *The Art of Not Being Governed: An Anarchist History of Upland Southeast Asia* (New Haven, CT: Yale University Press, 2009), 178–219.
4. See, for example, Huri İslamoğlu-İnan, *State and Peasant in the Ottoman Empire: Agrarian Power Relations and Regional Economic Development in Ottoman Anatolia during the Sixteenth Century* (Leiden: Brill, 1994); Bruce McGowan, *Economic Life in Ottoman Europe: Taxation, Trade, and the Struggle for Land, 1600–1800* (New York: Cambridge University Press, 1981); Amy Singer, *Palestinian Peasants and Ottoman Officials: Rural Administration around Sixteenth-Century Jerusalem* (New York: Cambridge University Press, 1994); Suraiya Faroqhi, *Towns and Townsmen of Ottoman Anatolia: Trade, Crafts, and Food Production in an Urban Setting, 1520–1650* (New York: Cambridge University Press, 1984); Beshara Doumani, *Rediscovering Palestine: Merchants and Peasants in Jabal Nablus, 1700–1900* (Berkeley: University of California Press, 1995); Nenad Moačanin, *Town and Country on the Middle Danube, 1526–1690* (Leiden: Brill, 2006); Kenneth M. Cuno, *The Pasha's Peasants: Land, Society, and Economy in Lower Egypt, 1740–1858* (New York: Cambridge University Press, 1992).
5. See, for example, Reşat Kasaba, *A Moveable Empire: Ottoman Nomads, Migrants, and Refugees* (Seattle: University of Washington Press, 2009), 13–52; Halil İnalcık, "'Arab' Camel Drivers in Western Anatolia in the Fifteenth Century," *Revue d'Histoire Maghrebine* 10, no. 31–32 (1983): 247–270; İnalcık, "The Yörüks: Their Origins, Expansion and Economic Role," in *Carpets of the Mediterranean Countries, 1400–1600*, ed. Robert Pinner and Walter B. Denny (London: Hali Magazine, 1986), 39–65; Suraiya Faroqhi, "Camels, Wagons, and the Ottoman State in the Sixteenth and Seventeenth Centuries," *International Journal of Middle East Studies* 14, no. 4 (1982): 523–539; Suavi Aydın and Oktay Özel, "Power Relations between State and Tribe in Ottoman Eastern Anatolia," *Bulgarian Historical Review* 34, no. 3–4 (2006): 51–67; Tom Sinclair, "The Ottoman Arrangements for the Tribal Principalities of the Lake Van Region of the Sixteenth Century," *International Journal of Turkish Studies* 9, no. 1–2 (2003): 119–143; Hakan Özoğlu, "State-Tribe Relations: Kurdish Tribalism in the 16th- and 17th-Century Ottoman Empire," *British Journal of Middle Eastern Studies* 23, no. 1 (1996): 5–27; Rhoads Murphey,

"The Resumption of Ottoman-Safavid Border Conflict, 1603–1638: Effects of Border Destabilization on the Evolution of State-Tribe Relations," in *Shifts and Drifts in Nomad-Sedentary Relations,* ed. Stefan Leder and Bernhard Streck (Wiesbaden: L. Reichert, 2005), 308–323.

6. Rudi Paul Lindner, *Nomads and Ottomans in Medieval Anatolia* (Bloomington: Research Institute for Inner Asian Studies, Indiana University, 1983), 65.
7. Roger Cribb, *Nomads in Archaeology* (New York: Cambridge University Press, 1991), 54–55; J. R. McNeill, *The Mountains of the Mediterranean World: An Environmental History* (New York: Cambridge University Press, 1992), 270, 279; Gábor Ágoston, "Where Environmental and Frontier Studies Meet: Rivers, Forests, Marshes and Forts along the Ottoman-Hapsburg Frontier in Hungary," in *The Frontiers of the Ottoman World*, ed. A. C. S. Peacock (Oxford: Oxford University Press, 2009), 73; Scott, *The Art of Not Being Governed*, 182–187.
8. Charles Julian Bishko, "The Castilian as Plainsman: The Medieval Ranching Frontier in La Mancha and Extremadura," in *The New World Looks at Its History*, ed. A. R. Lewis and T. F. McGann (Austin: University of Texas Press, 1963), 47–69; Carla Rahn Phillips and William D. Phillips Jr., *Spain's Golden Fleece: Wool Production and the Wool Trade from the Middle Ages to the Nineteenth Century* (Baltimore: Johns Hopkins University Press, 1997), 24–42.
9. Virginia DeJohn Anderson, *Creatures of Empire: How Domestic Animals Transformed Early America* (New York: Oxford University Press, 2004).
10. John F. Richards, *The Unending Frontier: An Environmental History of the Early Modern World* (Berkeley: University of California Press, 2003), 274–306.
11. Henry David Thoreau, *The Portable Thoreau*, ed. Jeffrey S. Cramer (New York: Penguin, 2012), 242.
12. Thoreau, *The Portable Thoreau*, 242.
13. Clive J. C. Phillips, *Principles of Cattle Production*, 3rd ed. (Boston: CABI, 2018); Thomas G. Field, *Scientific Farm Animal Production: An Introduction to Animal Science*, 10th ed. (Upper Saddle River, NJ: Prentice Hall, 2012).
14. Rada Dyson-Hudson and Neville Dyson-Hudson, "Nomadic Pastoralism," *Annual Review of Anthropology* 9 (1980): 17.
15. Phillips, *Principles of Cattle Production*, 101.
16. Peter Beaumont, Gerald H. Blake, and J. Malcolm Wagstaff, *The Middle East: A Geographical Study*, 2nd ed. (New York: Halsted Press, 1988), 50–112; Ewan W. Anderson, *The Middle East: Geography and Geopolitics* (London: Routledge, 2000), 30–54.
17. Dr. Leonhart Rauwolff, "Travels into the Eastern Countries," in *A Collection of Curious Travels and Voyages,* ed. John Ray (London: Royal Society, 1693), 1:144.
18. D. C. P. Thalen, *Ecology and Utilization of Desert Shrub Rangelands in Iraq* (The Hague: Dr. W. Junk, 1979), 24–25; Harry Wayne Springfield, *Forage Problems and Resources of Iraq* (Washington, DC: International Cooperation Administration,

1957), 4–5; E. R. Guest, "The Rustam Herbarium, Iraq: Part VI. General and Ecological Account," *Kew Bulletin* 8, no. 3 (1953): 396–401; Grahame Williamson, "Iraqi Livestock," *Empire Journal of Experimental Agriculture* 17 (1949): 50. The botanical terms for the plants mentioned are *Poa bulbosa* (bulbous bluegrass), *Artemisia herba-alba* (white wormwood), *Artemisia scoparia* (redstem wormwood), *Achillea* species (yarrow), *Hyparrhenia hirta* (thatching grass), *Medicago laciniate* (cutleaf medick), *Citrullus colocynthis* (colocynth), *Koeleria phleoides* (Mediterranean hairgrass), *Carex stenophylla* (needleleaf sedge), *Ranunculus asiaticus* (Asian buttercup), *Alhagi maurorum* (camelthorn), and *Astragalus* species (milkvetch).

19. Thalen, *Ecology and Utilization*, 22–23; Springfield, *Forage Problems*, 7–9; Guest, "The Rustam Herbarium," 398–399. The botanical names for the plants mentioned are *Salix acmophylla* (a willow species), *Populus euphratica* (Euphrates poplar), *Tamarix pentandra* (five-stamen tamarisk), *Glinus lotoides* (lotus sweetjuice), *Crypsis alopecuroides* (foxtail pricklegrass), *Polygonum salicifolium* (knotweed), *Salvinia natans* (floating watermoss), *Nymphoides indicum* (water snowflakes), *Phragmites communis* (common reed), *Typha angustata* (southern cattail).

20. Eyles Irwin, *A Series of Adventures in the Course of a Voyage Up the Red-Sea, on the Coasts of Arabia and Egypt; and of a Route through the Desarts of Thebais, in the Year 1777*, 3rd ed. (London: J. Dodsley, 1787), 2:312. For other early modern descriptions of the annual pastoral cycle in Iraq, see John Jackson, *Journey from India towards England in the Year 1797* (London: Printed for T. Cadell, Jun. and W. Davies, Strand, 1799), 17; Guillaume Antoine Olivier, *Voyage dans l'Empire Othoman, l'Égypte et la Perse: Fait par ordre du Gouvernement, pendant les six premières années de la République* (Paris: Chez H. Agasse, 1804), 4:383. For modern accounts, see D. Hywel Davies, "Observations on Land Use in Iraq," *Economic Geography* 33, no. 2 (1957): 125; Williamson, "Iraqi Livestock," 50–52; TNA, FO 922/74, I. Gillespie, "Survey of the Livestock Industry in Iraq" (May 1943): 10.

21. "Los valles en el invierno/las cumbres en el verano / como si fueran de nieve / blanquean con tus rebaños." Quoted in Phillips and Phillips Jr., *Spain's Golden Fleece*, 100. On classical transhumance in the Mediterranean basin, see Fernand Braudel, *The Mediterranean and the Mediterranean World in the Age of Phillip II*, trans. Siân Reynolds (New York: Harper & Raw, 1972), 1:85–102; McNeill, *The Mountains of the Mediterranean World*, 112–114.

22. George B. Cressey, *Crossroads: Land and Life in Southwest Asia* (Chicago: J. B. Lippincott, 1960), 16–17, 38.

23. David Gilmartin, *Blood and Water: The Indus River Basin in Modern History* (Oakland: University of California Press, 2015), 16.

24. Lajos Rácz, *The Steppe to Europe: An Environmental History of Hungary in the Traditional Age*, trans. Alan Campbell (Knapwell, UK: White Horse Press, 2013), 47–49.

25. William C. Brice, *South-West Asia* (London: University of London Press, 1966), 245; Hans J. Nissen and Peter Heine, *From Mesopotamia to Iraq: A Concise History,* trans. Hans J. Nissen (Chicago: University of Chicago Press, 2009), 4; J. N. Postgate, *Early Mesopotamia: Society and Economy at the Dawn of History* (New York: Routledge, 1994), xxi.
26. Rifaat Ali Abou-el-Haj, "The Social Uses of the Past: Recent Arab Historiography of Ottoman Rule," *International Journal of Middle East Studies* 14, no. 2 (1982): 195. On the high degree of interaction between nomads and settlers in West Asia in general, see J. R. McNeill, "The Eccentricity of the Middle East and North Africa's Environmental History," in *Water on Sand: Environmental Histories of the Middle East and North Africa,* ed. Alan Mikhail (New York: Oxford University Press, 2013), 33–41; M. B. Rowton, "Autonomy and Nomadism in Western Asia," *Orientalia* 42 (1973): 247–258; Rowton, "Urban Autonomy in a Nomadic Environment," *Journal of Near Eastern Studies* 32, no. 1–2 (1973): 201–215; Rowton, "Enclosed Nomadism," *Journal of the Economic and Social History of the Orient* 17, no. 1 (1974): 1–30; Thomas J. Barfield, *The Nomadic Alternative* (Englewood Cliffs, NJ: Prentice Hall, 1993), 93–94; Anatoly M. Khazanov, *Nomads and the Outside World,* trans. Julia Crookenden, 2nd ed. (Madison: University of Wisconsin Press, 1994), 62.
27. OA, MD 25/347 (18 L 981/10 February 1574).
28. Guest, "The Rustam Herbarium," 398–399. The botanical names for the plants mentioned are *Ranunculus* (buttercup), *Cyperus rotundus* (nutgrass), and *Lippia nodiflora* (turkey tangle fogfruit).
29. Edward Ives, *A Voyage from England to India, in the Year 1754, and an Historical Narrative of the Operations of the Squadron and Army in India, under the Command of Vice-Admiral Watson and Colonel Clive, in the Years 1755, 1756, 1757* (London: Printed for Edward and Charles Dilly, 1773), 259. Mobile pastoralists themselves managed water resources to enhance the production of pasture. See Emily Hammer, "Water Management by Mobile Pastoralists in the Middle East," in *Water and Power in Past Societies,* ed. Emily Holt (Albany: State University of New York Press, 2018), 63–88.
30. Thalen, *Ecology and Utilization,* 23; Springfield, *Forage Problems,* 8; Guest, "The Rustam Herbarium," 403; Gillespie, "Survey of the Livestock Industry," 3; Williamson, "Iraqi Livestock," 50; Robert McC. Adams, *Heartland of Cities: Surveys of Ancient Settlement and Land Use on the Central Floodplain of the Euphrates* (Chicago: University of Chicago Press, 1981), 142, 148. The botanical terms for the plants mentioned are *Melilotus indica* (annual yellow sweet clover), *Anagallis arvensis* (scarlet pimpernel), *Eragrostis cilianensis* (stinkgrass), and *Convolvulus* species (bindweed).
31. The longevity of these arrangements, regulated by the laws of Hammurabi (d. 1750 BC) and witnessed by twentieth-century ethnographers, is a testament to the convenient rewards they yielded to farmer and herder. See Postgate, *Early*

Mesopotamia, 158–159; Susan Pollock, *Ancient Mesopotamia: The Eden That Never Was* (Cambridge: Cambridge University Press, 1999), 70.

32. Edward Ochsenschlager, *Iraq's Marsh Arabs in the Garden of Eden* (Philadelphia: University of Pennsylvania Museum of Archaeology and Anthropology, 2004), 197; Williamson, "Iraqi Livestock," 50; Tate Paulette, "Pastoral Systems and Economies of Mobility," in *Models of Mesopotamian Landscapes: How Small-Scale Processes Contributed to the Growth of Early Civilizations*, ed. T. J. Wilkinson, McGuire Gibson, and Magnus Widell (Oxford: Archaeopress, 2013), 130–139.

33. Robert McC. Adams and Hans J. Nissen, *The Uruk Countryside: The Natural Setting of Urban Societies* (Chicago: University of Chicago Press, 1972), 73. For a similar observation, see Max von Oppenheim, *Die Beduinen*, vol. 1, *Die Beduinenstämme in Mesopotamien und Syrien* (Leipzig: Otto Harrassowitz, 1939), 22–23; Gillespie, "Survey of the Livestock Industry," 2.

34. Olivier, *Voyage dans l'Empire Othoman*, 4:391–392; J. Baillie Fraser, *Mesopotamia and Assyria from the Earliest Ages to the Present Time, with Illustrations of Their Natural History* (New York: Harper and Brothers, 1845), 320; Mahmud Shukri al-Alusi, *Akhbar Baghdad wa ma Jawaraha min al-Bilad*, ed. Imad Abdussalam Ra'uf (Beirut: al-Dar al-Arabiyya li'l-Mawsu'at, 2008), 350–351.

35. M. L. Ryder, *Sheep and Man* (London: Duckworth, 1983); Philip Armstrong, *Sheep* (London: Reaktion Books, 2016).

36. On the role of sheep in the rise of Sumerian civilization in the middle of the fourth millennium BC, see Guillermo Algaze, *Ancient Mesopotamia at the Down of Civilization: The Evolution of an Urban Landscape* (Chicago: University of Chicago Press, 2008), 77–92.

37. J. J. Finkelstein, "An Old Babylonian Herding Contract and Genesis 31:38," *Journal of the American Oriental Society* 88, no. 1 (1968), 30–36; Postgate, *Early Mesopotamia*, 159. Likewise, twentieth-century ethnographers mention that many villagers owned water buffalo but kept them in the care of specialized breeders to avoid the indignity of raising the animal and to protect their houses and date palms from the damage that it could cause. Breeders received the buffalo's dairy produce and half of the offspring in return for their service. S. M. Salim, *Marsh Dwellers of the Euphrates Delta* (London: Athlone Press, 1962), 91.

38. TKG.KK, TT 29, ff. 9v-28r, 101v-104r, 198v. For a comparable case of animal ownership as a capital investment, see Alan Mikhail, "Animals as Property in Early Modern Ottoman Egypt," *Journal of the Economic and Social History of the Orient* 53, no. 4 (2010): 621–652.

39. Phillips and Phillips Jr., *Spain's Golden Fleece*.

40. John A. Marino, *Pastoral Economics in the Kingdom of Naples* (Baltimore: Johns Hopkins University Press, 1988).

41. Erdinç Gülcü, "Osmanlı İdaresinde Bağdat (1534–1623)" (PhD diss., Fırat Üniversitesi, 1999), 310–311.

42. Katib Çelebi, *Cihannüma* (Istanbul: Darü't-Tıbaati'l-Amire, 1145/1732), 469.
43. Gülcü, "Osmanlı İdaresinde Bağdat," 341–343; Halil İnalcık, "The Ottoman State: Economy and Society, 1300–1600," in *An Economic and Social History of the Ottoman Empire, 1300–1914*, vol. 1, *1300–1600*, ed. İnalcık and Donald Quataert (New York: Cambridge University Press, 1994), 35; Kasaba, *A Moveable Empire*, 21.
44. Gülcü, "Osmanlı İdaresinde Bağdat," 302; Abbas al-Azzawi, *Mawsuʻat Ashaʾir al-Iraq* (Beirut: al-Dar al-Arabiyya li'l-Mawsuʻat, 2005), 2:151–154; Thomas T. Allsen, *The Royal Hunt in Eurasian History* (Philadelphia: University of Pennsylvania Press, 2006), 83. Other spellings for the Qara'ul could be Qaravul or Qaragöl.
45. Murtaza Nazmizade, *Gülşen-i Hulefa: Bağdat Tarihi, 762–1717*, ed. Mehmet Karataş (Ankara: Türk Tarih Kurumu, 2014), 302.
46. We know about the Qara Ulus judge through his correspondence with the Imperial Council in Istanbul, conveying the grievances of his constituents against corrupt Ottoman agents. For an early example, see OA, MAD 2775, 745 (16 C 973/8 January 1566). On Ottoman judges appointed to accompany other tribal groups, see Kasaba, *A Moveable Empire*, 24. On the Ottoman "sea judge," see Joshua M. White, *Piracy and Law in the Ottoman Mediterranean* (Stanford, CA: Stanford University Press, 2018), 91.
47. My understanding of aggregation as a state simplification strategy is informed by James C. Scott, *Seeing Like a State: How Certain Schemes to Improve the Human Condition Have Failed* (New Haven, CT: Yale University Press, 1998), 183–306.
48. TKG.KK, TT 29, ff. 36v-40v, 386v.
49. See, for example, OA, MD 24/894 (n.d.); TKG.KK, TT 29, ff. 196r–196v.
50. Ryder, *Sheep and Man*, 5.
51. Varda Kagan-Zur et al., eds., *Desert Truffles: Phylogeny, Physiology, Distribution and Domestication* (Heidelberg: Springer, 2014); Zachary Nowak, *Truffle: A Global History* (London: Reaktion Books, 2015); Neil MacFarquhar, "Beneath Desert Sands, an Eden of Truffles," *New York Times*, April 14, 2004, F1.
52. Ives, *A Voyage from England to India*, 233.
53. Abraham Parsons, *Travels in Asia and Africa* (London: Longman, Hurst, Rees, and Orme, 1808), 88.
54. OA, TT 534, 8.
55. Halil İnalcık, *The Ottoman Empire: The Classical Age, 1300–1600*, trans. Norman Itzkowitz and Colin Imber (New York: Praeger, 1973), 32.
56. See, for example, OA, MAD 2775, 745 (15 C 973/7 January 1566); OA, MAD 2775, 860 (4 B 973/26 January 1566); OA, MAD 2775, 1503 (27 L 973/17 May 1566); OA, MAD 2931, 104 (10 ZA 1095/19 October 1684).

CHAPTER 5

1. William Ainsworth, *Researches in Assyria, Babylonia, and Chaldæa* (London: John W. Parker, 1838), 131.

2. Ecologists have not been able to standardize the variegated terminology for wetlands across all the classification schemes that exist today. Paul A. Keddy, *Wetland Ecology: Principles and Conservation*, 2nd ed. (New York: Cambridge University Press, 2010), 4–5.
3. Hans J. Nissen and Peter Heine, *From Mesopotamia to Iraq: A Concise History*, trans. Hans J. Nissen (Chicago: University of Chicago Press, 2009), 143.
4. Peter Christensen, "Middle Eastern Irrigation: Legacies and Lessons," in *Transformations of Middle Eastern Natural Environments: Legacies and Lessons*, ed. Jeff Albert, Magnus Bernhardsson, and Roger Kenna (New Haven, CT: Yale School of Forestry and Environmental Studies, 1998), 15–16.
5. Muhammad Rashid al-Feel, *The Historical Geography of Iraq between the Mongolian and Ottoman Conquests, 1258–1534* (Najaf: al-Adab Press, 1965), 49–53; Faisal Husain, "In the Bellies of the Marshes: Water and Power in the Countryside of Ottoman Baghdad," *Environmental History* 19, no. 4 (2014): 642–645. Likewise, the collapse of the Sasanian Empire in the seventh century coincided with a period of wetland expansion in the region. See A. Asa Eger, "The Swamps of Home: Marsh Formation and Settlement in the Early Medieval Near East," *Journal of Near Eastern Studies* 70, no. 1 (2011): 57–62.
6. Paolo Squatriti, *Water and Society in Early Medieval Italy: AD 400-1000* (New York: Cambridge University Press, 1998), 67–76.
7. Jean Baptiste Louis Jacques Rousseau, *Description du Pachalik de Bagdad* (Paris: Treuttel et Würtz, 1809), 59.
8. For a more detailed hydrography of the sub-units that comprise this marsh area, see Ali al-Sharqi, "Al-Bata'ih al-Haliyya," *Lughat al-Arab* 7 (January 1927): 375–384; S. M. Salim, *Marsh Dwellers of the Euphrates Delta* (London: Athlone Press, 1962), 5–8; Wilfred Thesiger, *The Marsh Arabs* (New York: Penguin, 2007), 13–14; H. Partow, *The Mesopotamian Marshlands: Demise of an Ecosystem* (Nairobi: United Nations Environmental Program, 2001), 11–15.
9. Sipahizade Mehmed Efendi, "Evdahu'l-Mesalik ila Ma'rifeti'l-Büldan ve'l-Memalik," British Library, Add. 23381, ff. 12r, 14r, 33r; Aşık Mehmed, *Menazırü'l Avalim*, ed. Mahmut Ak (Ankara: Türk Tarih Kurumu, 2007), 2:259.
10. Feridun Bey, *Nuzhet-i esrarü'l-ahyar der-ahbar-ı sefer-i Sigetvar: Sultan Süleyman'ın son seferi*, ed. H. Ahmet Arslantürk, Günhan Börekçi, and Abdülkadir Özcan (Istanbul: Zeytinburnu Belediyesi, 2012), 399. Basra's Ottoman governor gives another estimate, informing Istanbul in 1565 that the southern marshes consisted of 300 rivers. OA, MD 5/353 (19 RA 973/14 October 1565).
11. TKG.KK, TT 29, ff. 30r–33r.
12. TKG.KK, TT 29, ff. 46r, 74v–75r, 235v–467v.
13. George B. Cressey, "The Shatt al-Arab Basin," *Middle East Journal* 12, no. 4 (1958): 448.
14. Thesiger, *The Marsh Arabs*, 75, 127; Gavin Young, *Return to the Marshes: Life with the Marsh Arabs of Iraq* (London: Collins, 1977), 35, 42; Gavin Maxwell, *A Reed*

Shaken by the Wind: Travels among the Marsh Arabs of Iraq (London: Eland, 2003), 60; Eger, "The Swamps of Home," 71.

15. W. Ross Cockrill, "The Working Buffalo," in *The Husbandry and Health of the Domestic Buffalo*, ed. Cockrill (Rome: Food and Agriculture Organization of the United Nations, 1974), 319.

16. OA, TT 282, 214–215.

17. Faisal H. Husain, "The Tigris-Euphrates Basin under Early Modern Ottoman Rule, c. 1534–1830" (PhD diss., Georgetown University, 2018), 212.

18. Pedro Teixeira, *The Travels of Pedro Teixeira*, trans. William F. Sinclair (London: Hakluyt Society, 1902), 29; Abraham Parsons, *Travels in Asia and Africa* (London: Longman, Hurst, Rees, and Orme, 1808), 157; Domenico Sestini, *Voyage de Constantinople à Bassora, en 1781, par le Tigre et l'Euphrate: Et Retour a Constantinople, en 1782, par le Désert et Alexandrie* (Paris: Chez Dupuis, 1798), 203.

19. OA, MD 3/1021 (3 § 967/29 April 1560).

20. See, for example, Faruk Tabak, *The Waning of the Mediterranean, 1550–1870: A Geohistorical Approach* (Baltimore: Johns Hopkins University Press, 2008), 18; Tom Nieuwenhuis, *Politics and Society in Early Modern Iraq: Mamluk Pashas, Tribal Shayks and Local Rule between 1802 and 1831* (The Hague: Martinus Nijhoff, 1982), 5. For a criticism of this portrayal, see Vittoria Di Palma, *Wasteland: A History* (New Haven, CT: Yale University Press, 2014), 84–127.

21. Fernand Braudel, *The Mediterranean and the Mediterranean World in the Age of Phillip II*, trans. Siân Reynolds (New York: Harper & Row, 1972), 1:63.

22. S. R. Christophers and H. E. Shortt, "Malaria in Mesopotamia," *Indian Journal of Medical Research* 8, no. 3 (1921): 508–552; A. R. Zahar, "Review of the Ecology of Malaria Vectors in the WHO Eastern Mediterranean Region," *Bulletin of the World Health Organization* 50 (1974): 431–432; Thesiger, *The Marsh Arabs*, 108–109.

23. T. J. Wilkinson, Louise Rayne, and Jaafar Jotheri, "Hydraulic Landscapes in Mesopotamia: The Role of Human Niche Construction," *Water History* 7, no. 4 (2015): 415.

24. Douglas J. Perkins et al., "The Global Burden of Severe Falciparum Malaria: An Immunological and Genetic Perspective on Pathogenesis," in *Dynamic Models of Infectious Diseases*, vol. 1, *Vector-Borne Diseases*, ed. V. Sree Hari Rao and Ravi Durvasula (New York: Springer, 2013), 231–283; Peter Perlmann and Marita Troye-Blomberg, "Malaria and Immune System in Humans," in *Malaria Immunology*, ed. Perlmann and Troye-Blomberg, 2nd ed. (Basel: Karger, 2002), 229–242; James L. A. Webb Jr., *The Long Struggle against Malaria in Tropical Africa* (New York: Cambridge University Press, 2014), 45–68.

25. Thesiger, *The Marsh Arabs*, 85, 108–109.

26. Fulanian, *The Marsh Arab: Haji Rikkan* (Philadelphia: J. B. Lippincott, 1928), 21. The actual authors are S. E. and M. E. Hedgecock, who used the pseudonym due to a prohibition against publishing books under their name during their tenure in uniform.

27. Edward Ochsenschlager, *Iraq's Marsh Arabs in the Garden of Eden* (Philadelphia: University of Pennsylvania Museum of Archaeology and Anthropology, 2004), 194; Young, *Return to the Marshes*, 42; Maxwell, *A Reed Shaken by the Wind*, 60, 66.
28. Young, *Return to the Marshes*, 165–167; Thesiger, *Marsh Arabs*, 108–109, 139, 187; Maxwell, *A Reed Shaken by the Wind*, 39–41, 171–179.
29. "Hükümname Mecmuası," Topkapı Sarayı Müzesi Kütüphanesi, Koğuşlar 888, fol. 56v (11 S 959/6 February 1552); TKG.KK, TT 29, fol. 75r. See also TKG.KK, TT 29, fol. 435v.
30. TKG.KK, TT 29, fol. 33r-v; OA, MAD 2926, 167 (27 L 1093/28 October 1682). One of the earliest and most detailed accounts of cultivation on marginal marsh areas is provided by J. Baillie Fraser, *Travels in Koordistan, Mesopotamia, & c., including an Account of Parts of Those Countries hitherto Unvisited by Europeans with Sketches of the Character and Manners of the Koordish and Arab Tribes* (London: Richard Bentley, 1840), 2:77–78. The account is based on the author's tour in Iraq in 1834–1835.
31. TKG.KK, TT 29, ff. 30r–33v, 46r, 74v–75r, 235v, 241r, 248r, 259r, 264r, 269v, 271v, 280v–281v, 292v–301v, 325v, 331r–331v, 345v, 356r, 359r, 361r–361v, 418r–425r, 435r–441r, 455v–467v; OA, MAD 2926, 167 (27 L 1093/28 October 1682); Piet Buringh, *Soils and Soil Conditions in Iraq* (Baghdad: Ministry of Agriculture, 1960), 151; Halil İnalcık, "The Ottoman State: Economy and Society, 1300–1600," in *An Economic and Social History of the Ottoman Empire, 1300–1914*, vol. 1, *1300–1600*, ed. İnalcık and Donald Quataert (New York: Cambridge University Press, 1994), 162–167; Tabak, *The Waning of the Mediterranean*, 275–282.
32. E. S. Drower, "The Arabs of the Hor Al Hawaiza," in *The Anthropology of Iraq*, ed. Henry Field, pt. 1, n. 2, *The Lower Euphrates-Tigris Region* (Chicago: Field Museum of Natural History, 1949), 372; Thesiger, *The Marsh Arabs*, 174–175; S. Haider, "Land Problems of Iraq" (PhD diss., University of London, 1942), 231–234; Salim, *Marsh Dwellers of the Euphrates Delta*, 85–87.
33. Te-Tzu Chang, "Rice," in *The Cambridge World History of Food*, ed. Kenneth F. Kiple and Kriemhild Coneè Ornelas (New York: Cambridge University Press, 2008), 132–149.
34. OA, TT 282; TKG.KK, TT 29. For evidence of rice fields with freehold status in Iraq, see Feridun Bey, *Nüzhet*, 399.
35. OA, MD 7/1312 (2 ZA 975/30 April 1568). The Ottoman regulation of rice cultivation in Anatolia and the Balkans is discussed in detail in Halil İnalcık, "Rice Cultivation and the Çeltük-Reaya System in the Ottoman Empire," *Turcica* 14 (1982): 59–141. From the fourteenth century, according to İnalcık, the Ottoman state pursued a highly interventionist approach toward rice cultivation. It directly monitored rice fields through inspectors, treasury agents, and tax farmers and relied on slave labor (*ortakcı-kul*) to meet the arduous demands of planting, irrigation, weeding, and canal maintenance. Due to constant complaints of abuse by

state agents and landholders and reports of peasant flight, a new labor system for state-controlled rice cultivation (*çeltükçi-reaya*) became dominant by the sixteenth century to sustain production with minimal interruptions. Overall, the new system improved labor conditions. It officially designated certain peasant groups as rice growers by sultanic decree, freed them from formal bondage, and granted them tax exemptions. Still, those peasants were legally required to grow rice as a public service for the state on a permanent basis. Ottoman archival records on Iraq that I have been able to examine show no traces of this centralized system of rice production. Imperial officials appear to have been less involved in the day-to-day business of rice cultivation than the way İnalcık portrays their counterparts in Anatolia and the Balkans.

36. Jean Baptiste Louis Jacques Rousseau, *Description du Pachalik de Bagdad* (Paris: Treuttel et Würtz, 1809), 61. For similar observations, see M. de Beauchamp, "Voyage de Bagdad à Bassora le long de l'Euphrate," *Le Journal des Sçavans* (1785): 294; Domenico Sestini, *Voyage de Constantinople à Bassora en 1781 par le Tigre et l'Euphrate, et retour à Contantinople en 1782, par le desert et Alexandrie* (Paris: Chez Dupuis, 1798), 173, 202–203, 221, 231.
37. Murtaza Nazmizade, *Gülşen-i Hulefa: Bağdat Tarihi, 762–1717*, ed. Mehmet Karataş (Ankara: Türk Tarih Kurumu, 2014), 376.
38. OA, TT 282; TKG.KK, TT 29; TNA, FO 922/74, I. Gillespie, "Survey of the Livestock Industry in Iraq" (May 1943).
39. The figure is inferred from Tavernier's observation that water buffalo herders paid 1.25 piaster for each animal per year, and as a result the Grand Signor (then Mehmed IV) collected from the water buffalo tax more than 180,000 piasters annually. Jean-Baptiste Tavernier, *Les Six Voyages de Jean Baptiste Tavernier* (Paris: G. Clouzier et C. Barbin, 1676), 1: 217.
40. OA, MAD 9947, 427 (4 ZA 1156/20 December 1743).
41. H. D. Kay, "Milk and Milk Production," in *The Husbandry and Health of the Domestic Buffalo*, 329–376; P. Mahadevan, "Distribution, Ecology and Adaptation," in *Buffalo Production*, ed. N. M. Tulloh and J. H. H. Holmes (Amsterdam: Elsevier, 1992), 1–12.
42. D. C. P. Thalen, *Ecology and Utilization of Desert Shrub Rangelands in Iraq* (The Hague: Dr. W. Junk, 1979), 22–23; Harry Wayne Springfield, *Forage Problems and Resources of Iraq* (Washington, DC: International Cooperation Administration, 1957), 7; Salim, *Marsh Dwellers of the Euphrates Delta*, 91–92; Maxwell, 65–67; Young, 167–170; F. I. El-Dessouky, "Iraq," in *Buffalo Production*, 86; Fraser, *Travels in Koordistan*, 2:42–43.
43. Per Olsson, Carl Folke, and Thomas Hahn, "Social-Ecological Transformation for Ecosystem Management: The Development of Adaptive Co-Management of a Wetland Landscape in Southern Sweden," *Ecology and Society* 9, no. 4 (2004).
44. Ochsenschlager, *Iraq's Marsh Arabs*, 194; Maxwell, *A Reed Shaken by the Wind*, 63–64; Young, *Return to the Marshes*, 41–42; Salim, *Marsh Dwellers of the Euphrates Delta*, 93.

45. Tavernier, *Les Six Voyages de Jean-Baptiste Tavernier*, 1:217.
46. Ochsenschlager, *Iraq's Marsh*, 194–195; Maxwell, *A Reed Shaken by the Wind*, 62–64; Young, *Return to the Marshes*, 41–42, 167–178; Kay, "Milk and Milk Production."
47. TKG.KK, TT 29, fol. 215v.
48. TKG.KK, TT 29, ff. 201v–215v; Erdinç Gülcü, "Osmanlı İdaresinde Bağdat (1534–1623)" (PhD diss., Fırat Üniversitesi, 1999), 334–335.
49. John F. Richards, *The Unending Frontier: An Environmental History of the Early Modern World* (Berkeley: University of California Press, 2003), 17–57, 169, 214–221; Raphaël Morera, "Environmental Change and Globalization in Seventeenth-Century France: Dutch Traders and the Draining of French Wetlands (Arles, Petit Poitou)," *International Review of Social History* 55, no. S18 (2010): 79–101; Vera S. Candiani, *Dreaming of Dry Land: Environmental Transformation in Colonial Mexico City* (Stanford, CA: Stanford University Press, 2014). For a survey of drainage projects in early modern Europe, see Salvatore Ciriacono, *Building on Water: Venice, Holland and the Construction of the European Landscape in Early Modern Times*, trans. Jeremy Scott (New York: Berghahn Books, 2006), 157–263.

PART III

1. Ali Yaycioglu, *Partners of the Empire: The Crisis of the Ottoman Order in the Age of Revolutions* (Stanford, CA: Stanford University Press, 2016), 1–2.

CHAPTER 6

1. Abbas al-Azzawi, *Mawsuʿat Tarikh al-Iraq Bayna Ihtilalayn* (Beirut: al-Dar al-ʿArabiyya li'l-Mawsuʿat, 2004), 5:178–182; Stephen Longrigg, *Four Centuries of Modern Iraq* (Oxford: Oxford University Press, 1925), 121–122; Clément Huart, *Histoire de Bagdad dans les temps modernes* (Paris: E. Leroux, 1901), 139–142; Mehmet Topal, "Bağdadlı Nazmi-Zade Murteza'nın, İcmal-i Sefer-i Nehr-i Ziyab Adlı Risalesine Göre Ziyab Seferi ve Nehirde Yapılan Düzenlemeler," in *Osmanlı Devleti'nde Nehirler ve Göller*, ed. Şakir Batmaz and Özen Tok (Kayseri: Not Yayınları, 2015), 1:71–85.
2. Philip Ball, *Nature's Patterns: A Tapestry in Three Parts*, vol. 2, *Flow* (New York: Oxford University Press, 2009), 11–12.
3. See, for example, Jaafar Jotheri, Mark B. Allen, and Tony J. Wilkinson, "Holocene Avulsions of the Euphrates River in the Najaf Area of Western Mesopotamia: Impacts on Human Settlement Patterns," *Geoarchaeology* 31, no. 3 (2016): 175–193; Carrie Hritz and T. J. Wilkinson, "Using Shuttle Radar Topography to Map Ancient Water Channels in Mesopotamia," *Antiquity* 80, no. 308 (2006): 415–424; Vanessa Mary An Heyvaert and Cecile Baeteman, "A Middle to Late Holocene Avulsion History of the Euphrates River: A Case Study from Tell ed-Der, Iraq, Lower Mesopotamia,"

Quaternary Science Reviews 27, no. 25–26 (2008): 2401–2410; Steven W. Cole and Hermann Gasche, "Second- and First-Millennium BC Rivers in Northern Babylonia," in *Changing Watercourses in Babylonia: Towards a Reconstruction of the Ancient Environment in Lower Mesopotamia*, ed. Hermann Gasche and Michel Tanret (Ghent: University of Ghent and the Oriental Institute of the University of Chicago, 1998), 1–64; Galina S. Morozova, "A Review of Holocene Avulsion of the Tigris and Euphrates Rivers and Possible Effects on the Evolution of Civilizations in Lower Mesopotamia," *Geoarchaeology* 20, no. 4 (2005): 401–423; McGuire Gibson, "Population Shift and the Rise of Mesopotamian Civilisation," in *The Exploration of Culture Change: Models in Prehistory*, ed. Colin Renfrew (London: Duckworth, 1973), 447–463.

4. My effort to integrate the human element into the study of river avulsion benefited most from Ling Zhang, *The River, the Plain, and the State: An Environmental Drama in Northern Song China, 1048–1128* (New York: Cambridge University Press, 2016).

5. Yasin al-Omari, "Al-Durr al-Maknun fi al-Ma'athir al-Madiya min al-Qurun," Bibliothèque Nationale de France, Arabe 4949, fol. 284r. See also Dina Rizk Khoury, *State and Provincial Society in the Ottoman Empire: Mosul, 1540–1834* (New York: Cambridge University Press, 1997), 35.

6. Silahdar Fındıklılı Mehmed Ağa, *Silahdar Tarihi* (Istanbul: Orhaniye Matbaası, 1928), 2:243.

7. Famine and inflation are recorded in Diyarbaır between 1687 and 1692. See OA, MAD 3871, 63 (22 ZA 1101/27 August 1690); OA, D.BŞM.DBH 14/54 (22 Ş 1104/28 April 1693). In Eskişehir and Akşehir, famine and inflation are recorded "a few years" before 1691 and 1693. See OA, İE.SH 1/76 (18 C 1102/19 March 1691); OA, AE.SAMD II 2/115 (10 Ş 1104/16 April 1693). In Seferihisar, famine and inflation are recorded from 1687 and was still in effect in 1691. See OA, İE.DH 9/880 (Evasıt C 1102/12–21 March 1691). Sources of the Dutch East India Company recorded inflation in Basra in February 1689. See Willem Floor, *The Persian Gulf: A Political and Economic History of Five Port Cities, 1500–1730* (Washington, DC: Mage, 2006), 538.

8. Rosanne D'Arrigo and Heidi M. Cullen, "A 350-Year (AD 1628–1980) Tree-Ring Record of Turkish Precipitation: Linkages to Tigris-Euphrates Streamflow and the NAO," *Dendrochronologia* 19 (2001): 169–177.

9. Pascal Flohr et al., "Late Holocene Droughts in the Fertile Crescent Recorded in a Speleothem from Northern Iraq," *Geophysical Research Letters* 44, no. 3 (2017): 1528–1536. It is true that there is always an uncertainty about the annual layer count of speleothems, more so than tree rings, for instance. But speleothem evidence still provides a more precise chronology than other geo-archival records like lake sediments, and the textual and dendrological evidence further strengthens the chronology of the Gejkar Cave's stalagmite. I am grateful to Dominik Fleitmann for educating me about speleothems as proxies for past environmental conditions.

10. J. Luterbacher et al., "The Late Maunder Minimum (1675–1715)—A Key Period for Studying Decadal Scale Climatic Change in Europe," *Climatic Change* 49, no. 4 (2001): 441–462; John A. Eddy, "Solar History and Human Affairs," *Human Ecology* 22 (1994): 23–35; Christian Pfister, "Spatial Patterns of Climatic Change in Europe A.D. 1675 to 1715," in *Climatic Trends and Anomalies in Europe 1675–1715*, ed. Burkhard Frenzel, Pfister, and Birgit Gläser (New York: G. Fischer, 1994), 287–316; Sam White, *The Climate of Rebellion in the Early Modern Ottoman Empire* (New York: Cambridge University Press, 2011), 133, 215–222; John F. Richards, *The Unending Frontier: An Environmental History of the Early Modern World* (Berkeley: University of California Press, 2003), 66–67; John L. Brooke, *Climate Change and the Course of Global History: A Rough Journey* (New York: Cambridge University Press, 2014), 175–176; Geoffrey Parker, *Global Crisis: War, Climate Change and Catastrophe in the Seventeenth Century* (New Haven, CT: Yale University Press, 2013), 13–17.

11. For Russia and Crimea, see Andrei O. Selivanov, "Global Climate Changes and Humidity Variations over East Europe and Asia by Historical Data," in *Global Precipitations and Climate Change*, ed. Michel Desbois and Françoise Désalmand (New York: Springer-Verlag, 1994), 88; Yevgeny P. Borisenkov, "Climatic and Other Natural Extremes in the European Territory of Russia in the Late Maunder Minimum (1675–1715)," *Climatic Trends*, 88; for India, Brian M. Fagan, *Floods, Famines, and Emperors: El Niño and the Fate of Civilizations* (New York: Basic Books, 2009), 8–9; for Iceland, Astrid E. J. Ogilvie, "Documentary Records of Climate from Iceland during the Late Maunder Minimum Period A.D. 1675 to 1715 with Reference to the Isotopic Record from Greenland," *Climatic Trends*, 17–18; for Switzerland, Christian Pfister, "Switzerland: The Time of Icy Winters and Chilly Springs," *Climatic Trends*, 218–219.

12. Murtaza Nazmizade, *Gülşen-i Hulefa: Bağdat Tarihi, 762–1717*, ed. Mehmet Karataş (Ankara: Türk Tarih Kurumu, 2014), 309–310.

13. Ahmad Ghurabzade, "Uyun Akhbar al-A'yan bi-man Mada fi Salif al-Asr wa-l-Zaman," British Library, Add. 23309, ff. 271v–272r.

14. For the relationship between stream power and sedimentation in the Mesopotamian context, see T. J. Wilkinson and Carrie Hritz, "Physical Geography, Environmental Change and the Role of Water," in *Models of Mesopotamian Landscapes: How Small-Scale Processes Contributed to the Growth of Early Civilizations*, ed. Wilkinson, McGuire Gibson, and Magnus Widell (Oxford: Archaeopress, 2013), 20.

15. Andrew S. Goudie, *Arid and Semi-arid Geomorphology* (New York: Cambridge University Press, 2013), 204–245; M. G. Ionides, *The Régime of the Rivers, Euphrates and Tigris* (London: E. & F. N. Spon, 1937).

16. William B. Bull, "Threshold of Critical Power in Streams," *Geological Society of America Bulletin* 90, no. 5 (1979): 453–464.

17. Robert J. Naiman, "Animal Influences on Ecosystem Dynamics," *BioScience* 38, no. 11 (1988): 750–762.

18. Raşid Mehmed Efendi, *Tarih-i Raşid ve Zeyli*, ed. Abdülkadir Özcan, Yunus Uğur, Baki Çakır, and Ahmet Zeki İzgöer (Istanbul: Klasik, 2013) 1:605.
19. Michael Church et al., "Rivers," in *Geomorphology and Global Environmental Change*, ed. Olav Slaymaker, Thomas Spencer, and Christine Embleton-Hamann (New York: Cambridge University Press, 2009), 109–112.
20. Piet Buringh, *Soils and Soil Conditions in Iraq* (Baghdad: Ministry of Agriculture, 1960), 144–148; T. J. Wilkinson, Louise Rayne, and Jaafar Jotheri, "Hydraulic Landscapes in Mesopotamia: The Role of Human Niche Construction," *Water History* 7, no. 4 (2015): 397–418.
21. Nazmizade, *Gülşen-i Hulefa*, 367–368; OA, D.BŞM.BGH 1/27; OA, D.BŞM. BGH 1/29. On nodes of avulsion, see T. J. Wilkinson, *Archaeological Landscapes of the Near East* (Tucson: University of Arizona Press, 2003), 82–85.
22. The interplay between recurring drought, rural flight to the cities, and disease outbreaks manifested itself across the Ottoman Empire during the height of the Little Ice Age in the late sixteenth and throughout the seventeenth centuries. See White, *The Climate of Rebellion*, 249–275. On the epidemiological consequences of urbanization in early modern Ottoman history more generally, see Nükhet Varlık, "Conquest, Urbanization and Plague Networks in the Ottoman Empire, 1453–1600," in *The Ottoman World*, ed. Christine Woodhead (New York: Routledge, 2012), 251–263; Sam White, "Rethinking Disease in Ottoman History," *International Journal of Middle East Studies* 42, no. 4 (2010): 560–561.
23. Nazmizade, *Gülşen-i Hulefa*, 325–326; Ghurabzade, "Akhbar," fol. 272v.
24. Nazmizade, *Gülşen-i Hulefa*, 328.
25. Mary Elizabeth Wilson, "The Power of Plague," *Epidemiology* 6, no. 4 (1995): 459.
26. The onset of the Little Ice Age crisis in the 1590s offered similar opportunities for the incursion of pastoral tribes onto Ottoman settlement centers. White, *The Climate of Rebellion*, 229–248.
27. Nazmizade, *Gülşen-i Hulefa*, 329–330; Raşid, *Tarih-i Raşid*, 1:420; Defterdar Sarı Mehmed Paşa, *Zübdet-i Vekayiat: Tahlil ve Metin, 1066–1116/1656–1704*, ed. Abdülkadir Özcan (Ankara: Türk Tarih Kurumu, 1995), 417, 454–55.
28. OA, MD 104/204 (Evasıt L 1103/26 June-5 July 1692); OA, MAD 18540, 23 (10 L 1103/25 June 1692). See also OA, MD 105/393 (Evahir CA 1106/7–16 January 1695).
29. OA, D.BŞM.BGH 1/17; OA, D.BŞM.BGH 1/27. For the losses recorded in 1693 (56,395 guruş), see OA, MAD 18537, 2.
30. OA, D.BŞM.BGH 16735; OA, MAD 9891, 14–17 (1 RA 1114/26 July 1702).
31. OA, MD 105/394 (Evahir CA 1106/7–16 January 1695).
32. Caroline Finkel, *Osman's Dream: The Story of the Ottoman Empire, 1300–1923* (New York: Basic Books, 2007), 289–321; Virginia Aksan, *Ottoman Wars, 1700–1870: An Empire Besieged* (New York: Routledge, 2007), 18–36.
33. OA, MD 104/201 (Evail L 1103/16–25 June 1692); OA, MAD 18540, 24–25 (10 L 1103/25 June 1692); Nazmizade, *Gülşen-i Hulefa*, 268.

34. OA, MD 104/218 (Evasıt L 1103/26 June–4 July 1692); OA, MD 104/225 (Evasıt L 1103/26 June–4 July 1692); OA, MD 104/473–474 (Evasıt R 1103/31 December 1691–9 January 1692); OA, MD 104/686–688 (Evasıt C 1104/17–26 February 1693); OA, MD 104/690–691 (Evasıt C 1104/17–26 February 1693); OA, MD 105/395 (Evahir CA 1106/9–18 November 1694); OA, MD 111/532–537 (Evail CA 1111/24 October–2 November 1699); OA, MD 111/1195 (Evail ZA 1110/1–10 May 1699); OA, MD 111/1196 (Evasıt ZA 1110/11–20 May 1699).

35. OA, MD 104/235 (Evahir L 1103/6–15 July 1692); OA, MD 105/393–395 (Evahir CA 1106/7–16 January 1695); OA, MD 106/563 (Evail ZA 1106/13–22 June 1695); OA, MD 106/575 (Evasıt ZA 1106/23 June–2 July 1695); OA, MD 106/1281 (Evail R 1107/8–17 November 1695); OA, MD 111/224 (Evahir S 1111/17–26 August 1699); OA, MD 111/1196 (Evasıt ZA 1111/1–10 May 1700); OA, OA, MAD 18540, 23 (10 L 1103/25 June 1692); Raşid, *Tarih-i Raşid*, 1:444–445, 1:516; Nazmizade, *Gülşen-i Hulefa*, 330–335; Defterdar, *Zübdet*, 675.

36. Longrigg, *Four Centuries of Modern Iraq*, 95; Silahdar Fındıklılı Mehmed Ağa, "Cild-i Salis Tarih-i Fındıklılı," Topkapı Sarayı Müzesi Kütüphanesi, Emanet Hazinesi 1413, fol. 153v.

37. Silahdar, "Cild-i Salis," fol. 179r; Nazmizade, *Gülşen-i Hulefa*, 370–371; Nazmizade, "İcmal-i Sefer-i Nehr-i Ziyab," Süleymaniye Kütüphanesi, Esad Efendi 2062/4, ff. 75r–91r.

38. Aksan, *Ottoman Wars*, 18–36; Finkel, *Osman's Dream*, 318–328.

39. Raşid, *Tarih-i Raşid*, 1:597–601; Defterdar, *Zübdet*, 706–713, 720–722; *Anonim Osmanlı Tarihi, 1099–1116/1688–1704*, ed. Abdülkadir Özcan (Ankara: Türk Tarih Kurumu, 2000), 154–160; al-Azzawi, *Tarikh al-Iraq*, 5:163–177.

40. Silahdar, "Cild-i Salis," ff. 177v, 185v; Nazmizade, *Gülşen-i Hulefa*, 367–368, 383–384; Nazmizade, "İcmal-i Sefer-i Nehr-i Ziyab," ff. 75v, 86v.

41. Raşid, *Tarih-i Raşid*, 1:605.

42. OA, D.BŞM.MSH 3/118 (Evasıt B 1110/13–22 January 1699); OA, D.BŞM.BGH 1/17; Silahdar, "Cild-i Salis," ff. 177v–179v; Nazmizade, *Gülşen-i Hulefa*, 367–370; Nazmizade, "İcmal-i Sefer-i Nehr-i Ziyab," ff. 75v–76r.

43. According to an accounting ledger, the Imperial Council allocated 1,409 purses (*kise*) for the damming of the Dhiyab Canal. TSMA 708 (1113/1701–1702). For cash funds and provisions allocated for the engineering expedition, see OA, D.BŞM 7653/81 (1113/1701–1702).

44. Nazmizade, "Tarih-i Seferü'l-Basra," Süleymaniye Kütüphanesi, Esad Efendi 2062/3, fol. 56v.

45. White, *The Climate of Rebellion*, 15–51.

46. OA, MAD 966, 192–215 (20 M 1113–17 M 1114/27 June 1701–13 June 1702); OA, MAD 3595, 6 (13 L 1113/12 March 1702); OA, MAD 3595, 24 (24 L 1113/23 March 1702); OA, MAD 3595, 72–73 (28 ZA 1113/26 April 1702); OA, MAD 3595, 73–74 (1 Z 1113/29 April 1702); OA, MAD 3595, 101–102 (25 Z 1113/23 May 1702); OA, MD 111/2261 (Evahır S 1113/27 July–5 August 1701); OA, MD 111/2349 (Evail R

1113/4–13 September 1701); OA, MAD 2510, 142 (20 ZA 1114/7 April 1703); OA, MAD 2510, 155 (12 S 1115/27 June 1703); OA, D.BŞM 7651/51 (20 M 1113/27 June 1701); OA, AE.SMST.II 116/12627 (14 CA 1113); Silahdar, "Cild-i Salis," ff. 179r–v; Nazmizade, *Gülşen-i Hulefa*, 370–371; Nazmizade, "İcmal-i Sefer-i Nehr-i Ziyab," fol. 78r. Some of the timber obtained from Maraş turned out to be ill-suited to dam construction. See OA, D.BŞM 7654/71 (20 B 1114/10 December 1702).

47. OA, MAD 966, 192–215 (20 M 1113–17 M 1114/27 June 1701–13 June 1702); Silahdar, "Cild-i Salis," ff. 180v–181v; Nazmizade, *Gülşen-i Hulefa*, 370–371; Nazmizade, "İcmal," ff. 80r–82v.
48. Silahdar, "Cild-i Salis," ff. 183r–v; Nazmizade, *Gülşen-i Hulefa*, 379–380; Nazmizade, "İcmal-i Sefer," ff. 83v–84r.
49. Nazmizade, "İcmal-i Sefer," fol. 83v.
50. Suraiya Faroqhi, *The Ottoman Empire and the World around It* (London: I. B. Tauris, 2007), 108; Rhoads Murphey, *Ottoman Warfare, 1500–1700* (New Brunswick, NJ: Rutgers University Press, 1999), 90–93.
51. Silahdar, "Cild-i Salis," ff. 183v–184r; Nazmizade, *Gülşen-i Hulefa*, 380–382; Nazmizade, "İcmal-i Sefer," ff. 84r–85r; Suraiya Faroqhi, "Crisis and Change, 1590–1699," in *An Economic and Social History of the Ottoman Empire*, vol. 2, *1600–1914*, ed. Halil İnalcık and Donald Quataert (Cambridge: Cambridge University Press, 1997), 483. The slaughtering of sheep to mark the completion of water control projects has ancient roots in Mesopotamia. Evidence of the practice is found in documents of the Third Dynasty of Ur III at the end of the third millennium BC. See Stephanie Rost, "Watercourse Management and Political Centralization in Third-Millennium B.C. Southern Mesopotamia: A Case Study of the Umma Province of the Ur III Period (2112–2004 B.C.)" (PhD diss., Stony Brook University, 2015), 116.
52. Silahdar, "Cild-i Salis," ff. 184r–185r; Nazmizade, *Gülşen-i Hulefa*, 381–388; Nazmizade, "İcmal-i Sefer," ff. 85r–88v.
53. Muhammad Rida al-Shabibi, "Al-Rumahiyya," *Lughat al-Arab* 1 (1913): 461–465.
54. TKG.KK, TT 29, ff. 313v–314r.
55. Nazmizade, *Gülşen-i Hulefa*, 186.
56. OA, TT 1028, 121; TKG.KK, TT 228, 25v–26r; TKG.KK, TT 29, 309v–314v; OA, D.BŞM.BGH 1/17.
57. Muhammad al-Maqdisi, *Ahsan al-Taqasim fi Ma'rifat al-Aqalim* (Leiden: Brill, 1906), 228.
58. Steven W. Cole, "Marsh Formation in the Borsippa Region and the Course of the Lower Euphrates," *Journal of Near Eastern Studies* 53, no. 2 (1994): 81–109.
59. P. Mahadevan, "Distribution, Ecology and Adaptation," in *Buffalo Production*, ed. N. M. Tulloh and J. H. H. Holmes (Amsterdam: Elsevier, 1992), 1–12.
60. TKG.KK, TT 29, fol. 304v.
61. Ernest Gellner, "Tribalism and the State in the Middle East," in *Tribes and State Formation in the Middle East*, ed. Philip S. Khoury and Joseph Kostiner

(Berkeley: University of California Press, 1990), 111. See also William Irons, "Nomadism as a Political Adaptation: The Case of the Yomut Turkmen," *American Ethnologist* 1, no. 4 (1974): 635–658; Philip Burnham, "Spatial Mobility and Political Centralization in Pastoral Societies," in *Pastoral Production and Society*, ed. L'Equipe écologie et anthropologie des societies pastorales (New York: Cambridge University Press, 1979), 349–360.

62. Quoted in Cemal Kafadar, *Between Two Worlds: The Construction of the Ottoman State* (Berkeley: University of California Press, 1995), 118.
63. Al-Azzawi, *Tarikh al-Iraq*, 5:160.
64. OA, D.BŞM.BGH 1/17; OA, D.BŞM.BGH 1/27; OA, D.BŞM.BGH 1/29; OA, D.BŞM.BGH 1/56; OA, D.BŞM.BGH 1/59; OA, D.BŞM.BGH 16735.
65. Carsten Niebuhr, *Reisebeschreibung nach Arabien und andern umliegenden Ländern* (Copenhagen: N. Möller, 1778), 2:252–253.
66. Jean Baptiste Louis Jacques Rousseau, *Description du Pachalik de Bagdad* (Paris: Treuttel et Würtz, 1809), 60.
67. White, *The Climate of Rebellion*, 123–225; Parker, *Global Crisis*.
68. For a deep history of the Tigris and Euphrates as integral components of Iraq's recurring political crises, see M. B. Rowton, "The Role of the Watercourses in the Growth of Mesopotamian Civilization," in *Alter Orient und Altes Testament*, ed. Kurt Bergerhof, Manfried Dietrich, and Oswald Loretz (Neukirchen-Vluyn, Germany: Butzon & Bercker, 1969), 307–316. On the role of the Nile in Egypt's experience with disaster during the eighteenth century, see Alan Mikhail, "The Nature of Plague in Late Eighteenth-Century Egypt," *Bulletin of the History of Medicine* 82, no. 2 (2008): 249–275; Mikhail, "Ottoman Iceland: A Climate History," *Environmental History* 20, no. 2 (2015): 262–284.

CHAPTER 7

1. Suraiya Faroqhi, *Approaching Ottoman History: An Introduction to the Sources* (Cambridge: Cambridge University Press, 1999), 214–215; Ali Yaycıoğlu, "Provincial Power-Holders and the Empire in the Late Ottoman World: Conflict or Partnership?" in *The Ottoman World*, ed. Christine Woodhead (New York: Routledge, 2012), 446.
2. Yaycıoğlu, "Provincial Power-Holders," 447–448; Tolga U. Esmer, "Economies of Violence, Banditry and Governance in the Ottoman Empire around 1800," *Past and Present* 224, no. 1 (2014): 163–199; Fikret Adanır, "Semi-Autonomous Provincial Forces in the Balkans and Anatolia," in *The Cambridge History of Turkey*, vol. 3, *The Later Ottoman Empire, 1603–1839*, ed. Suraiya N. Faroqhi (Cambridge: Cambridge University Press, 2006), 157–185.
3. Dina Rizk Khoury, "The Ottoman Centre versus Provincial Power-Holders: An Analysis of the Historiography," in *The Later Ottoman Empire*, 154; Margaret L. Meriwether, "Urban Notables and Rural Resources in Aleppo, 1770–1830," *International Journal of Turkish Studies* 4 (1987): 55–73.

4. Mehmet Genç, "A Study of the Feasibility of Using Eighteenth-Century Ottoman Financial Records as an Indicator of Economic Activity," in *The Ottoman Empire and the World-Economy*, ed. Huri İslamoğlu-İnan (Cambridge: Cambridge University Press, 1987), 345–373; Ariel Salzmann, "An Ancien Régime Revisited: 'Privatization' and Political Economy in the Eighteenth-Century Ottoman Empire," *Politics & Society* 21, no. 4 (1993): 393–423; Yaycıoğlu, "Provincial Power-Holders," 444, 448.

5. Canay Şahin, "The Economic Power of Anatolian *Ayan*s in the Late Eighteenth Century: The Case of the Caniklizades," *International Journal of Turkish Studies* 11, no. 1/2 (2005): 28–47; Ali Yaycioglu, *Partners of the Empire: The Crisis of the Ottoman Order in the Age of Revolutions* (Stanford, CA: Stanford University Press, 2016), 84–89; Jane Hathaway, *The Arab Lands under Ottoman Rule, 1516–1800* (Abingdon: Routledge, 2013), 87–90; Dina Rizk Khoury, *State and Provincial Society in the Ottoman Empire: Mosul, 1540–1834* (Cambridge: Cambridge University Press, 1997), 75–108.

6. "Regional governance regimes" is a term used by sociologist Karen Barkey, which she defines as "networks of large patriarchal families who established themselves around one or two leaders; developed their resources and influence through multiple state and nonstate activities and positions; extended their networks to incorporate clients, whether lesser notables or peasants; and both in their local rule and in their understanding of their legitimacy mimicked the ruling household of the sultan." Barkey, *Empire of Difference: The Ottomans in Comparative Perspective* (New York: Cambridge University Press, 2008), 242.

7. OA, D.BŞM.BGH 1/17. For more evidence on the financial strength of the Ahşamat and Cemmasat during the seventeenth century, see OA, D.BŞM 976; TSMA 5852; TSMA 9508.

8. OA, MD 112/1672 (Evahir S 1114/17 July 1702). In the seventeenth century, Ottoman bureaucrats treated the Ahşamat and Cemmasat as one group, often referring to them simply as the Ahşamat, as the scribe of this document did.

9. OA, MD 112/1672 (Evahir S 1114/17 July 1702).

10. Mehmed Hurşid Paşa, *Seyahatname-i Hudud*, ed. Alaattin Eser (Istanbul: Simurg, 1997), 164–165.

11. Murtaza Nazmizade, *Gülşen-i Hulefa: Bağdat Tarihi, 762–1717*, ed. Mehmet Karataş (Ankara: Türk Tarih Kurumu, 2014), 369.

12. Nazmizade, *Gülşen-i Hulefa*, 368–369; Hamid al-Saʿdun, *Imarat al-Muntafiq wa Atharuha fi Tarikh al-Iraq wa al-Mantiqa al-Iqlimiyya, 1546–1918* (Amman: Dar Waʾil, 1999), 64–66.

13. Nazmizade, *Gülşen-i Hulefa*, 369; Max von Oppenheim, *Die Beduinen*, vol. 1, *Die Beduinenstämme in Mesopotamien und Syrien* (Leipzig: Otto Harrassowitz, 1939), 131–144; Tom Nieuwenhuis, *Politics and Society in Early Modern Iraq: Mamluk Pashas, Tribal Shayks and Local Rule Between 1802 and 1831* (The Hague: Martinus Nijhoff, 1982), 124–125.

14. Abbas al-Azzawi, *Mawsu'at Tarikh al-Iraq Bayna Ihtilalayn* (Beirut: al-Dar al-Arabiyya li'l-Mawsu'at, 2004), 5:161–162.
15. The chronicles of Iraq mention more confrontations between Shammar tribes and Ottoman governors in Iraq in 1705, 1706, 1725, 1733, and 1744. See Nazmizade, *Gülşen-i Hulefa*, 398; Abdulrahman al-Suwaydi, *Hadiqat al-Zawra fi Sirat al-Wuzara*, Imad Abdussalam Ra'uf (Baghdad: Al-Majma al-Ilmi, 2003), 68–75, 99–110, 259–278; al-Azzawi, *Mawsu'at Tarikh al-Iraq*, 5:197–199, 5:250–251, 5:284–285; 5:316. A member of the Shammar tribe, Ghazi al-Yawar, would become Iraq's first interim president after the fall of the Baathist regime in 2003.
16. Nazmizade, *Gülşen-i Hulefa*, 367–368.
17. Al-Suwaydi, *Hadiqat al-Zawra*, 88–89. See also Nazmizade, *Gülşen-i Hulefa*, 368.
18. Silahdar Fındıklılı Mehmet Ağa, "Cild-i Salis Tarih-i Fındıklılı," Topkapı Sarayı Müzesi Kütüphanesi, Emanet Hazinesi 1413, ff. 177v–178v; Nazmizade, *Gülşen-i Hulefa*, 368–369; Hammud al-Sa'idi, *Dirasat an Asha'ir al-Iraq: Al-Khaza'il* (Najaf: Matba'at al-Adab, 1974), 12–27.
19. TKG.KK, TT 29, ff. 304r–467r.
20. Edward Ochsenschlager, *Iraq's Marsh Arabs in the Garden of Eden* (Philadelphia: University of Pennsylvania Museum of Archaeology and Anthropology, 2004), 27; TNA, FO 922/74, I. Gillespie, "Survey of the Livestock Industry in Iraq," (May 1943): 1.
21. R. B. Griffiths, "Parasites and Parasitic Diseases," in *The Husbandry and Health of the Domestic Buffalo*, ed. W. Ross Cockrill (Rome: Food and Agriculture Organization of the United Nations, 1974), 236–275.
22. S. M. Salim, *Marsh Dwellers of the Euphrates Delta* (London: Athlone Press, 1962), 138–140.
23. Nazmizade, *Gülşen-i Hulefa*, 234; Yasin al-Omari, "Al-Durr al-Maknun fi al-Ma'athir al-Madiya min al-Qurun," Bibliothèque Nationale de France, Arabe 4949, fol. 253r; al-Sa'idi, *Al-Khaza'il*, 8–11; Stephen Longrigg, *Four Centuries of Modern Iraq* (Oxford: Oxford University Press, 1925), 82; Mustafa Naima, *Tarih-i Naima* (Istanbul: Darü't-Tıbaati'l-Amire, 1147/1734–1735), 2:6; Max von Oppenheim, *Die Beduinen*, vol. 3, *Die Beduinenstämme in Nord—und Mittelarabien und Irak* (Leipzig: Otto Harrassowitz Wiesbaden, 1952), 313; Selim Güngörürler, "Diplomacy and Political Relations between the Ottoman Empire and Safavid Iran, 1639–1722" (PhD diss., Georgetown University, 2016), 128.
24. Al-Suwaydi, *Hadiqat al-Zawra*, 88–95.
25. "Annual Administration Report, Shamiyah Division, from 1st January to 31st December 1918," in *Iraq Administration Reports 1914–1932* (Slough: Archive Editions, 1992), 2:66. For a fuller account of the Khaza'il's story, see al-Sa'idi, *Al-Khaza'il*; von Oppenheim, *Die Beduinenstämme in Nord—und Mittelarabien und Irak*, 313–321; Faisal Husain, "In the Bellies of the Marshes: Water and Power in the Countryside of Ottoman Baghdad," *Environmental History* 19, no. 4 (2014): 638–664.

26. Kamal S. Salibi, "Middle Eastern Parallels: Syria-Iraq-Arabia in Ottoman Times," *Middle Eastern Studies* 15, no. 1 (1979): 70–81; Hala Fattah, *The Politics of Regional Trade in Iraq, Arabia, and the Gulf, 1745–1900* (Albany: State University of New York Press, 1997), 28–31.
27. Nazmizade, *Gülşen-i Hulefa*, 394; al-Suwaydi, *Hadiqat al-Zawra*, 68–75. The two tribes in question were Al Shahwan and Al Ghurayr, both of whom belong to the Shammar confederation. See Abbas al-Azzawi, *Mawsuʿat Ashaʾir al-Iraq* (Beirut: al-Dar al-Arabiyya liʾl-Mawsuʿat, 2005), 1:254.
28. Al-Suwaydi, *Hadiqat al-Zawra*, 77.
29. OA, MD 114-1/1661 (Evail C 1116/1–9 October 1704); al-Suwaydi, *Hadiqat al-Zawra*, 67–78.
30. Hanna Batatu, *The Old Social Classes and the Revolutionary Movements of Iraq* (Princeton, NJ: Princeton University Press, 1978), 219; Nieuwenhuis, *Politics and Society*, 81; Ali Shakir Ali, *Tarikh al-Iraq fi al-Ahd al-Othmani, 1638–1750 Miladiyya/ 1048–1174 Hijriyya: Dirasa fi Ahwalihi al-Siyasiyya* (Nineveh: Maktabat 30 Tammuz, 1985), 85–100. For slightly different calculations of how many Ottoman governors served in Baghdad between 1638 and 1704, see Carsten Niebuhr, *Reisebeschreibung nach Arabien und andern umliegenden Ländern* (Copenhagen: N. Möller, 1778), 2:309–311; Habib K. Chiha, *La Province de Bagdad: Son Passé, Son Présent, Son Avenir* (Cairo: El-Maaref, 1908), 39–42.
31. Thomas Lier, *Haushalte und Haushaltspolitik in Bagdad, 1704–1831* (Würzburg: Ergon, 2004), 15–64.
32. On the Suez and Danube admiralties, see Cengiz Orhonlu, "Hint Kaptanlığı ve Piri Reis," *Belleten* 134 (1970): 235–254; Rossitsa Gradeva, "War and Peace along the Danube: Vidin at the End of the Seventeenth Century," *Oriente Moderno* 81, no. 1 (2001): 163; Gradeva, *War and Peace in Rumeli: 15th to Beginning of 19th Century* (Istanbul: Isis Press, 2008), 78–81; Svetlana Ivanova, "Ali Pasha: Sketches from the Life of a Kapudan Pasha on the Danube," in *The Kapudan Pasha: His Office and His Domain*, ed. Elizabeth Zachariadou (Rethymnon: Crete University Press, 2002), 325–345.
33. Feridun Bey, *Nüzhet-i esrarüʾl-ahyar der-ahbar-ı sefer-i Sigetvar: Sultan Süleymanʾın son seferi*, ed. H. Ahmet Arslantürk, Günhan Börekçi, and Abdülkadir Özcan (Istanbul: Zeytinburnu Belediyesi, 2012), 385–387.
34. Ali Yılmaz, "XVI. Yüzyılda Birecik Sancağı" (PhD diss., İstanbul Üniversitesi, 1996), 200.
35. OA, İE.DH 20/1831 (Evahir Ş 1111/11–20 February 1700); Ivanova, "Ali Pasha," 336. Some of Aşçızade's financial dealings with the Ottoman state in Vidin are highlighted in OA, İE.ML 59/5587 (1 C 1107/6 January 1696); OA, İE.BH 7/632 (15 N 1111/6 March 1700); OA, MAD 9885, 218–219 (23 N 1111/14 March 1700); OA, MD 111/1194 (Evail ZA 1111/11–20 March 1700); OA, MD 111/1199 (Evahir M 1112/8–17 July 1700). The Ottoman fleet in the Danube basin dates back to the early fifteenth century and was a major exporter of skilled labor and naval technology to

the Shatt Fleet from its establishment in the sixteenth century. See Gábor Ágoston, "Where Environmental and Frontier Studies Meet: Rivers, Forests, Marshes and Forts along the Ottoman-Hapsburg Frontier in Hungary," in *The Frontiers of the Ottoman World*, ed. A. C. S. Peacock (Oxford: Oxford University Press, 2009), 58–60; Colin Imber, "The Navy of Süleyman the Magnificent," *Archivum Ottomanicum* 6 (1980): 275–277; Gradeva, *War and Peace in Rumeli*, 301–323; Gradeva, "War and Peace Along the Danube," 163–165.

36. Güngörürler, "Diplomacy and Political Relations," 345–350. Aşçızade's name remained closely associated with the Shatt admiralty after his death. At least two of his successors were close family members. Following his death, Istanbul tapped Aşçızade's nephew to the post in 1706, a position he held for about six years until he was sacked due to his incompetence and neglect. Half a century later in 1763, one of Aşçızade's grandsons, Salih Bey, was promoted to the post to reform what was at the time a mismanaged fleet. See OA, C.BH 1/7 (CA 1177/November–December 1763); OA, AE.SMST.III 297/23775 (24 B 1177/28 January 1764).

37. Hasan Pasha's expeditions and his reign more generally are best documented by al-Suwaydi, *Hadiqat al-Zawra*, 67–222.

38. Lier, *Haushalte und Haushaltspolitik in Bagdad*, 32–38; Nieuwenhuis, *Politics and Society*, 13–15; John R. Perry, "The Mamluk Paşalık of Baghdad and Ottoman-Iranian Relations in the Late Eighteenth Century," in *Studies on Ottoman Diplomatic History*, ed. Sinan Kuneralp (Istanbul: Isis Press, 1987), 59–70; Longrigg, *Four Centuries of Modern Iraq*, 163–165.

39. Lier, *Haushalte und Haushaltspolitik in Bagdad*, 41; Nieuwenhuis, *Politics and Society*, 76–78.

40. Lier, *Haushalte und Haushaltspolitik in Bagdad*, 62, 132–133.

41. Abraham Parsons, *Travels in Asia and Africa* (London: Longman, Hurst, Rees, and Orme, 1808), 159. See also John Taylor, *Travels from England to India, In the Year 1789* (London: S. Low, 1799), 1:270.

42. Süheyla Yenidünya, "XIX. Yüzyıl Başlarında Osmanlı Devleti'nin Bağdat'ta Kölemen Hakimiyetini Kaldırma Teşebbüsleri," *Türk Dünyası Araştırmaları* 182 (2009): 1–46.

43. Khoury, *State and Provincial Society*, 71, 208–209; Percy Kemp, "Mosul and Mosuli Historians of the Jalili Era (1726–1834)" (PhD diss., University of Oxford, 1979), 76–94; Stefan Winter, "The Other *Nahdah*: The Bedirxans, the Millis and the Tribal Roots of Kurdish Nationalism in Syria," *Oriente Moderno* 86, no. 3 (2006): 461–474; Ariel Salzmann, *Tocqueville in the Ottoman Empire: Rival Paths to the Modern State* (Brill: Leiden, 2004), 134; Nazmizade, *Gülşen-i Hulefa*, 408; Lier, *Haushalte und Haushaltspolitik in Bagdad*, 59–64; Nieuwenhuis, *Politics and Society*, 24, 30, 32–33, 41–42, 104–106; Longrigg, *Four Centuries of Modern Iraq*, 210–212.

44. OA, HAT 397/20896 (1225/1810–1811).

45. Lier, *Haushalte und Haushaltspolitik in Bagdad*, 59–64; Nieuwenhuis, *Politics and Society*, 24.
46. Dina Riz Khoury, "The Introduction of Commercial Agriculture in the Province of Mosul and Its Effects on the Peasantry, 1750–1850," in *Landholding and Commercial Agriculture in the Middle East*, ed. Çağlar Keyder and Faruk Tabak (Albany: State University of New York Press, 1991), 157–158; Kemp, "Mosul and Mosuli Historians," 67, 78–80. In fact, two sons of the founder of the Jalili dynasty contracted to supply Baghdad with grain. See Hathaway, *The Arab Lands under Ottoman Rule*, 94.
47. The timber-provisioning process is described in detail by the British resident in Baghdad during his visit to Süleymaniye in the spring of 1820. See Claudius James Rich, *Narrative of a Residence in Koordistan, and on the Site of Ancient Nineveh* (London: James Duncan, 1836), 1:104–107. See also Kemp, "Mosul and Mosuli Historians," 67.
48. Nazmizade, *Gülşen-i Hulefa*, 404.
49. OA, HAT 397/20896 (1225/1810–1811).
50. Thabit A. J. Abdullah, *Merchants, Mamluks, and Murder: The Political Economy of Trade in Eighteenth-Century Basra* (Albany: State University of New York Press, 2001), 80.
51. IOR/G/29/21, Henry Moore to the Court of Directors, fol. 1v (16 January 1774).
52. J. R. Perry, "The Banu Ka'b: An Ambitious Brigand State in Khuzistan," *Le Monde Iranien et l'Islam* 1 (1971): 131–152; Willem Floor, "The Rise and Fall of the Banu Ka'b: A Borderer State in Southern Khuzestan," *Iran* 44 (2006): 277–315.
53. Patricia Risso, *Oman and Muscat: An Early Modern History* (London: Croom Helm, 1986), 39–93.
54. Zaki Saleh, *Mesopotamia (Iraq), 1600–1914: A Study in British Foreign Affairs* (Baghdad: Al-Ma'aref Press, 1957), 71.
55. The promotion, according to a British historian, was to protect the East India Company in Iraq from "the jealous bickering of the Frenchman." Longrigg, *Four Centuries of Modern Iraq*, 188. See also IOR/L/P&S/20/C236, Jerome A. Saldanha, "Précis of Turkish Arabia Affairs, 1801–1905," ff. 57r–64r.
56. Abdullah, *Merchants, Mamluks, and Murder*, 51.
57. IOR/L/P&S/20/C91, John Gordon Lorimer, "Gazetteer of the Persian Gulf, 'Oman, and Central Arabia," 1219, 1252–1257.
58. IOR/G/29/21, Henry Moore to the Court of Directors, fol. 4r (16 January 1774); Risso, *Oman and Muscat*, 89. The carrying capacity of each ketch is based on the eyewitness account of Abraham Parsons in 1774, who noted that the two ketches combined had 230 "Turks and Arabs" on board. Parsons, *Travels in Asia and Africa*, 181.
59. Parsons, *Travels in Asia and Africa*, 152, 166.
60. OA, BŞM.BSH 3/106 (19 S 1194/25 February 1780); OA, C.BH 68/3245 (9 RA 1194/15 March 1780); OA, C.BH 230/10714 (9 RA 1194/15 March 1780). See also

IOR/G/29/21, William D. Latouche to the Court of Directors, ff. 388r–390r (5 December 1780).
61. Saldanha, "Précis," fol. 49v; Lorimer, "Gazetteer," 1278.
62. OA, C.AS 304/12580 (18 N 1199/26 July 1785).
63. OA, HAT 93/3799 (17 M 1217/20 May 1802); OA, HAT 241/13544 (n.d.).
64. Robert Stewart Castlereagh, *Correspondence, Despatches, and Other Papers of Viscount Castlereagh*, ed. Charles William Vane (London: William Shoberl, 1851), 5:188–189.
65. Lier, *Haushalte und Haushaltspolitik in Bagdad*, 144.
66. OA, C.AS 304/12580 (18 N 1199/26 July 1785); OA, C.AS 460/19189 (25 B 1211/24 January 1797); OA, C.AS 400/16521 (16 R 1210/30 October 1795); OA, C.DH 214/10661 (19 S 1211/24 August 1796); OA, HAT 391/20761 (5 Ş 1249/18 December 1833); Castlereagh, *Correspondence*, 175–176, 185–188; Saldanha, "Précis," fol. 56r; IOR/G/29/21, William Digges Latouche to the Court of Directors, fol. 397r (23 January 1781); IOR/G/29/21, William Digges Latouche to the Court of Directors, fol. 75v (26 September 1781); Longrigg, *Four Centuries of Modern Iraq*, 254.
67. Ahmed Cevdet, *Tarih-i Cevdet* (Istanbul: Matbaʻa-i Amire, 1288/1871–1872), 8:6. Cited in M. Şükrü Hanioğlu, *A Brief History of the Late Ottoman Empire* (Princeton, NJ: Princeton University Press, 2008), 14.
68. OA, HAT 396/20879 (15 B 1225/16 August 1810); Lier, *Haushalte und Haushaltspolitik in Bagdad*, 162.
69. On the introduction of steam navigation to the Tigris and Euphrates, see Cengiz Orhonlu and Turgut Işıksal, "Osmanlı Devrinde Nehir Nakliyatı Hakkında Araştırmalar: Dicle ve Fırat Nehirlerinde Nakliyat," *Tarih Dergisi* 13 (1962–1963): 100–102; Camille Lyans Cole, "Precarious Empires: A Social and Environmental History of Steam Navigation on the Tigris," *Journal of Social History* 50, no. 1 (2016): 74–101; J. P. Parry, "Steam Power and the British Influence in Baghdad, 1820–1860," *Historical Journal* 56, no. 1 (2013): 145–173; John S. Guest, *The Euphrates Expedition* (London: Kegan Paul International, 1992); M. E. Yapp, "The Euphrates Expedition," in *The Islamic World from Classical to Modern Times: Essays in Honor of Bernard Lewis*, ed. C. E. Bosworth, Charles Issawi, Roger Savory, and A. L. Udovitch (Princeton, NJ: Darwin Press, 1989), 891–915; Ebubekir Ceylan, *The Ottoman Origins of Modern Iraq: Political Reform, Modernization and Development in the Nineteenth-Century Middle East* (New York: I. B. Tauris, 2011), 189–199.
70. OA, MAD 7915, 393 (10 N 1112/18 February 1701); OA, MAD 5433 (6 N 1112/14 February 1701).
71. OA, D.BŞM.TRE 14694 (Evahir Z 1146/25 May–4 June 1734); OA, MAD 9934, 95, 167–168, 197, 246, 262 (25 N 1146/1 March 1734); OA, MAD 9934, 173–174, 177–178, 309 (9 L 1146/15 March 1734); OA, C.AS 922/39875 (Evahir Z 1146/25 May–4 June 1734); OA, C.BH 193/9048 (18 Z 1147/11 May 1735); OA, C.AS 1104/48809 (22 CA 1147/20 October 1734); OA, C.BH 85/4078 (15 Ş 1147/10 January

1735); OA, C.BH 193/9047 (18 CA 1147/16 October 1734); OA, C.BH 222/10313 (26 RA 1156/20 May 1743); OA, MAD 9947, 134 (3 R 1156/27 May 1743); OA, MAD 9947, 375 (15 N 1156/2 November 1743); OA, MAD 9948, 14 (14 S 1157/29 March 1744); OA, MAD 9952, 237 (12 N 1157/18 October 1744); OA, C.BH 238/11063 (12 N 1157/19 October 1744); OA, AE.SMHD.I 181/14113 (2 C 1157/13 July 1744); OA, AE.SMHD.I 129/9464 (23 CA 1157/4 July 1744); OA, D.BŞM 2855 (1158–1159/1745–1746); OA, D.BŞM.TRE 14800 (9 R 1158/11 May 1745); OA, C.AS 709/29745 (10 R 1158/12 May 1745); OA, AE.SMHD.I 233/18716 (9 RA1158/11 April 1745); OA, HAT 7/223 (11 CA 1159/1 June 1746).
72. OA, HAT 6/212 (23 R 1191/31 May 1777).
73. Virginia H. Aksan, "Feeding the Ottoman Troops on the Danube, 1768–1774," *War & Society* 13, no. 1 (1995): 1–14.
74. For the financial crisis of the late eighteenth century, see Caroline Finkel, *Osman's Dream: The Story of the Ottoman Empire, 1300–1923* (New York: Basic Books, 2005), 387–389. For the period's military crisis, see Gábor Ágoston, "Military Transformation in the Ottoman Empire and Russia, 1500–1800," *Kritika: Explorations in Russian and Eurasian History* 12, no. 2 (2011): 281–319. For concurrent efforts to modernize Ottoman land forces, see Virginia H. Aksan, *Ottoman Wars, 1700–1870: An Empire Besieged* (London: Routledge, 2013), 180–213; Yaycioglu, *Partners of the Empire*, 40–63. For the modernization of the Ottoman navy in this period, see Tuncay Zorlu, *Innovation and Empire in Turkey: Sultan Selim III and the Modernisation of the Ottoman Navy* (New York: Tauris Academic Studies, 2008).
75. OA, C.BH 115/5553 (16 L 1234/8 August 1819); OA, C.BH 155/7387 (28 Safar 1233/7 January 1818); OA, C.BH 191/8950 (22 RA 1236/28 December 1820).
76. OA, MD 179/430 (Evail Zilkade 1195/19 October 1781); Lier, *Haushalte und Haushaltspolitik in Bagdad*, 122.
77. Khaled Fahmy, "The Era of Muhammad ʿAli, 1805–1848," in *The Cambridge History of Egypt*, vol. 2, *Modern Egypt from 1517 to the End of the Twentieth Century*, ed. M. W. Daly (Cambridge: Cambridge University Press, 1998), 139–179.
78. On the foreign policies of Osman Pazvantoğlu and Ali Pasha, see Yaycioglu, *Partners of the Empire*, 103–106.

CONCLUSION

1. For comparable analyses from the modern and ancient periods, see Keiko Kiyotaki, *Ottoman Land Reform in the Province of Baghdad* (Leiden: Brill, 2019), 115; Piotr Steinkeller, "City and Countryside in Third-Millennium Southern Babylonia," in *Settlement and Society: Essays Dedicated to Robert McCormick Adams*, ed. Elizabeth C. Stone (Los Angeles: Cotsen Institute of Archaeology, 2007), 185–211.
2. Dina Rizk Khoury, "The Ottoman Centre versus Provincial Power-Holders: An Analysis of the Historiography," in *The Cambridge History of Turkey*, vol. 3, *The*

Later Ottoman Empire, 1603–1839, ed. Suraiya N. Faroqhi (Cambridge: Cambridge University Press, 2006), 155–156; Thomas Lier, *Haushalte und Haushaltspolitik in Bagdad, 1704–1831* (Würzburg: Ergon, 2004), 215–220; Tom Nieuwenhuis, *Politics and Society in Early Modern Iraq: Mamluk Pashas, Tribal Shayks and Local Rule Between 1802 and 1831* (The Hague: Martinus Nijhoff, 1982), 76–81; Percy Kemp, "Mosul and Mosuli Historians of the Jalili Era (1726–1834)" (PhD diss., Oxford University, 1979), 77.

3. Ottoman environmental policies during the nineteenth and early twentieth centuries are documented in Camille Lyans Cole, "Precarious Empires: A Social and Environmental History of Steam Navigation on the Tigris," *Journal of Social History* 50, no. 1 (2016): 74–101; Cole, "Controversial Investments: Trade and Infrastructure in Ottoman-British Relations in Iraq, 1861–1918," *Middle Eastern Studies* 54, no. 5 (2018): 744–768; Rhoads Murphey, "The Ottoman Centuries in Iraq: Legacy or Aftermath? A Survey Study of Mesopotamian Hydrology and Ottoman Irrigation Projects," *Journal of Turkish Studies* 11 (1987): 24–27; Isacar A. Bolaños, "The Ottomans during the Global Crises of Cholera and Plague: The View from Iraq and the Gulf," *International Journal of Middle East Studies* 51, no. 4 (2019): 603–620; Cengiz Orhonlu and Turgut Işıksal, "Osmanlı Devrinde Nehir Nakliyatı Hakkinda Araştırmalar: Dicle ve Fırat Nehirlerinde Nakliyat," *Tarih Dergisi* 13 (1962–1963): 100–102; Ebubekir Ceylan, *The Ottoman Origins of Modern Iraq: Political Reform, Modernization and Development in the Nineteenth-Century Middle East* (New York: I. B. Tauris, 2011), 189–201; Kiyotaki, *Ottoman Land Reform*, 59–107. On Ottoman policies toward nomadic populations during the same period, see Samuel Dolbee, "The Locust and the Starling: People, Insects, and Disease in the Late Ottoman Jazira and After, 1860–1940" (PhD diss., New York University, 2017), 104–208; Selim Deringil, "'They Live in a State of Nomadism and Savagery': The Late Ottoman Empire and the Post-Colonial Debate," *Comparative Studies in Society and History* 45, no. 2 (2003): 311–342; Reşat Kasaba, *A Moveable Empire: Ottoman Nomads, Migrants, and Refugees* (Seattle: University of Washington Press, 2009), 84–122; Ceylan, *The Ottoman Origins of Modern Iraq*, 132–152. I am grateful to Samuel Dolbee for sharing his dissertation with me.

4. For a comprehensive history of the Tigris-Euphrates basin in the twentieth century, see Dale Stahl, "The Two Rivers: Water, Development and Politics in the Tigris-Euphrates Basin, 1920–1975" (PhD diss., Columbia University, 2014). On GAP, see Brahma Chellaney, *Water, Peace, and War: Confronting the Global Water Crisis* (Lanham, MD: Rowman and Littlefield, 2013), 197–202; Leila M. Harris, "State as Socionatural Effect: Variable and Emergent Geographies of the State in Southeastern Turkey," *Comparative Studies of South Asia, Africa and the Middle East* 32, no. 1 (2012): 25–39; Harris, "Water and Conflict Geographies of the Southeastern Anatolia Project," *Society and Natural Resources* 15, no. 8 (2002): 743–759; Nilgun Harmancioğlu, Necdet Alpaslan, and Eline Boelee, *Irrigation, Health and Environment: A Review of Literature from Turkey* (Colombo, Sri

Lanka: International Water Management Institute, 2001), 11–20. On Hasankeyf, see Julia Harte, "New Dam in Turkey Threatens to Flood Ancient City and Archaeological Sites," *National Geographic*, February 21, 2014; Ali Kucukgocmen, "'History Disappears' as Dam Waters Flood Ancient Turkish Town," Reuters, February 25, 2020, https://www.reuters.com/article/us-turkey-dam/history-disappears-as-dam-waters-flood-ancient-turkish-town-idUSKBN20J1TW. I am grateful to Debra Javeline for directing me to some of these references and for educating me about the global water crisis today.

5. Harte, "New Dam in Turkey"; Federico Borsari and Irene Pasqua, eds., *MENA's Fertile Crescent in the Time of Dry Geopolitics* (Milan: Istituto Per Gli Studi Di Politica Internazionale, 2020); Pieter-Jan Dockx, *Water Scarcity in Iraq: From Inter-Tribal Conflict to International Disputes* (New Delhi: Institute of Peace and Conflict Studies, 2019); M. Nouar Shamout and Glada Lahn, *The Euphrates in Crisis: Channels of Cooperation for a Threatened River* (London: Chatham House, 2015); Peter Beaumont, "Restructuring of Water Usage in the Tigris-Euphrates Basin: The Impact of Modern Water Management Policies," in *Transformations of Middle Eastern Natural Environments: Legacies and Lessons*, ed. Jeff Albert, Magnus Bernhardsson, and Roger Kenna (New Haven, CT: Yale School of Forestry and Environmental Studies, 1998), 168–186; Campbell Robertson, "Iraq, a Land between 2 Rivers, Suffers as One of Them Dwindles," *New York Times*, July 14, 2009, A1. I am grateful to Julia Harte for taking the time to discuss with me her trip along the Tigris River in 2013, documenting the region's rapid environmental and social transformations for *National Geographic*.

6. Jennifer Hattam, "Dammed, Dirty, Drained by War: Can Iraq's Tigris River Be Restored?" *Christian Science Monitor*, October 17, 2013.

7. "A Mesopotamian Odyssey," *The Economist*, October 22, 2013.

8. *New Oxford American Dictionary*, s.v. "rival," accessed June 30, 2020, https://www.lexico.com/en/definition/rival.

Bibliography

ARCHIVAL SOURCES

Osmanlı Arşivi (Istanbul)

Ali Emiri
　Abdülhamid I: 92/6294, 230/15255, 281/18900, 325/22163, 325/22164, 325/22165
　Ahmed I: 1/33
　Ahmed II: 2/115, 10/983, 21/2231, 21/2244, 21/2252, 22/2266
　Ahmed III: 3/221, 4/277, 20/1864, 23/2146, 28/2710, 32/3048, 43/4218, 51/5150, 58/5791, 69/6921, 85/8567, 88/8829, 91/9069, 95/9361, 106/10451, 106/10452, 106/10453, 118/11598, 126/12348, 126/12443, 131/12837, 140/13613, 146/14315, 156/15350, 159/15597, 161/15809, 165/16168, 165/16169, 176/17090, 176/17112, 179/17428, 181/17643, 186/18038, 194/18761, 198/19169, 210/20291, 210/20292, 213/20609, 218/21002
　İbrahim: 5/560
　Mahmud I: 85/5750, 129/9464, 181/14113, 192/14977, 233/18716
　Mahmud II: 2/177, 3/313, 6/619, 9/900, 17/1929, 23/2579, 95/11151, 96/11302, 100/11727
　Mustafa II: 2/171, 12/1115, 12/1148, 13/1225, 16/1567, 83/8940, 96/10361, 116/12627
　Mustafa III: 130/10098, 131/10183, 174/13634, 189/14887, 297/23775
　Osman III: 58/4192, 69/5223, 86/6647
　Selim III: 84/5094, 132/8022
　Süleyman II: 23/2324

Bab-ı Defteri Başmuhasebe Kalemi
　Defterler: 976, 2855, 3383
　Evrak: 7651/36, 7651/51, 7653/81, 7654/47, 7654/71

Bağdat Hazinesi
　Defterler: 16735
　Evrak: 1/17, 1/25, 1/26, 1/27, 1/29, 1/56, 1/59, 2/28, 3/16, 3/52

Basra Hazinesi
　Evrak: 3/106

Diyarbakır Hazinesi
　Evrak: 14/54

Musul Hazinesi
　Evrak: 1/88, 1/190, 2/103, 2/149, 2/191, 3/118, 4/184, 6/196, 8/1, 8/19, 8/33, 8/36, 8/41, 8/44

Tersane-i Amire Eminliği
　Defterler: 14598, 14694, 14800
　Evrak: 3/8

Cevdet
　Adliye: 85/5107
　Askeriye: 71/3355, 124/5554, 304/12580, 400/16521, 460/19189, 572/24061, 709/29745, 786/33308, 922/39875, 954/41442, 977/42578, 1104/48809
　Bahriye: 1/7, 10/914, 50/2377, 66/3118, 68/3245, 83/3978, 85/4078, 90/4305, 104/5049, 115/5553, 149/7106, 155/7387, 191/8950, 193/9047, 193/9048, 222/10313, 230/10714, 238/11063
　Dahiliye: 31/1544, 214/10661
　Evkaf: 51/2522, 63/3117, 122/6078, 244/12198, 249/12464
　Hariciye: 68/3397
　Maliye: 522/21342, 753/30679, 522/21342, 221/9191
　Maarif: 1/42, 169/8420
　Nafia: 38/1891

Hatt-ı Hümayun: 6/212, 7/223, 93/3799, 241/13544, 391/20761, 396/20879, 397/20896

İbnülemin
　Askeriye: 54/4873, 78/7027, 81/7334
　Bahriye: 7/632
　Dahiliye: 9/880, 15/1395, 20/1831

Evkaf: 1/29, 1/39, 2/196, 2/205, 6/628, 8/964, 12/1467, 14/1607, 15/1818, 16/1884, 17/2061, 17/2079, 21/2506, 23/2696, 23/2731, 24/2883, 28/3215, 30/3424, 30/3504, 32/3717, 34/3897, 38/4386, 39/4429, 45/5064, 46/5222, 46/5232, 48/5341, 49/5448, 51/5617, 52/5734, 52/5781, 53/5830, 56/6192, 59/6446, 59/6448, 61/6690

Hatt-ı Hümayun: 4/348

Maliye: 5/353, 56/5309, 59/5587, 107/10135

Sıhhiye: 1/76

Tevcihat: 15/1717, 23/2462

Kamil Kepeci: 79

Maliyeden Müdevver Defterler: 46, 75, 966, 975, 2510, 2737, 2775, 2915, 2926, 2931, 2933, 3134, 3242, 3595, 3871, 4117, 4879, 5433, 7915, 9885, 9891, 9934, 9947, 9948, 9952, 9968, 9970, 10151, 18537, 18540

Mühimme Defterleri: 3, 5, 6, 7, 10, 12, 16, 19, 22, 24, 25, 26, 27, 30, 32, 33, 38, 40, 46, 53, 69, 78, 87, 104, 105, 106, 108, 111, 112, 114–1, 179

Tapu Tahrir Defterleri: 184, 200, 276, 282, 386, 397, 496, 501, 534, 582, 660, 667, 1028, 1073

Tapu ve Kadastro Genel Müdürlüğü, Kuyud-ı Kadime Arşivi (Ankara)

Tapu Tahrir Defterleri: 29, 30, 228, 379, 386

Topkapı Sarayı Müzesi Arşivi (Istanbul)

Defterler: 708, 5852, 9508
Evrak: 301/14, 643/72, 813/44, 890/32, 1046/44, 1067/55

India Office Records (London)

East India Company Factory Records: 29/21
Political and Secret Department Records: 9/98, 20/C91, 20/C236

The National Archives (London)

Foreign Office: 922/74

MANUSCRIPT LIBRARIES

İstanbul Üniversitesi Nadir Eserler Kütüphanesi (Istanbul)

Matrakçı Nasuh. "Beyan-ı Menazil-i Sefer-i Irakeyn." T.5964.

Süleymaniye Kütüphanesi (Istanbul)

Murtaza Nazmizade. "Tarih-i Seferü'l-Basra." Esad Efendi 2062/3.
Murtaza Nazmizade. "İcmal-i Sefer-i Nehr-i Ziyab." Esad Efendi 2062/4.

Topkapı Sarayı Müzesi Kütüphanesi (Istanbul)

"Dastan-i Sultan Süleyman." Revan Köşkü 1286.
"Hükümname Mecmuası." Koğuşlar 888.
Silahdar Fındıklılı Mehmed Ağa. "Cild-i Salis Tarih-i Fındıklılı." Emanet Hazinesi 1413.

Österreichische Nationalbibliothek (Vienna)

Matrakçı Nasuh. "Tarih-i Al-i Osman." Cod. Mixt. 339 Han.

British Library (London)

Ahmad Ghurabzade. "Uyun Akhbar al-A'yan bi-man Mada fi Salif al-'Asr wa-l-Zaman." Add. 23309.
"Correspondence, mainly of Sir Harford Jones, with 1st and 2nd Viscounts Melville: 1785-1820." Add. 41767.
Lokman bin Hüseyin. "Mücmel-ul-Tumar." Or. 1135.
Sipahizade Mehmed Efendi. "Evdahu'l-Mesalik ila Ma'rifeti'l-Büldan ve'l-Memalik." Add. 23381.

Bibliothèque Nationale de France (Paris)

al-Omari, Yasin Ibn Khayrallah. "Al-Durr al-Maknun fi al-Ma'athir al-Madiya min al-Qurun." Arabe 4949.

Qatar National Library (Doha)

HC.MAP 1.

James Ford Bell Library, University of Minnesota (Minneapolis)

Bembo, Ambrosio. "Viaggio e giornale per parte dell' Asia di quattro anni incicra fatto." Bell 1676 fBe.

PUBLISHED PRIMARY SOURCES

Akgündüz, Ahmed. *Osmanlı Kanunnameleri ve Hukuki Tahlilleri*. 9 vols. Istanbul: FEY Vakfı, 1990–1996.

al-Alusi, Mahmud Shukri. *Akhbar Baghdad wa ma Jawaraha min al-Bilad*. Edited by Imad Abdussalam Ra'uf. Beirut: al-Dar al-Arabiyya li'l-Mawsuʿat, 2008.

Ainsworth, William Francis. *Researches in Assyria, Babylonia, and Chaldea*. London: John W. Parker, 1838.

"Annual Administration Report, Shamiyah Division, from 1st January to 31st December 1918." In *Iraq Administration Reports 1914–1932*. 10 vols. Slough, UK: Archive Editions, 1992.

Anonim Osmanlı Tarihi, 1099–1116/168–1704. Edited by Abdülkadir Özcan. Ankara: Türk Tarih Kurumu, 2000.

Aşık Mehmed. *Menazırü'l-Avalim*. Edited by Mahmut Ak. 3 vols. Ankara: Türk Tarih Kurumu, 2007.

Beawes, William. "Remarks and Occurrences in a Journey from Aleppo to Bassora, by the Way of the Desert." In *The Desert Route to India: Being the Journals of Four Travellers by the Great Desert Caravan Route between Aleppo and Basra, 1745–1751*, edited by Douglas Carruthers, 1–40. London: Hakluyt Society, 1929.

Cameron, Verney Lovett. *Our Future Highway*. 2 vols. London: Macmillan, 1880.

Carmichael, John. "A Journey from Aleppo to Basra in 1751." In *The Desert Route to India: Being the Journals of Four Travellers by the Great Desert Caravan Route between Aleppo and Basra, 1745–1751*, edited by Douglas Carruthers, 129–179. London: Hakluyt Society, 1929.

Cartwright, John. "Observations of Master John Cartwright in his Voyage from Aleppo to Hispaan, and backe againe: published by himselfe, and here contracted." In *Hakluytus Posthumus: Or Purchas His Pilgrimes: Contayning a History of the World in Sea Voyages and Lande Travells by Englishmen and Others*, edited by Samuel Purchas, 8:482–523. Glasgow: James MacLehose and Sons, 1905.

Castlereagh, Robert Stewart. *Correspondence, Despatches, and Other Papers of Viscount Castlereagh*. Edited by Charles William Vane. 12 vols. London: William Shoberl, 1851.

Cevdet, Ahmed. *Tarih-i Cevdet*. Istanbul: Matbaʿa-i Amire, 1288/1871–1872.

de Beauchamp, M. "Voyage de Bagdad à Bassora le long de l'Euphrate." *Le Journal des Sçavans* (1785): 285–303.

Defterdar Sarı Mehmed Paşa. *Zübdet-i Vekayiat: Tahlil ve Metin, 1066–1116/1656–1704*. Edited by Abdülkadir Özcan. Ankara: Türk Tarih Kurumu, 1995.

de Thévenot, Jean. *The Travels of Monsieur de Thévenot into the Levant in Three Parts*. Translated by Archibald Lovell. 3 pts. London: Printed by H. Clark, H. Faithorne, J. Adamson, C. Skegnes, and T. Newborough, 1687.

de Thévenot, Jean. *Suite du Voyage de Mr. de Thévenot au Levant*. Paris: Angot, 1689.

della Valle, Pietro. *Viaggi di Pietro della valle il Pellegrino*. 4 vols. Rome: n.p., 1650.

della Valle, Pietro. *The Travels of Sig. Pietro della Valle, A Noble Roman, into East-India and Arabia Deserta, In Which, the Several Countries, Together with the Customs,*

Manners, Traffique, and Rites both Religious and Civil, of Those Oriental Princes and Nations, Are Faithfully Described. Translated by G. Havers. London: Printed by J. Macock, 1665.

Eldred, M. John. "The Voyage of M. John Eldred to Tripolis in Syria by Sea, and from thence by Land and River to Babylon, and Balsara, Anno 1583." In *The Principal Navigations, Voyages, Traffiques & Discoveries of the English Nations*, edited by Richard Hakluyt, 3:321–328. London: J. M. Dent, 1907.

Evliya Çelebi. *Evliya Çelebi Seyahatnamesi: Topkapı Sarayı Bağdat 304 Yazmasının Transkripsiyonu-Dizini*. Edited by Yücel Dağlı and Seyit Ali Kahraman. 7 vols. Istanbul: Yapı Kredi Yayınları, 2001.

Feridun Bey. *Nüzhet-i esrarü'l-ahyar der-ahbar-ı sefer-i Sigetvar: Sultan Süleyman'ın son seferi*. Edited by H. Ahmet Arslantürk, Günhan Börekçi, and Abdülkadir Özcan. Istanbul: Zeytinburnu Belediyesi, 2012.

Fraser, J. Baillie. *Travels in Koordistan, Mesopotamia, & c., including an Account of Parts of Those Countries hitherto Unvisited by Europeans with Sketches of the Character and Manners of the Koordish and Arab Tribes.* 2 vols. London: Richard Bentley, 1840.

Fraser, J. Baillie. *Mesopotamia and Assyria from the Earliest Ages to the Present Time, With Illustrations of Their Natural History*. New York: Harper and Brothers, 1845.

Fulanian. *The Marsh Arab: Haji Rikkan*. Philadelphia: J. B. Lippincott, 1928.

Godinho, Manuel. *Relação do novo caminho que fez por terra e mar: vindo da India para Portugal, no anno de 1663.* Lisbon: Sociedade propagadora dos conhecimentos uteis, 1842.

Herodotus. *The Landmark Herodotus: The Histories*. Edited by Robert B. Strassler. Translated by Andrea L. Purvis. New York: Anchor Books, 2009.

Hurşid Paşa, Mehmed. *Seyahatname-i Hudud*. Edited by Alaattin Eser. Istanbul: Simurg, 1997.

Irwin, Eyles. *A Series of Adventures in the Course of a Voyage Up the Red-Sea, on the Coasts of Arabia and Egypt; and of a Route through the Desarts of Thebais, in the Year 1777.* 3rd ed. 2 vols. London: Printed for J. Dodsley, 1787.

Ives, Edward. *A Voyage from England to India, in the Year 1754, and an Historical Narrative of the Operations of the Squadron and Army in India, under the Command of Vice-Admiral Watson and Colonel Clive, in the Years 1755, 1756, 1757*. London: Printed for Edward and Charles Dilly, 1773.

Jackson, John. *Journey from India, Towards England, in the Year 1797*. London: Printed for T. Cadell, Jun. and W. Davies, Strand, 1799.

Katib Çelebi. *Cihannüma*. Istanbul: Darü't-Tıbaati'l-Amire, 1145/1732.

Katib Çelebi. *Fezleke-i Katib Çelebi*. 2 vols. Istanbul: Ceride-i Havadis Matbaası, 1286/1869.

Katib Çelebi. *Tuhfetü'l-Kibar fi Esfari'l-Bihar*. Istanbul: Matbaa-i Bahriye, 1329/1911.

Manwaring, George. *The Three Brothers; or, The Travels and Adventures of Sir Anthony, Sir Robert, and Sir Thomas Sherley, in Persia, Russia, Turkey, Spain, etc.* London: Printed for Hurst, Robinson, & Co., 1825.

al-Maqdisi, Muhammad. *Ahsan al-Taqasim fi Ma'rifat al-Aqalim*. Leiden: Brill, 1906.

Matrakçı Nasuh. *Beyan-i Menazil-i Sefer-i Irakeyn-i Sultan Süleyman Han*. Edited by Hüseyin G. Yurdaydın. Ankara: Türk Tarih Kurumu, 1976.

Murchio, Vincenzo Maria. *Il viaggio all' Indie Orientali*. Rome: Filippomaria Mancini, 1672.

Naima, Mustafa. *Tarih-i Naima*. 2 vols. Istanbul: Darü't-Tıbaati'l-Amire, 1147/1734–1735.

Nazmizade, Murtaza. *Gülşen-i Hulefa*. Edited by Mehmet Karataş. Ankara: Türk Tarih Kurumu, 2014.

Newberie, John. "Two Voyages of Master John Newberie, One into the Holy Land; The Other to Balsara, Ormus, Persia, and Backe Thorow Turkie." In *Hakluytus Posthumus; or, Purchas His Pilgrimes: Contayning a History of the World in Sea Voyages and Lande Travells by Englishmen and Others*, edited by Samuel Purchas, 8:449–481. Glasgow: James MacLehose and Sons, 1905.

Niebuhr, Carsten. *Reisebeschreibung nach Arabien und andern umliegenden Ländern*. 2 vols. Copenhagen: N. Möller, 1774–1778.

Olivier, Guillaume Antoine. *Voyage dans l'Empire Othoman, l'Égypte et la Perse: Fait par ordre du Gouvernement, pendant les six premières années de la République*. 6 vols. Paris: Chez H. Agasse, 1801–1807.

Parsons, Abraham. *Travels in Asia and Africa*. London: Longman, Hurst, Rees, and Orme, 1808.

Peçevi İbrahim. *Tarih-i Peçevi*. 2 vols. Istanbul: Matbaa-i Amire, 1283/1866–1867.

Porter, Sir Robert Ker. *Travels in Georgia, Persia, Armenia, Ancient Babylonia, & c. & c. during the Years 1817, 1818, 1819, and 1820*. 2 vols. London: Longman, Hurst, Rees, Orme, and Brown, 1821–1822.

Raşid Mehmed Efendi. *Tarih-i Raşid ve Zeyli*. Edited by Abdülkadir Özcan, Yunus Uğur, Baki Çakır, Ahmet Zeki İzgöer. 3 vols. Istanbul: Klasik, 2013.

Rauwolff, Dr. Leonhart. "Travels into the Eastern Countries." Vol. 1, *A Collection of Curious Travels and Voyages*, edited by John Ray. London: Royal Society, 1693.

Rich, Claudius James. *Narrative of a Residence in Koordistan, and on the Site of Ancient Nineveh; with Journal of a Voyage Down the Tigris to Bagdad and an Account of a Visit to Shirauz and Persepolis*. 2 vols. London: Juames Duncan, 1836.

Rousseau, Jean Baptiste Louis Jacques. *Description du Pachalik de Bagdad*. Paris: Treuttel et Würtz, 1809.

Ryley, J. Horton. *Ralph Fitch, England's Pioneer to India and Burma; His Companions and Contemporaries, with His Remarkable Narrative Told in His Own Words*. London: T. F. Unwin, 1899.

Sestini, Domenico. *Voyage de Constantinople à Bassora en 1781 par le Tigre et l'Euphrate, et retour à Contantinople en 1782, par le desert et Alexandrie*. Paris: Chez Dupuis, 1798.

Silahdar Fındıklılı Mehmet Ağa. *Silahdar Tarihi*. 2 vols. Istanbul: Orhaniye Matbaası, 1928.

al-Suwaydi, Abdulrahman. *Hadiqat al-Zawra fi Sirat al-Wuzara*. Edited by Imad Abdussalam Ra'uf. Baghdad: Manshurat al-Majma al-Ilmi, 2003.

Tavernier, Jean-Baptiste. *Les Six Voyages de Jean Baptiste Tavernier*. 2 vols. Paris: G. Clouzier et C. Barbin, 1676.
Taylor, John. *Travels from England to India, In the Year 1789*. London: S. Low, 1799.
Teixeira, Pedro. *The Travels of Pedro Teixeira*. Translated by William F. Sinclair. London: Hakluyt Society, 1902.

UNPUBLISHED SECONDARY SOURCES

Dolbee, Samuel. "The Locust and the Starling: People, Insects, and Disease in the Late Ottoman Jazira and After, 1860–1940." PhD diss., New York University, 2017.
al-Duri, Khidr. "Society and Economy of Iraq under the Seljuqs (1055–1160 A.D.)." PhD diss., University of Pennsylvania, 1970.
Gülcü, Erdinç. "Osmanlı İdaresinde Bağdat (1534–1623)." PhD diss., Fırat Üniversitesi, 1999.
Güngörürler, Selim. "Diplomacy and Political Relations between the Ottoman Empire and Safavid Iran, 1639–1722." PhD diss., Georgetown University, 2016.
Haider, S. "Land Problems of Iraq." PhD diss., University of London, 1942.
Husain, Faisal H. "The Tigris-Euphrates Basin under Early Modern Ottoman Rule, c. 1534–1830." PhD diss., Georgetown University, 2018.
Kemp, Percy. "Mosul and Mosuli Historians of the Jalili Era (1726–1834)." PhD diss., University of Oxford, 1979.
Murphey, Rhoads. "The Functioning of the Ottoman Army under Murad IV (1623–1639/1032/1049): Key to the Understanding of the Relationship between Center and Periphery in Seventeenth-Century Turkey." PhD diss., University of Chicago, 1979.
Pournelle, Jennifer R. "Marshland of Cities: Deltaic Landscapes and the Evolution of Early Mesopotamian Civilization." PhD diss., University of California, San Diego, 2003.
Rost, Stephanie. "Watercourse Management and Political Centralization in Third-Millennium B.C. Southern Mesopotamia: A Case Study of the Umma Province of the Ur III Period (2112–2004 B.C.)." PhD diss., Stony Brook University, 2015.
Stahl, Dale. "The Two Rivers: Water, Development and Politics in the Tigris-Euphrates Basin, 1920–1975." PhD diss., Columbia University, 2014.
Williams, Elizabeth Rachel. "Cultivating Empires: Environment, Expertise, and Scientific Agriculture in Late Ottoman and French Mandate Syria." PhD diss., Georgetown University, 2015.
Yılmaz, Ali. "XVI. Yüzyılda Birecik Sancağı." PhD diss., İstanbul Üniversitesi, 1996.

PUBLISHED SECONDARY SOURCES

Abdullah, Thabit A. J. *Merchants, Mamluks, and Murder: The Political Economy of Trade in Eighteenth-Century Basra*. Albany: State University of New York Press, 2001.

Abdulhusayn, Al Tuʿma. *Bughyat an-Nubala fi Tarikh Karbala.* Baghdad: Matbaʿat al-Irshad, 1966.

Abou-el-Haj, Rifaat Ali. "The Social Uses of the Past: Recent Arab Historiography of Ottoman Rule." *International Journal of Middle East Studies* 14, no. 2 (1982): 185–201.

Abu-Lughod, Janet L. *Before European Hegemony: The World System A.D. 1250–1350.* New York: Oxford University Press, 1989.

Adams, Robert McC. *Land behind Baghdad: A History of Settlement on the Diyala Plains.* Chicago: University of Chicago Press, 1965.

Adams, Robert McC. "Historic Patterns of Mesopotamian Irrigation Agriculture." In *Irrigation's Impact on Society,* edited by Adams, Theodore E. Downing, and McGuire Gibson, 1–5. Tucson: University of Arizona Press, 1974.

Adams, Robert McC. *Heartland of Cities: Surveys of Ancient Settlement and Land Use on the Central Floodplain of the Euphrates.* Chicago: University of Chicago Press, 1981.

Adams, Robert McC. "Designed Flexibility in a Sewn Boat of the Western Indian Ocean." In *Sewn Plank Boats: Archaeological and Ethnographic Papers Based on Those Presented to a Conference at Greenwich in November, 1984,* edited by Sean McGrail and Eric Kentley, 289–302. Oxford: BAR, 1985.

Adams, Robert McC. "Intensified Large-Scale Irrigation as an Aspect of Imperial Policy: Strategies of Statecraft on the Late Sasanian Mesopotamian Plain." In *Agricultural Strategies,* edited by Joyce Marcus and Charles Stanish, 17–37. Los Angeles: Cotsen Institute of Archaeology, 2006.

Adams, Robert McC., and Hans J. Nissen. *The Uruk Countryside: The Natural Setting of Urban Societies.* Chicago: University of Chicago Press, 1972.

Adanır, Fikret. "Semi-Autonomous Provincial Forces in the Balkans and Anatolia." In *The Later Ottoman Empire, 1603–1839.* Vol. 3 of *The Cambridge History of Turkey,* edited by Suraiya N. Faroqhi, 157–185. Cambridge: Cambridge University Press, 2006.

Ágoston, Gábor. "The Costs of the Ottoman Fortress-System in Hungary in the Sixteenth and Seventeenth Centuries." In *Ottomans, Hungarians, and Habsburgs in Central Europe: The Military Confines in the Era of Ottoman Conquest,* edited by Pál Fodor and Géza Dávid, 195–228. Leiden: Brill, 2000.

Ágoston, Gábor. "A Flexible Empire: Authority and Its Limits on the Ottoman Frontiers." *International Journal of Turkish Studies* 9, no. 1/2 (2003): 15–31.

Ágoston, Gábor. *Guns for the Sultan: Military Power and the Weapons Industry in the Ottoman Empire.* New York: Cambridge University Press, 2005.

Ágoston, Gábor. "Where Environmental and Frontier Studies Meet: Rivers, Forests, Marshes and Forts along the Ottoman-Hapsburg Frontier in Hungary." In *The Frontiers of the Ottoman World,* edited by A. C. S. Peacock, 57–79. Oxford: Oxford University Press, 2009.

Ágoston, Gábor. "Military Transformation in the Ottoman Empire and Russia, 1500–1800." *Kritika: Explorations in Russian and Eurasian History* 12, no. 2 (2011): 281–319.

Aksan, Virginia H. "Feeding the Ottoman Troops on the Danube, 1768–1774." *War & Society* 13, no. 1 (1995): 1–14.

Aksan, Virginia H. *Ottoman Wars, 1700–1870: An Empire Besieged*. London: Routledge, 2013.

Aksan, Virginia H. "What's Up in Ottoman Studies?" *Journal of the Ottoman and Turkish Studies Association* 1, no. 1/2 (2014): 3–21.

Algaze, Guillermo. *Ancient Mesopotamia at the Dawn of Civilization: The Evolution of an Urban Landscape*. Chicago: University of Chicago Press, 2008.

Ali, Ali Shakir. *Tarikh al-Iraq fi al-Ahd al-Othmani, 1638–1750 Miladiyya/1048–1174 Hijriyya: Dirasa fi Ahwalihi al-Siyasiyya*. Nineveh: Maktabat 30 Tammuz, 1985.

Allan, J. David, and María M. Castillo. *Stream Ecology: Structure and Function of Running Waters*. 2nd ed. Dordrecht: Springer, 2007.

Allsen, Thomas T. *The Royal Hunt in Eurasian History*. Philadelphia: University of Pennsylvania Press, 2006.

Altaweel, Mark. "Simulating the Effects of Salinization on Irrigation Agriculture in Southern Mesopotamia." In *Models of Mesopotamian Landscapes: How Small-Scale Processes Contributed to the Growth of Early Civilizations*, edited by T. J. Wilkinson, McGuire Gibson, and Magnus Widell, 219–238. Oxford: Archaeopress, 2013.

Altaweel, Mark. "Southern Mesopotamia: Water and the Rise of Urbanism." *Wiley Interdisciplinary Reviews: Water* 6, no. 4 (2019).

Altaweel, Mark, Anke Marsh, Jaafar Jotheri, Carrie Hritz, Dominik Fleitmann, Stephanie Rost, Stephen F. Lintner, McGuire Gibson, Matthew Bosomworth, Matthew Jacobson, Eduardo Garzanti, Mara Limonta, and Giuditta Radeff. "New Insights on the Role of Environmental Dynamics Shaping Southern Mesopotamia: From the Pre-Ubaid to the Early Islamic Period." *Iraq* 81 (2019): 23–46.

Anderson, Ewan W. *The Middle East: Geography and Geopolitics*. London: Routledge, 2000.

Anderson, Virginia DeJohn. *Creatures of Empire: How Domestic Animals Transformed Early America*. New York: Oxford University Press, 2004.

Armstrong, Philip. *Sheep*. London: Reaktion Books, 2016.

Ashtor, E. *A Social and Economic History of the Near East in the Middle Ages*. Berkeley: University of California Press, 1976.

Ataman, Bekir Kemal. "Ottoman Demographic History (14th–17th Centuries): Some Considerations." *Journal of the Economic and Social History of the Orient* 35, no. 2 (1992): 187–198.

Axworthy, Michael. "Nader Shah and Persian Naval Expansion in the Persian Gulf, 1700–1747." *Journal of the Royal Asiatic Society* 21, no. 1 (2011): 31–39.

Aydın, Suavi, and Oktay Özel. "Power Relations Between State and Tribe in Ottoman Eastern Anatolia." *Bulgarian Historical Review* 34, no. 3–4 (2006): 51–67.

al-Azzawi, Abbas. *Mawsuʻat Tarikh al-Iraq Bayna Ihtilalayn*. 8 volumes. Beirut: al-Dar al-ʻArabiyya li'l-Mawsuʻat, 2004.

al-Azzawi, Abbas. *Mawsuʿat Ashaʾir al-Iraq.* 4 vols. Beirut: al-Dar al-Arabiyya li'l-Mawsuʿat, 2005.

Ball, Philip. *Life's Matrix: A Biography of Water.* New York: Farrar, Straus and Giroux, 2001.

Ball, Philip. *Flow.* Vol. 2 of *Nature's Patterns: A Tapestry in Three Parts.* New York: Oxford University Press, 2009.

Ballato, Paolo, Cornelius E. Uba, Angela Landgraf, Manfred R. Strecker, Masafumi Sudo, Daniel F. Stockli, Anke Friedrich, and Saeid H. Tabatabaei. "Arabia-Eurasia Continental Collision: Insights from Late Tertiary Foreland-Basin Evolution in the Alborz Mountains, Northern Iran." *Geological Society of America Bulletin* 123, no. 1/2 (2011): 106–131.

Barfield, Thomas J. *The Nomadic Alternative.* Englewood Cliffs, NJ: Prentice Hall, 1993.

Barkan, Ömer Lutfi. "Essai sur les données statistiques des registres de recensement dans l'Empire ottoman aux XVe et XVIe siècles." *Journal of the Economic and Social History of the Orient* 1, no. 1 (1957): 9–36.

Barkan, Ömer Lutfi. "Research on the Ottoman Fiscal Surveys." In *Studies in the Economic History of the Middle East: From the Rise of Islam to the Present Day*, edited by M. A. Cook, 163–171. London: Oxford University Press, 1970.

Barkey, Karen. *Empire of Difference: The Ottomans in Comparative Perspective.* New York: Cambridge University Press, 2008.

Başarır, Özlem. "Diyarbekir Voyvodası Mustafa Ağa'nın Terekesi Üzerıne Bazı Düşünceler." *Bilig* 65 (2013): 23–46.

Batatu, Hanna. *The Old Social Classes and the Revolutionary Movements of Iraq.* Princeton, NJ: Princeton University Press, 1978.

Beaumont, Peter. "Restructuring of Water Usage in the Tigris-Euphrates Basin: The Impact of Modern Water Management Policies." In *Transformations of Middle Eastern Natural Environments: Legacies and Lessons*, edited by Jeff Albert, Magnus Bernhardsson, and Roger Kenna, 168–186. New Haven, CT: Yale School of Forestry and Environmental Studies, 1998.

Beaumont, Peter, Gerald H. Blake, and J. Malcolm Wagstaff. *The Middle East: A Geographical Study.* 2nd ed. New York: Halsted Press, 1988.

Bishko, Charles Julian. "The Castilian as Plainsman: The Medieval Ranching Frontier in La Mancha and Extremadura." In *The New World Looks at Its History*, edited A. R. Lewis and T. F. McGann, 47–69. Austin: University of Texas Press, 1963.

Boesch, Hans H. "El-ʿIraq." *Economic Geography* 15, no. 4 (1939): 325–361.

Bolaños, Isacar A. "The Ottomans during the Global Crises of Cholera and Plague: The View from Iraq and the Gulf." *International Journal of Middle East Studies* 51, no. 4 (2019): 603–620.

Borisenkov, Yevgeny P. "Climatic and other Natural Extremes in the European Territory of Russia in the Late Maunder Minimum (1675–1715)." In *Climatic Trends and Anomalies in Europe 1675–1715*, edited by Burkhard Frenzel, Christian Pfister, and Birgit Gläser, 83–94. New York: G. Fischer, 1994.

Borsari, Federico, and Irene Pasqua, eds. *MENA's Fertile Crescent in the Time of Dry Geopolitics*. Milan: Istituto Per Gli Studi Di Politica Internazionale, 2020.

Bostan, İdris. *Osmanlı Bahriye Teşkilatı: XVII. Yüzyılda Tersane-i Amire*. Ankara: Türk Tarih Kurumu, 1992.

Bostan, İdris. "Birecik." *Türkiye Diyanet Vakfı Ansiklopedisi* 6 (1992): 187–189.

Bostan, İdris. *Kürekli ve Yelkenli Osmanlı Gemileri*. Istanbul: Bilge, 2005.

Bostan, İdris. "Imperial Arsenal." *Encyclopaedia of Islam*. 3rd ed. Leiden: Brill Online, 2016.

Boyar, Ebru. "Ottoman Expansion in the East." In *The Cambridge History of Turkey*. Vol. 2 of *The Ottoman Empire as a World Power, 1453–1603*, edited by Suraiya N. Faroqhi and Kate Fleet, 74–140. Cambridge: Cambridge University Press, 2006.

Braudel, Fernand. *The Mediterranean and the Mediterranean World in the Age of Phillip II*. Translated by Siân Reynolds. 2 vols. New York: Harper & Row, 1972.

Brice, William C. *South-West Asia*. London: University of London Press, 1966.

Brooke, John L. *Climate Change and the Course of Global History: A Rough Journey*. New York: Cambridge University Press, 2014.

Brummett, Palmira. *Ottoman Seapower and Levantine Diplomacy in the Age of Discovery*. Albany: State University of New York Press, 1994.

Bull, William B. "Threshold of Critical Power in Streams." *Geological Society of America Bulletin* 90, no. 5 (1979): 453–464.

Burak, Guy. *The Formation of Islamic Law: The Hanafi School in the Early Modern Ottoman Empire*. New York: Cambridge University Press, 2015.

Buringh, Piet. *Soils and Soil Conditions in Iraq*. Baghdad: Ministry of Agriculture, 1960.

Burke, Kenneth. *The Philosophy of Literary Form: Studies in Symbolic Action*. Baton Rouge: Louisiana State University Press, 1941.

Burnham, Philip. "Spatial Mobility and Political Centralization in Pastoral Societies." In *Pastoral Production and Society*, edited by L'Equipe écologie et anthropologie des societies pastorales, 349–360. New York: Cambridge University Press, 1979.

Campopiano, Michele. "State, Land Tax and Agriculture in Iraq from the Arab Conquest to the Crisis of the Abbasid Caliphate (Seventh-Tenth Centuries)." *Studia Islamica* 3 (2012): 5–50.

Campopiano, Michele. "Cooperation and Private Enterprise in Water Management in Iraq: Continuity and Change between the Sasanian and Early Islamic Periods (Sixth to Tenth Centuries)." *Environment and History* 23, no. 3 (2017): 385–407.

Campbell, Brian. *Rivers and the Power of Ancient Rome*. Chapel Hill: University of North Carolina Press, 2012.

Candiani, Vera S. *Dreaming of Dry Land: Environmental Transformation in Colonial Mexico City*. Stanford, CA: Stanford University Press, 2014.

Casale, Giancarlo. "The Ethnic Composition of Ottoman Ship Crews and the 'Rumi Challenge' to Portuguese Identity." *Medieval Encounters* 13 (2007): 122–144.

Casale, Giancarlo. *The Ottoman Age of Exploration*. New York: Oxford University Press, 2010.

Casale, Giancarlo. "The Islamic Empires of the Early Modern World." In *The Construction of a Global World, 1400–1800*, edited by Jerry H. Bentley, Sanjay Subrahmanyam, and Merry E. Wiesner-Hanks, 323–344. Vol. 6 of *The Cambridge World History*. Cambridge: Cambridge University Press, 2015.

Casson, Lionel. *Ships and Seamanship in the Ancient World.* Princeton, NJ: Princeton University Press, 1971.

Ceylan, Ebubekir. *The Ottoman Origins of Modern Iraq: Political Reform, Modernization and Development in the Nineteenth-Century Middle East.* New York: I. B. Tauris, 2011.

Chang, Te-Tzu. "Rice." In *The Cambridge World History of Food*, edited by Kenneth F. Kiple and Kriemhild Coneè Ornelas, 132–149. New York: Cambridge University Press, 2008.

Charles, M. P. "Onions, Cucumbers and the Date Palm: An Introduction to the Cultivation of Alliaceae, Cucurbitaceae and Fruit Trees in Modern Iraq." *Bulletin on Sumerian Agriculture* 3 (1987): 1–21.

Charles, M. P. "Irrigation in Lowland Mesopotamia." *Bulletin on Sumerian Agriculture* 4 (1988): 1–39.

Chase, Kenneth. *Firearms: A Global History to 1700*. New York: Cambridge University Press, 2003.

Chellaney, Brahma. *Water, Peace, and War: Confronting the Global Water Crisis.* Lanham, MD: Rowman and Littlefield, 2013.

Chiha, Habib K. *La Province de Bagdad: Son Passé, Son Présent, Son Avenir.* Cairo: El-Maaref, 1908.

Chorley, Richard J., and Barbara A. Kennedy. *Physical Geography: A Systems Approach.* London: Prentice-Hall International, 1971.

Christensen, Peter. *The Decline of Iranshahr: Irrigation and Environments in the History of the Middle East, 500 B.C. to A.D. 1500.* Copenhagen: Museum Tusculanum Press, 1993.

Christensen, Peter. "Middle Eastern Irrigation: Legacies and Lessons." In *Transformations of Middle Eastern Natural Environments: Legacies and Lessons*, edited by Jeff Albert, Magnus Bernhardsson, and Roger Kenna, 15–30. New Haven, CT: Yale School of Forestry and Environmental Studies, 1998.

Christophers, S. R., and H. E. Shortt. "Malaria in Mesopotamia." *Indian Journal of Medical Research* 8, no. 3 (1921): 508–552.

Church, Michael, Tim P. Burt, Victor J. Galay, and G. Mathias Kondolf. "Rivers." In *Geomorphology and Global Environmental Change*, edited by Olav Slaymaker, Thomas Spencer, and Christine Embleton-Hamann, 98–129. New York: Cambridge University Press, 2009.

Cipolla, Carlo M. *Guns, Sails and Empires: Technological Innovation and the Early Phases of European Expansion, 1400–1700.* New York: Pantheon Books, 1965.

Ciriacono, Salvatore. *Building on Water: Venice, Holland and the Construction of the European Landscape in Early Modern Times.* Translated by Jeremy Scott. New York: Berghahn Books, 2006.

Çizakça, Murat. "The Ottoman Empire: Recent Research on Shipping and Shipbuilding in the Sixteenth to Nineteenth Centuries." In *Maritime History at the Crossroads: A Critical Review of Recent Historiography*, edited by Frank Broeze, 213–228. St. John's, Canada: International Maritime Economic History Association, 1995.

Cockrill, W. Ross. "The Working Buffalo." In *The Husbandry and Health of the Domestic Buffalo*, edited by W. Ross Cockrill, 313–328. Rome: Food and Agriculture Organization of the United Nations, 1974.

Cole, Camille Lyans. "Precarious Empires: A Social and Environmental History of Steam Navigation on the Tigris." *Journal of Social History* 50, no. 1 (2016): 74–101.

Cole, Camille Lyans. "Controversial Investments: Trade and Infrastructure in Ottoman-British Relations in Iraq, 1861–1918." *Middle Eastern Studies* 54, no. 5 (2018): 744–768.

Cole, Steven W. "Marsh Formation in the Borsippa Region and the Course of the Lower Euphrates." *Journal of Near Eastern Studies* 53, no. 2 (1994): 81–109.

Cole, Steven W., and Hermann Gasche. "Second- and First-Millennium BC Rivers in Northern Babylonia." In *Changing Watercourses in Babylonia: Towards a Reconstruction of the Ancient Environment in Lower Mesopotamia*, edited by Hermann Gasche and Michel Tanret, 1–64. Ghent: University of Ghent and the Oriental Institute of the University of Chicago, 1998.

Colten, Craig E. "Fluid Geographies: Urbanizing River Basins." In *Urban Rivers: Remaking Rivers, Cities, and Space in Europe and North America*, edited by Stéphane Castonguay and Matthew Evenden, 201–218. Pittsburgh: University of Pittsburgh Press, 2012.

Contenau, Georges. *Everyday Life in Babylon and Assyria*. Translated by K. R. and A. R. Maxwell-Hyslop. New York: St. Martin's Press, 1954.

Cook, Michael. "The Long-Term Geopolitics of the Pre-Modern Middle East." *Journal of the Royal Asiatic Society* 26, no.1/2 (2016): 33–41.

Cressey, George B. "The Shatt al-Arab Basin." *Middle East Journal* 12, no. 4 (1958): 448–460.

Cressey, George B. *Crossroads: Land and Life in Southwest Asia*. Chicago: J. B. Lippincott, 1960.

Cribb, Roger. *Nomads in Archaeology*. New York: Cambridge University Press, 1991.

Cunha, João Teles e. "The Portuguese Presence in the Persian Gulf." In *The Persian Gulf in History*, edited by Lawrence G. Potter, 207–234. New York: Palgrave Macmillan, 2009.

Cuno, Kenneth M. *The Pasha's Peasants: Land, Society, and Economy in Lower Egypt, 1740–1858*. New York: Cambridge University Press, 1992.

Coşgel, Metin M. "Ottoman Tax Registers (*Tahrir Defterleri*)." *Historical Methods* 37, no. 2 (2004): 87–100.

Cvetkova, Bistra. "Early Ottoman Tahrir Defters as a Source on the History of Bulgaria and the Balkans." *Archivum Ottomanicum* 8 (1983): 133–213.

Dağlı, Murat. "The Limits of Ottoman Pragmatism." *History and Theory* 52, no. 2 (2013): 194–213.

Damluji, Mona, Arbella Bet-Shlimon, Alda Benjamen, Saleem Al-Baloly, Haytham Bahoora, Caecilia Pieri, Bridget L. Guarasci, Zainab Saleh, and Peter Sluglett. "Roundtable: Perspectives on Researching Iraq Today." *Arab Studies Journal* 23, no. 1 (2015): 236–265.

Darling, Linda T. *Revenue-Raising and Legitimacy: Tax Collection and Finance Administration in the Ottoman Empire, 1560–1660.* Leiden: E. J. Brill, 1996.

Darling, Linda T. "Political Change and Political Discourse in the Early Modern Mediterranean World." *Journal of Interdisciplinary History* 38, no. 4 (2008): 505–531.

Darling, Linda T. *A History of Social Justice and Political Power in the Middle East: The Circle of Justice from Mesopotamia to Globalization.* New York: Routledge, 2012.

D'Arrigo, Rosanne, and Heidi M. Cullen. "A 350-Year (AD 1628–1980) Tree-Ring Record of Turkish Precipitation: Linkages to Tigris-Euphrates Streamflow and the NAO." *Dendrochronologia* 19 (2001): 169–177.

Davies, D. Hywel. "Observations on Land Use in Iraq." *Economic Geography* 33, no. 2 (1957): 122–134.

Demos, John. *Circles and Lines: The Shape of Life in Early America.* Cambridge, MA: Harvard University Press, 2004.

Deringil, Selim. "'They Live in a State of Nomadism and Savagery': The Late Ottoman Empire and the Post-Colonial Debate." *Comparative Studies in Society and History* 45, no. 2 (2003): 311–342.

El-Dessouky, F. I. "Iraq." In *Buffalo Production*, edited by N. M. Tulloh and J. H. H. Holmes, 81–94. Amsterdam: Elsevier, 1992.

Di Palma, Vittoria. *Wasteland: A History.* New Haven, CT: Yale University Press, 2014.

Dinç, Fasih. "Osmanlı Diyarbakır'ında Keleğin Yapımı ve Kullanımı." In *Osmanlıdan Günümüze Diyarbakır*, edited by İbrahim Özcoşar, Ali Karakaş, Mustafa Öztürk, and Ziya Polat, 61–100. Istanbul: Ensar, 2018.

Dockx, Pieter-Jan. *Water Scarcity in Iraq: From Inter-Tribal Conflict to International Disputes.* New Delhi: Institute of Peace and Conflict Studies, 2019.

Doumani, Beshara. *Rediscovering Palestine: Merchants and Peasants in Jabal Nablus, 1700–1900.* Berkeley: University of California Press, 1995.

Doumanis, Nicholas. "Durable Empire: State Virtuosity and Social Accommodation in the Ottoman Mediterranean." *Historical Journal* 49, no. 3 (2006): 953–966.

Dowson, V. H. W. *Dates and Date Cultivation of the Iraq.* 3 pts. Cambridge: W. Heffer and Sons, 1921–1923.

Drower, E. S. "The Arabs of the Hor Al Hawaiza." In *The Lower Euphrates-Tigris Region*, edited by Henry Field, 368–406. Part 1, number 2 of *The Anthropology of Iraq*. Chicago: Field Museum of Natural History, 1949.

Dunne, Thomas, and Luna B. Leopold. *Water in Environmental Planning.* San Francisco: W. H. Freeman, 1978.

Dyson-Hudson, Rada, and Neville Dyson-Hudson. "Nomadic Pastoralism." *Annual Review of Anthropology* 9 (1980): 15–61.

Eddy, John A. "Solar History and Human Affairs." *Human Ecology* 22 (1994): 23–35.

Eger, A. Asa. "The Swamps of Home: Marsh Formation and Settlement in the Early Medieval Near East." *Journal of Near Eastern Studies* 70, no. 1 (2011): 55–79.

Erder, Leila. "The Measurement of Preindustrial Population Changes: The Ottoman Empire from the 15th to the 17th Centuries." *Middle Eastern Studies* 11, no. 3 (1975): 284–301.

Esmer, Tolga U. "Economies of Violence, Banditry and Governance in the Ottoman Empire around 1800." *Past and Present* 224, no. 1 (2014): 163–199.

Fagan, Brian M. *Floods, Famines, and Emperors: El Niño and the Fate of Civilizations.* New York: Basic Books, 2009.

Fahmy, Khaled. "The Era of Muhammad ʿAli, 1805–1848." In *Modern Egypt from 1517 to the End of the Twentieth Century.* Vol. 2 of *The Cambridge History of Egypt*, edited by M. W. Daly, 139–179. Cambridge: Cambridge University Press, 1998.

Northedge, Alastair, T. J. Wilkinson, and Robin Falkner. "Survey and Excavations at Samarra 1989." *Iraq* 50 (1990): 121–147.

Faroqhi, Suraiya. "Camels, Wagons, and the Ottoman State in the Sixteenth and Seventeenth Centuries." *International Journal of Middle East Studies* 14, no. 4 (1982): 523–539.

Faroqhi, Suraiya. *Towns and Townsmen of Ottoman Anatolia: Trade, Crafts, and Food Production in an Urban Setting, 1520–1650.* New York: Cambridge University Press, 1984.

Faroqhi, Suraiya. "Crisis and Change, 1590–1699." In *1600–1914*, edited by Halil İnalcık and Donald Quataert, 411–636. Vol. 2 of *An Economic and Social History of the Ottoman Empire.* Cambridge: Cambridge University Press, 1997.

Faroqhi, Suraiya. *Approaching Ottoman History: An Introduction to the Sources.* Cambridge: Cambridge University Press, 1999.

Faroqhi, Suraiya. *The Ottoman Empire and the World Around It.* London: I. B. Tauris, 2007.

Faroqhi, Suraiya. *Artisans of Empire: Crafts and Craftspeople Under the Ottomans.* New York: I. B. Tauris, 2009.

Fattah, Hala. *The Politics of Regional Trade in Iraq, Arabia, and the Gulf, 1745–1900.* Albany: State University of New York Press, 1997.

al-Feel, Muhammad Rashid. *The Historical Geography of Iraq between the Mongolian and Ottoman Conquests, 1258–1534.* Najaf: al-Adab Press, 1965.

Field, Thomas G. *Scientific Farm Animal Production: An Introduction to Animal Science.* 10th ed. Upper Saddle River, NJ: Prentice Hall, 2012.

Finkel, Caroline. *Osman's Dream: The Story of the Ottoman Empire, 1300–1923.* New York: Basic Books, 2007.

Finkelstein, J. J. "An Old Babylonian Herding Contract and Genesis 31:38." *Journal of the American Oriental Society* 88, no. 1 (1968), 30–36.

Flohr, Pascal, Dominik Fleitmann, Eduardo Zorita, Aleksey Sadekov, Hai Cheng, Matt Bosomworth, Lawrence Edwards, Wendy Matthews, and Roger Matthews. "Late Holocene Droughts in the Fertile Crescent Recorded in a Speleothem from Northern Iraq." *Geophysical Research Letters* 44, no. 3 (2017): 1528–1536.

Floor, Willem. "The Iranian Navy in the Persian Gulf during the Eighteenth Century." *Iranian Studies* 20, no. 1 (1987): 31–53.

Floor, Willem. *The Persian Gulf: A Political and Economic History of Five Port Cities, 1500–1730*. Washington, DC: Mage, 2006.

Floor, Willem. "The Rise and Fall of the Banu Kaʿb: A Borderer State in Southern Khuzestan." *Iran* 44 (2006): 277–315.

Floor, Willem. "Dutch Relations with the Persian Gulf." In *The Persian Gulf in History*, edited by Lawrence G. Potter, 235–259. New York: Palgrave Macmillan, 2009.

Flores, Dan. "Place: An Argument for Bioregional History." *Environmental History Review* 18, no. 4 (1994): 1–18.

Frey, Perry A., and George H. Reed. "The Ubiquity of Iron." *ACS Chemical Biology* 7, no. 9 (2012): 1474–1476.

Gabriel, Richard A., and Karen S. Metz. *From Sumer to Rome: The Military Capabilities of Ancient Armies*. New York: Greenwood Press, 1991.

Gellner, Ernest. "Tribalism and the State in the Middle East." In *Tribes and State Formation in the Middle East*, edited by Philip S. Khoury and Joseph Kostiner, 109–126. Berkeley: University of California Press, 1990.

Genç, Mehmet. "A Study of the Feasibility of Using Eighteenth-Century Ottoman Financial Records as an Indicator of Economic Activity." In *The Ottoman Empire and the World-Economy*, edited by Huri İslamoğlu-İnan, 345–373. Cambridge: Cambridge University Press, 1987.

Genç, Mehmet. "Osmanlı İktisadi Dünya Görüşünün İlkeleri." *Sosyoloji Dergisi* 3, no. 1 (1989): 175–185.

Gibson, McGuire. "Population Shift and the Rise of Mesopotamian Civilisation." In *The Exploration of Culture Change: Models in Prehistory*, edited by Colin Renfrew, 447–463. London: Duckworth, 1973.

Gilmartin, David. *Blood and Water: The Indus River Basin in Modern History*. Oakland: University of California Press, 2015.

Glete, Jan. *Navies and Nations: Warships, Navies, and State Building in Europe and America, 1500–1860*. 2 vols. Stockholm: Almqvist and Wiksell International, 1993.

Goffman, Daniel, and Christopher Stroop. "Empire as Composite: The Ottoman Polity and the Typology of Dominion." In *Imperialisms: Historical and Literary Investigations, 1500–1900*, edited by Balachandra Rajan and Elizabeth Sauer, 129–145. New York: Palgrave Macmillan, 2004.

Goudie, Andrew S. *Arid and Semi-arid Geomorphology*. New York: Cambridge University Press, 2013.

Goody, Jack. *Metals, Culture and Capitalism: An Essay on the Origins of the Modern World*. New York: Cambridge University Press, 2012.

Göyünç, Nejat. "Diyarbekir Beylerbeyiliği'nin İlk İdari Taksimatı." *Tarih Dergisi* 23 (1969): 23–34.
Göyünç, Nejat. "'Hane' Deyimi Hakkında." *Tarih Dergisi* 32 (1979): 331–348.
Göyünç, Nejat. "Dicle ve Fırat Nehirlerinde Nakliyat." *Belleten* 65, no. 243 (2001): 655–660.
Gradeva, Rossitsa. "War and Peace Along the Danube: Vidin at the End of the Seventeenth Century." *Oriente Moderno* 81, no. 1 (2001): 149–175.
Gradeva, Rossitsa. *War and Peace in Rumeli: 15th to Beginning of 19th Century.* Istanbul: Isis Press, 2008.
Graff, Gerald, and Cathy Birkenstein. *"They Say/I Say": The Moves That Matter in Academic Writing.* 3rd ed. New York: W. W. Norton, 2014.
Grant, Jonathan. "Rethinking the Ottoman 'Decline': Military Technology Diffusion in the Ottoman Empire, Fifteenth to Eighteenth Centuries." *Journal of World History* 10, no. 1 (1999): 179–201.
Greco, Angela. "The Taming of the Wilderness: Marshes as an Economic Resource in 3rd Millennium BC Southern Mesopotamia." *Water History* 12, no. 1 (2020): 23–38.
Greene, Molly. "An Islamic Experiment? Ottoman Land Policy on Crete." *Mediterranean Historical Review* 11, no. 1 (1996): 60–78.
Griffiths, R. B. "Parasites and Parasitic Diseases." In *The Husbandry and Health of the Domestic Buffalo*, edited by W. Ross Cockrill, 236–275. Rome: Food and Agriculture Organization of the United Nations, 1974.
Guest, E. R. "The Rustam Herbarium, Iraq: Part VI. General and Ecological Account." *Kew Bulletin* 8, no. 3 (1953): 396–401.
Guest, John S. *The Euphrates Expedition.* London: Kegan Paul International, 1992.
Guilmartin, John F. Jr. *Galleons and Galleys.* London: Cassel, 2002.
Gradeva, Rossitsa. "The Military Revolution in Warfare at Sea during the Early Modern Era: Technological Origins, Operational Outcomes and Strategic Consequences." *Journal of Maritime Research* 13, no. 2 (2011): 129–137.
Gündüz, Ahmet. *Osmanlı İdaresinde Musul (1523–1639).* Elazığ: Fırat Üniversitesi Basımevi, 2003.
Güney, Emrullah. "Dicle Irmağında Kelek Taşımacılığı." *Coğrafya Araştırmaları* 2, no. 2 (1990): 323–328.
Gümüşçü, Osman. "The Ottoman *Tahrir Defters* as a Source for Historical Geography." *Belleten* 72, no. 265 (2008): 911–941.
Hammer, Emily Louise, and Benjamin S. Arbuckle. "10,000 Years of Pastoralism in Anatolia: A Review of Evidence for Variability in Pastoral Lifeways." *Nomadic Peoples* 21, no. 2 (2017): 214–267.
Hammer, Emily Louise, and Benjamin S. Arbuckle. "Water Management by Mobile Pastoralists in the Middle East." In *Water and Power in Past Societies*, edited by Emily Holt, 63–88. Albany: State University of New York Press, 2018.
Hanioğlu, M. Şükrü. *A Brief History of the Late Ottoman Empire.* Princeton, NJ: Princeton University Press, 2008.

Harmancioğlu, Nilgun, Necdet Alpaslan, and Eline Boelee. *Irrigation, Health and Environment: A Review of Literature from Turkey*. Colombo, Sri Lanka: International Water Management Institute, 2001.

Harris, Leila M. "Water and Conflict Geographies of the Southeastern Anatolia Project." *Society and Natural Resources* 15, no. 8 (2002): 743–759.

Harris, Leila M. "State as Socionatural Effect: Variable and Emergent Geographies of the State in Southeastern Turkey." *Comparative Studies of South Asia, Africa and the Middle East* 32, no. 1 (2012): 25–39.

Harris, Leila M., and Samer Alatout. "Negotiating Hydro-Scales, Forging States: Comparison of the Upper Tigris/Euphrates and Jordan River Basins." *Political Geography* 29, no. 3 (2010): 148–156.

Harte, Julia. "New Dam in Turkey Threatens to Flood Ancient City and Archaeological Sites." *National Geographic*, February 21, 2014.

Hathaway, Jane. *The Arab Lands under Ottoman Rule, 1516–1800*. Abingdon, UK: Routledge, 2013.

Hattam, Jennifer. "Dammed, Dirty, Drained by War: Can Iraq's Tigris River Be Restored?" *Christian Science Monitor*, October 17, 2013.

Headrick, Daniel R. *Power over Peoples: Technology, Environments, and Western Imperialism, 1400 to the Present*. Princeton, NJ: Princeton University Press, 2010.

Hegyi, Klára. "The Ottoman Networks of Fortresses in Hungary." In *Ottomans, Hungarians, and Habsburgs in Central Europe: The Military Confines in the Era of Ottoman Conquest*, edited by Pál Fodor and Géza Dávid, 163–193. Leiden: Brill, 2000.

Hess, Andrew C. "The Ottoman Conquest of Egypt (1517) and the Beginning of the Sixteenth-Century World War." *International Journal of Middle East Studies* 4, no. 1 (1973): 55–76.

Heyvaert, Vanessa Mary An, and Cecile Baeteman. "A Middle to Late Holocene Avulsion History of the Euphrates River: A Case Study from Tell ed-Der, Iraq, Lower Mesopotamia." *Quaternary Science Reviews* 27, no. 25–26 (2008): 2401–2410.

Heywood, Colin. "Between Historical Myth and 'Mythohistory': The Limits of Ottoman History." *Byzantine and Modern Greek Studies* 12 (1988): 315–345.

Hodder, Ian. *The Present Past: An Introduction to Anthropology for Archaeologists*. Barnsley, UK: Pen & Sword Archaeology, 2012.

Hodgson, Marshall G. S. *The Venture of Islam: Conscience and History in a World Civilization*. 3 vols. Chicago: University of Chicago Press, 1974.

Hourani, George. *Arab Seafaring in the Indian Ocean in Ancient and Early Medieval Times*. Princeton, NJ: Princeton University Press, 1995.

Howard-Johnston, James. "The Two Great Powers in Late Antiquity: A Comparison." In *States, Resources, and Armies*, edited by Averil Cameron, 157–226. Vol. 3 of *The Byzantine and Early Islamic Near East*. Princeton, NJ: Darwin Press, 1992.

Hritz, Carrie, and T. J. Wilkinson. "Using Shuttle Radar Topography to Map Ancient Water Channels in Mesopotamia." *Antiquity* 80, no. 308 (2006): 415–424.

Huart, Clément. *Histoire de Bagdad dans les temps modernes.* Paris: E. Leroux, 1901.

Hunt, Robert C. "Size and the Structure of Authority in Canal Irrigation Systems." *Journal of Anthropological Research* 44, no. 4 (1988): 335–355.

Husain, Faisal H. "In the Bellies of the Marshes: Water and Power in the Countryside of Ottoman Baghdad." *Environmental History* 19, no. 4 (2014): 638–664.

İlhan, M. Mehdi. "The Katif District (*Liva*) during the First Few Years of Ottoman Rule: A Study of the 1551 Ottoman Cadastral Survey." *Belleten* 51, no. 200 (1987): 781–798.

İlhan, M. Mehdi. "The Process of Ottoman Cadastral Surveys during the Second Half of the Sixteenth Century: A Study Based on Documents from Muhimme Defters." *A. D. Xenopol* 24, no. 1 (1987): 17–25.

İlhan, M. Mehdi. "XVI. Yüzyılın İlk Yarısında Diyarbakır Şehrinin Nüfusu ve Vakıfları: 1518 ve 1540 Tarihli Tapu Tahrir Defterlerinden Notlar." *Tarih Araştırmaları Dergisi* 16, no. 27 (1994): 45–113.

Imber, Colin. "The Persecution of the Ottoman Shiʿites according to the Mühimme Defterleri, 1565–1585." *Der Islam* 56 (1979): 245–273.

Imber, Colin. "The Navy of Süleyman the Magnificent." *Archivum Ottomanicum* 6 (1980): 211–282.

İnalcık, Halil. "Ottoman Methods of Conquest." *Studia Islamica* 2 (1954): 103–129.

İnalcık, Halil. *The Ottoman Empire: The Classical Age, 1300–1600.* Translated by Norman Itzkowitz and Colin Imber. New York: Praeger, 1973.

İnalcık, Halil. "The Question of the Emergence of the Ottoman State." *International Journal of Turkish Studies* 2 (1980): 71–79.

İnalcık, Halil. "Rice Cultivation and the Çeltük-Reaya System in the Ottoman Empire." *Turcica* 14 (1982): 59–141.

İnalcık, Halil. "'Arab' Camel Drivers in Western Anatolia in the Fifteenth Century." *Revue d'Histoire Maghrebine* 10, no. 31/32 (1983): 247–270.

İnalcık, Halil. "The Yörüks: Their Origins, Expansion and Economic Role." In *Carpets of the Mediterranean Countries, 1400–160*, edited by Robert Pinner and Walter B. Denny, 39–65. London: Hali Magazine, 1986.

İnalcık, Halil. "Osman Ghazi's Siege of Nicaea and the Battle of Bapheus." In *The Ottoman Emirate (1300–1389)*, edited by Elizabeth Zachariadou, 77–99. Rethymnon: Crete University Press, 1993.

İnalcık, Halil. "The Ottoman State: Economy and Society, 1300–1600." In *1300–1600*, edited by Halil İnalcık and Donald Quataert, 9–409. Vol. 1 of *An Economic and Social History of the Ottoman Empire, 1300–1914.* New York: Cambridge University Press, 1994.

İnalcık, Halil. "Weights and Measures." In *1300–1600*, edited by Halil İnalcık and Donald Quataert, xxxvii–xliv. Vol. 1 of *An Economic and Social History of the Ottoman Empire, 1300–1914.* New York: Cambridge University Press, 1994.

Irons, William. "Nomadism as a Political Adaptation: The Case of the Yomut Turkmen." *American Ethnologist* 1, no. 4 (1974): 635–658.

Ionides, M. G. *The Régime of the Rivers, Euphrates and Tigris.* London: E. and F. N. Spon, 1937.

Işıksal, Turgut. "Gunpowder in Ottoman Documents of the Last Half of the 16th Century." *International Journal of Turkish Studies* 2, no. 2 (1981–1982): 81–91.

İslamoğlu-İnan, Huri. *State and Peasant in the Ottoman Empire: Agrarian Power Relations and Regional Economic Development in Ottoman Anatolia during the Sixteenth Century.* Leiden: Brill, 1994.

Ivanova, Svetlana. "Ali Pasha: Sketches from the Life of a Kapudan Pasha on the Danube." In *The Kapudan Pasha: His Office and His Domain*, edited by Elizabeth Zachariadou, 325–345. Rethymnon: Crete University Press, 2002.

Jacobsen, Thorkild, and Robert M. Adams. "Salt and Silt in Ancient Mesopotamian Agriculture." *Science* 128, no. 3334 (1958): 1251–1258.

Jotheri, Jaafar. "Recognition Criteria for Canals and Rivers in the Mesopotamian Floodplain." In *Water Societies and Technologies from the Past and Present*, edited by Yijie Zhuang and Mark Altaweel, 111–126. London: UCL Press, 2018.

Jotheri, Jaafar, Mark B. Allen, and Tony J. Wilkinson. "Holocene Avulsions of the Euphrates River in the Najaf Area of Western Mesopotamia: Impacts on Human Settlement Patterns." *Geoarchaeology* 31, no. 3 (2016): 175–193.

Kafadar, Cemal. *Between Two Worlds: The Construction of the Ottoman State.* Berkeley: University of California Press, 1995.

Kagan-Zur, Varda, Nurit Roth-Bejerano, Yaron Sitrit, and Asunción Morte, eds. *Desert Truffles: Phylogeny, Physiology, Distribution and Domestication.* Heidelberg: Springer, 2014.

Káldy-Nagy, J. "The Administration of the *Sanjaq* Registrations in Hungary." *Acta Orientalia Academiae Scientiarum Hungaricae* 21, no. 2 (1968): 181–223.

Karakaya-Stump, Ayfer. *The Kizilbash-Alevis in Ottoman Anatolia: Sufism, Politics and Community.* Edinburgh: Edinburgh University Press, 2019.

Karamustafa, Ahmet T. "Military, Administrative, and Scholarly Maps and Plans." In *Cartography in the Traditional Islamic and South Asian Societies*, edited by J. B. Harley and David Woodward, 209–227. Vol. 2, bk. 1 of *The History of Cartography*. Chicago: University of Chicago Press, 1987.

Kasaba, Reşat. *A Moveable Empire: Ottoman Nomads, Migrants, and Refugees.* Seattle: University of Washington Press, 2009.

Kay, H. D. "Milk and Milk Production." In *The Husbandry and Health of the Domestic Buffalo*, edited by W. Ross Cockrill, 329–376. Rome: Food and Agriculture Organization of the United Nations, 1974.

Keddy, Paul A. *Wetland Ecology: Principles and Conservation.* 2nd ed. New York: Cambridge University Press, 2010.

Kennedy, Hugh. "The Feeding of the Five Hundred Thousand: Cities and Agriculture in Early Islamic Mesopotamia." *Iraq* 73 (2011): 177–199.

Khazanov, Anatoly M. *Nomads and the Outside World.* Translated by Julia Crookenden. 2nd ed. Madison: University of Wisconsin Press, 1994.

Khoury, Dina Rizk. "The Introduction of Commercial Agriculture in the Province of Mosul and Its Effects on the Peasantry, 1750–1850." In *Landholding and Commercial Agriculture in the Middle East*, edited by Çağlar Keyder and Faruk Tabak, 155–171. Albany: State University of New York Press, 1991.

Khoury, Dina Rizk. *State and Provincial Society in the Ottoman Empire: Mosul, 1540–1834*. New York: Cambridge University Press, 1997.

Khoury, Dina Rizk. "The Ottoman Centre versus Provincial Power-Holders: An Analysis of the Historiography." In *The Later Ottoman Empire, 1603–1839*, edited by Suraiya N. Faroqhi, 135–156. Vol. 3 of *The Cambridge History of Turkey*. Cambridge: Cambridge University Press, 2006.

Kiyotaki, Keiko. *Ottoman Land Reform in the Province of Baghdad*. Leiden: Brill, 2019.

Koshnaw, Renas I., Daniel F. Stockli, and Fritz Schlunegger. "Timing of the Arabia-Eurasia Continental Collision—Evidence from Detrital Zicron U-Pb Geochronology of the Red Bed Series Strata of the Northwest Zagros Hinterland, Kurdistan Region of Iraq." *Geology* 47, no. 1 (2018): 47–50.

Kucukgocmen, Ali. "'History Disappears' as Dam Waters Flood Ancient Turkish Town." Reuters, February 25, 2020, https://www.reuters.com/article/us-turkey-dam/history-disappears-as-dam-waters-flood-ancient-turkish-town-idUSKBN20J1TW.

Kunt, İ. Metin. "An Ottoman Imperial Campaign: Suppressing the Marsh Arabs, Central Power and Peripheral Rebellion in the 1560s." *Journal of Ottoman Studies* 43 (2014): 1–18.

Kurşun, Zekeriya. "Does the Qatar Map of the Tigris and Euphrates belong to Evliya Çelebi?" *Osmanlı Araştırmaları* 39 (2012): 1–15.

Laiou, Sophia. "The Levends of the Sea in the Second Half of the 16th Century: Some Considerations." *Archivum Ottomanicum* 23 (2005–2006): 233–247.

Landry, Donna. *Noble Brutes: How Eastern Horses Transformed English Culture*. Baltimore: Johns Hopkins University Press, 2009.

Lapidus, Ira M. "Tribes and State Formation in Islamic History." In *Tribes and State Formation in the Middle East*, edited by Philip S. Khoury and Joseph Kostiner, 25–47. Berkeley: University of California Press, 1990.

Lee, John S. "Postwar Pines: The Military and the Expansion of State Forests in Post-Imjin Korea, 1598–1684." *Journal of Asian Studies* 77, no. 2 (2018): 319–332.

Lee, Wayne E. *Waging War: Conflict, Culture, and Innovation in World History*. New York: Oxford University Press, 2016.

Lees, Susan H., and Daniel G. Bates. "The Origins of Specialized Nomadic Pastoralism: A Systemic Model." *American Antiquity* 39, no. 2 (1974): 187–193.

Lewis, Bernard. "The Ottoman Archives as a Source for the History of the Arab Lands." *Journal of the Royal Asiatic Society* 83, no. 3–4 (1951): 139–155.

Khoury, Dina Rizk. "Studies in the Ottoman Archives—I." *Bulletin of the School of Oriental and African Studies* 16, no. 3 (1954): 469–501.

Khoury, Dina Rizk. "The Mongols, the Turks and the Muslim Polity." *Transactions of the Royal Historical Society* 18 (1968): 49–68.

Lier, Thomas. *Haushalte und Haushaltspolitik in Bagdad, 1704–1831.* Würzburg: Ergon, 2004.

Lindner, Rudi Paul. *Nomads and Ottomans in Medieval Anatolia.* Bloomington: Research Institute for Inner Asian Studies, Indiana University, 1983.

Lindner, Rudi Paul. *Explorations in Ottoman Prehistory.* Ann Arbor: University of Michigan Press, 2007.

Linton, Jamie. "Is the Hydrologic Cycle Sustainable? A Historical-Geographical Critique of a Modern Concept." *Annals of the Association of American Geographers* 98, no. 3 (2008): 630–649.

Liverani, Mario. *The Ancient Near East: History, Society and Economy.* Translated by Soraia Tabatabai. New York: Routledge, 2014.

Longrigg, Stephen. *Four Centuries of Modern Iraq.* Oxford: Oxford University Press, 1925.

Lowry, Heath W. *Studies in Defterology: Ottoman Society in the Fifteenth and Sixteenth Centuries.* Istanbul: Isis Press, 1992.

Lowry, Heath W. *The Nature of the Early Ottoman State.* Albany: State University of New York Press, 2003.

Luterbacher, J., R. Rickli, E. Xoplaki, C. Tinguely, C. Beck, C. Pfister, and H. Wanner. "The Late Maunder Minimum (1675–1715)—A Key Period for Studying Decadal Scale Climatic Change in Europe." *Climatic Change* 49, no. 4 (2001): 441–462.

Lytle, David A., and N. LeRoy Poff. "Adaptation to Natural Flow Regimes." *Trends in Ecology and Evolution* 19, no. 2 (2004): 94–100.

MacFarquhar, Neil. "Beneath Desert Sands, an Eden of Truffles." *New York Times*, April 14, 2004.

Mahadevan, P. "Distribution, Ecology and Adaptation." In *Buffalo Production*, edited by N. M. Tulloh and J. H. H. Holmes, 1–12. Amsterdam: Elsevier, 1992.

al-Mahbuba, Ja'far. *Madhi an-Najaf wa Hadhiruha.* 2nd ed. 3 vols. Beirut: Dar al-Adwa, 1986.

Marino, John A. *Pastoral Economics in the Kingdom of Naples.* Baltimore: Johns Hopkins University Press, 1988.

Mason, Laura. *Pine.* London: Reaktion Books, 2013.

Masters, Bruce. "Aleppo: The Ottoman Empire's Caravan City." In *The Ottoman City between East and West: Aleppo, Izmir, and Istanbul*, edited by Edhem Eldem, Daniel Goffman, and Bruce Masters, 17–78. Cambridge: Cambridge University Press, 1999.

Matthee, Rudi. "Unwalled Cities and Restless Nomads: Firearms and Artillery in Safavid Iran." In *Safavid Persia: The History and Politics of an Islamic Society*, edited by Charles Melville, 389–416. New York: I. B. Tauris, 1996.

Matthee, Rudi. "The Portuguese Presence in the Persian Gulf: An Overview." In *Imperial Crossroads: The Great Powers and the Persian Gulf*, edited by Jeffrey R. Macris and Saul Kelly, 3–11. Annapolis, MD: Naval Institute Press, 2012.

Matthee, Rudi. "The Safavid Economy as Part of the World Economy." In *Iran and the World in the Safavid Age*, edited by Willem Floor and Edmund Herzig, 31–79. New York: I. B. Tauris, 2012.

Matthews, Roger. *The Archaeology of Mesopotamia: Theories and Approaches*. New York: Routledge, 2003.

Maxwell, Gavin. *A Reed Shaken by the Wind: Travels among the Marsh Arabs of Iraq*. London: Eland, 2003.

McGowan, Bruce. *Economic Life in Ottoman Europe: Taxation, Trade, and the Struggle for Land, 1600–1800*. New York: Cambridge University Press, 1981.

McNeill, J. R. *The Mountains of the Mediterranean World: An Environmental History*. New York: Cambridge University Press, 1992.

McNeill, J. R. "Woods and Warfare in World History." *Environmental History* 9, no. 3 (2004): 388–410.

McNeill, J. R. "The Eccentricity of the Middle East and North Africa's Environmental History." In *Water on Sand: Environmental Histories of the Middle East and North Africa*, edited by Alan Mikhail, 27–50. New York: Oxford University Press, 2013.

McNeill, J. R. "Peak Document and the Future of History." *American Historical Review* 125, no. 1 (2020): 1–18.

McNeill, J. R., and William H. McNeill. *The Human Web: A Bird's-Eye View of World History*. New York: W. W. Norton, 2003.

McNeill, William H. *The Age of Gunpowder Empires, 1450–1800*. Washington, DC: American Historical Association, 1989.

Meriwether, Margaret L. "Urban Notables and Rural Resources in Aleppo, 1770–1830." *International Journal of Turkish Studies* 4 (1987): 55–73.

"A Mesopotamian Odyssey." *The Economist*, October 22, 2013.

Mikhail, Alan. "The Nature of Plague in Late Eighteenth-Century Egypt." *Bulletin of the History of Medicine* 82, no. 2 (2008): 249–275.

Mikhail, Alan. "Animals as Property in Early Modern Ottoman Egypt." *Journal of the Economic and Social History of the Orient* 53, no. 4 (2010): 621–652.

Mikhail, Alan. *Nature and Empire in Ottoman Egypt: An Environmental History*. New York: Cambridge University Press, 2011.

Mikhail, Alan. "Ottoman Iceland: A Climate History." *Environmental History* 20, no. 2 (2015): 262–284.

Mikhail, Alan, and Christine M. Philliou. "The Ottoman Empire and the Imperial Turn." *Comparative Studies in Society and History* 54, no. 4 (2012): 721–745.

Mitchell, Raoul C. "Instability of the Mesopotamian Plains." *Bulletin de la Société de Géographie d'Égypte* 31 (1958): 127–140.

Moačanin, Nenad. *Town and Country on the Middle Danube, 1526–1690*. Leiden: Brill, 2006.

Molle, François. "River-Basin Planning and Management: The Social Life of a Concept." *Geoforum* 40, no. 3 (2009): 484–494.

Morera, Raphaël. "Environmental Change and Globalization in Seventeenth-Century France: Dutch Traders and the Draining of French Wetlands (Arles, Petit Poitou)." *International Review of Social History* 55, no. S18 (2010): 79–101.

Morozova, Galina S. "A Review of Holocene Avulsion of the Tigris and Euphrates Rivers and Possible Effects on the Evolution of Civilizations in Lower Mesopotamia." *Geoarchaeology* 20, no. 4 (2005): 401–423.

Moorey, P. R. S. *Ancient Mesopotamian Materials and Industries: The Archaeological Evidence*. Oxford: Clarendon Press, 1994.

Muckelroy, Keith. *Maritime Archaeology*. New York: Cambridge University Press, 1978.

Mumford, Lewis. *The City in History: Its Origins, Its Transformations, and Its Prospects*. New York: Harcourt Brace Jovanovich, 1961.

Murphey, Rhoads. "The Ottoman Centuries in Iraq: Legacy or Aftermath? A Survey Study of Mesopotamian Hydrology and Ottoman Irrigation Projects." *Journal of Turkish Studies* 11 (1987): 17–29.

Murphey, Rhoads. "Ottoman Census Methods in the Mid-Sixteenth Century: Three Case Histories." *Studia Islamica* 71 (1990): 115–126.

Murphey, Rhoads. *Ottoman Warfare, 1500–1700*. New Brunswick, NJ: Rutgers University Press, 1999.

Murphey, Rhoads. "The Resumption of Ottoman-Safavid Border Conflict, 1603–1638: Effects of Border Destablization on the Evolution of State-Tribe Relations." In *Shifts and Drifts in Nomad-Sedentary Relations*, edited by Stefan Leder and Bernhard Streck, 308–323. Wiesbaden: L. Reichert, 2005.

Murra, John V. "'El Archipiélago Vertical' Revisited." In *Andean Ecology and Civilization: An Interdisciplinary Perspective on Andean Ecological Complementarity: Papers from Wenner-Gren Foundation for Anthropological Research Symposium*, edited by Shozo Masuda, Izumi Shimada, and Craig Morris, 3–13. Tokyo: University of Tokyo Press, 1985.

Murra, John V. "The Limits and Limitations of the 'Vertical Archipelago' in the Andes." In *Andean Ecology and Civilization: An Interdisciplinary Perspective on Andean Ecological Complementarity: Papers from the Wenner-Gren Foundation for Anthropological Research Symposium*, edited by Shozo Masuda, Izumi Shimada, and Craig Morris, 15–20. Tokyo: University of Tokyo Press, 1985.

Murray, Andrew. *Ship-Building in Iron and Wood*. Edinburgh: Adam and Charles Black, 1863.

Muscolino, Micah S. *The Ecology of War in China: Henan Province, the Yellow River, and Beyond, 1938–1950*. New York: Cambridge University Press, 2015.

Naiman, Robert J. "Animal Influences on Ecosystem Dynamics." *BioScience* 38, no. 11 (1988): 750–762.

Nasrallah, Nawal. *Dates: A Global History*. London: Reaktion Books, 2011.

Necipoğlu, Gülru. *The Age of Sinan: Architectural Culture in the Ottoman Empire*. Princeton, NJ: Princeton University Press, 2005.

Nieuwenhuis, Tom. *Politics and Society in Early Modern Iraq: Mamluk Pashas, Tribal Shayks and Local Rule between 1802 and 1831*. The Hague: Martinus Nijhoff, 1982.
Nissen, Hans J., and Peter Heine. *From Mesopotamia to Iraq: A Concise History*. Translated by Hans J. Nissen. Chicago: University of Chicago Press, 2009.
Nowak, Zachary. *Truffle: A Global History*. London: Reaktion Books, 2015.
Ochsenschlager, Edward. "Village Weavers: Ethnoarchaeology at al-Hiba." *Bulletin on Sumerian Agriculture* 7 (1993): 43–62.
Ochsenschlager, Edward. *Iraq's Marsh Arabs in the Garden of Eden*. Philadelphia: University of Pennsylvania Museum of Archaeology and Anthropology, 2004.
Ogilvie, Astrid E. J. "Documentary Records of Climate from Iceland during the Late Maunder Minimum Period A.D. 1675 to 1715 with Reference to the Isotopic Record from Greenland." In *Climatic Trends and Anomalies in Europe 1675–1715*, edited by Burkhard Frenzel, Christian Pfister, and Birgit Gläser, 9–22. New York: G. Fischer, 1994.
Öğüt, Tahir. *18–19 Yüzyıllarda Birecik Sancağında İktisadi ve Sosyal Yapı*. Ankara: Türk Tarih Kurumu, 2013.
Olsson, Per, Carl Folke, and Thomas Hahn. "Social-Ecological Transformation for Ecosystem Management: The Development of Adaptive Co-Management of a Wetland Landscape in Southern Sweden." *Ecology and Society* 9, no. 4 (2004).
Orhonlu, Cengiz. "Hint Kaptanlığı ve Piri Reis." *Belleten* 134 (1970): 235–254.
Orhonlu, Cengiz, and Turgut Işıksal. "Osmanlı Devrinde Nehir Nakliyatı Hakkinda Araştırmalar: Dicle ve Fırat Nehirlerinde Nakliyat." *Tarih Dergisi* 13 (1962–1963): 79–102.
Osmanlı Arşivi Daire Başkanlığı. *Başbakanlık Osmanlı Arşivi Rehberi*. Istanbul: Seçil Ofset, 2017.
Özbaran, Salih. "XVI. Yüzyılda Basra Körfezi Sahillerinde Osmanlılar Basra Beylerbeyliğinin Kuruluşu." *Tarih Dergisi* 25 (1971): 51–73.
Özbaran, Salih. "The Ottoman Turks and the Portuguese in the Persian Gulf, 1534–1581." *Journal of Asian History* 6, no. 1 (1972): 45–87.
Özbaran, Salih. *The Ottoman Response to European Expansion: Studies on Ottoman-Portuguese Relations in the Indian Ocean and Ottoman Administration in the Arab Lands during the Sixteenth Century*. Istanbul: Isis Press, 1994.
Özbaran, Salih. "Ottoman Naval Policy in the South." In *Süleyman the Magnificent and His Age: The Ottoman Empire in the Early Modern World*, edited by Metin Kunt and Christine Woodhead, 55–70. New York: Longman, 1995.
Özbaran, Salih. *Ottoman Expansion toward the Indian Ocean in the 16th Century*. Istanbul: Istanbul Bilgi University Press, 2009.
Özel, Oktay. *The Collapse of Rural Order in Ottoman Anatolia: Amasya, 1576–1643*. Leiden: Brill, 2016.
Özoğlu, Hakan. "State-Tribe Relations: Kurdish Tribalism in the 16th- and 17th-Century Ottoman Empire." *British Journal of Middle Eastern Studies* 23, no. 1 (1996): 5–27.

Parker, Geoffrey. *The Military Revolution: Military Innovation and the Rise of the West, 1500–1800*. 2nd ed. New York: Cambridge University Press, 1996.

Parker, Geoffrey. *Global Crisis: War, Climate Change and Catastrophe in the Seventeenth Century*. New Haven, CT: Yale University Press, 2013.

Parry, J. P. "Steam Power and the British Influence in Baghdad, 1820–1860." *Historical Journal* 56, no. 1 (2013): 145–173.

Partow, H. *The Mesopotamian Marshlands: Demise of an Ecosystem*. Nairobi: United Nations Environmental Program, 2001.

Paulette, Tate. "Pastoral Systems and Economies of Mobility." In *Models of Mesopotamian Landscapes: How Small-Scale Processes Contributed to the Growth of Early Civilizations*, edited by T. J. Wilkinson, McGuire Gibson, and Magnus Widell, 130–139. Oxford: Archaeopress, 2013.

Peacock, A. C. S. "Introduction: The Ottoman Empire and Its Frontiers." In *The Frontiers of the Ottoman World*, edited by Peacock, 1–27. Oxford: Oxford University Press, 2009.

Pearson, M. N. *The Portuguese in India*. New York: Cambridge University Press, 1987.

Perkins, Douglas J., Tom Were, Samuel Anyona, James B. Hittner, Prakasha Kempaiah, Gregory C. Davenport, and John Michael Ong'echa. "The Global Burden of Severe Falciparum Malaria: An Immunological and Genetic Perspective on Pathogenesis." In *Vector-Borne Diseases*, edited by V. Sree Hari Rao and Ravi Durvasula, 231–283. Vol. 1 of *Dynamic Models of Infectious Diseases*. New York: Springer, 2013.

Perlmann, Peter, and Marita Troye-Blomberg. "Malaria and Immune System in Humans." In *Malaria Immunology*, 2nd ed, edited by P. Perlmann and M. Troye-Blomberg, 229–242. Basel: Krager, 2002.

Perry, John R. "The Banu Kaʿb: An Ambitious Brigand State in Khuzistan." *Le Monde Iranien et l'Islam* 1 (1971): 131–152.

Perry, John R. "The Mamluk Paşalık of Baghdad and Ottoman-Iranian Relations in the Late Eighteenth Century." In *Studies on Ottoman Diplomatic History*, edited by Sinan Kuneralp, 59–70. Istanbul: Isis Press, 1987.

Pfister, Christian. "Spatial Patterns of Climatic Change in Europe A.D. 1675 to 1715." In *Climatic Trends and Anomalies in Europe 1675–1715*, edited by Burkhard Frenzel, Christian Pfister, and Birgit Gläser, 287–316. New York: G. Fischer, 1994.

Pfister, Christian. "Switzerland: The Time of Icy Winters and Chilly Springs." In *Climatic Trends and Anomalies in Europe 1675–1715*, edited by Burkhard Frenzel, Christian Pfister, and Birgit Gläser, 205–224. New York: G. Fischer, 1994.

Phillips, Carla Rahn, and William D. Phillips Jr. *Spain's Golden Fleece: Wool Production and the Wool Trade from the Middle Ages to the Nineteenth Century*. Baltimore: Johns Hopkins University Press, 1997.

Phillips, Clive J. C. *Principles of Cattle Production*. 3rd ed. Boston: CABI, 2018.

Pitcher, Donald Edgar. *An Historical Geography of the Ottoman Empire from Earliest Times to the End of the Sixteenth Century, with Detailed Maps to Illustrate the Expansion of the Sultanate*. Leiden: E. J. Brill, 1972.

Pollock, Susan. *Ancient Mesopotamia: The Eden That Never Was*. Cambridge: Cambridge University Press, 1999.

Popenoe, Paul B. *Date Growing in the Old World and the New*. Altadena: West India Gardens, 1913.

Pournelle, Jennifer R. "KLM to CORONA: A Bird's-Eye View of Cultural Ecology and Early Mesopotamian Urbanization." In *Settlement and Society: Essays Dedicated to Robert McCormick Adams*, edited by Elizabeth C. Stone, 29–62. Los Angeles: Cotsen Institute of Archaeology, 2007.

Pournelle, Jennifer R., and Guillermo Algaze. "Travels in Edin: Deltaic Resilience and Early Urbanism in Greater Mesopotamia." In *Preludes to Urbanism: The Late Chalcolithic of Mesopotamia*, edited by Augusta McMahon and Harriet E. W. Crawford, 7–34. Cambridge: McDonald Institute for Archaeological Research, 2014.

Pritchard, Sara B. *Confluence: The Nature of Technology and the Remaking of the Rhône*. Cambridge, MA: Harvard University Press, 2011.

Postgate, J. N. "Notes on Fruit in the Cuneiform Sources." *Bulletin on Sumerian Agriculture* 3 (1987): 115–144.

Postgate, J. N. *Early Mesopotamia: Society and Economy at the Dawn of History*. New York: Routledge, 1994.

Potache, Dejanirah. "The Commercial Relations between Basrah and Goa in the Sixteenth Century." *Studia* 48 (1989): 145–161.

Potts, Daniel T. *Mesopotamian Civilization: The Material Foundations*. London: Athlone Press, 1997.

Potts, Daniel T. "Tigris River." *Encyclopaedia Iranica*. New York: Iranica Online, 2017.

Powers, W. L. "Soil and Land-Use Capabilities in Iraq: A Preliminary Report." *Geographical Review* 44, no. 3 (1954): 373–380.

Pul, Ayşe. "Osmanlı Tuna Donanmasının Üstüaçık Gemileri." *Tarih Okulu Dergisi* 18 (2014): 285–317.

Radkau, Joachim. *Wood: A History*. Translated by Patrick Camiller. Cambridge: Polity Press, 2012.

Rácz, Lajos. *The Steppe to Europe: An Environmental History of Hungary in the Traditional Age*. Translated by Alan Campbell. Knapwell, UK: White Horse Press, 2013.

Rieber, Alfred J. *The Struggle for the Eurasian Borderlands: From the Rise of Early Modern Empires to the End of the First World War*. New York: Cambridge University Press, 2014.

Richards, John F. *The Unending Frontier: An Environmental History of the Early Modern World*. Berkeley: University of California Press, 2003.

Risso, Patricia. *Oman and Muscat: An Early Modern History*. London: Croom Helm, 1986.

Ritter, Hellmut. "Autographs in Turkish Libraries." *Oriens* 6, no. 1 (1953): 65.

Roaf, Michael. *Cultural Atlas of Mesopotamia and the Ancient Near East.* New York: Facts On File, 1996.

Roberts, Michael. *Essays in Swedish History.* Minneapolis: University of Minnesota Press, 1967.

Robertson, Campbell. "Iraq, a Land between 2 Rivers, Suffers as One of Them Dwindles." *New York Times,* July 14, 2009.

Rogers, Clifford J. ed. *The Military Revolution Debate: Readings on the Military Transformation of Early Modern Europe.* Boulder, CO: Westview, 1995.

Rost, Stephanie. "Water Management in Mesopotamia from the Sixth till the First Millennium B.C." *Wiley Interdisciplinary Reviews: Water* 4, no. 5 (2017).

Rost, Stephanie, and Abdulamir Hamdani. "Traditional Dam Construction in Modern Iraq: A Possible Analogy for Ancient Mesopotamian Irrigation Practices." *Iraq* 73 (2011): 201–220.

Rowton, M. B. "The Role of the Watercourses in the Growth of Mesopotamian Civilization." In *Alter Orient und Altes Testament,* edited by Kurt Bergerhof, Manfried Dietrich, and Oswald Loretz, 307–316. Neukirchen-Vluyn, Germany: Butzon and Bercker, 1969.

Rowton, M. B. "Autonomy and Nomadism in Western Asia." *Orientalia* 42 (1973): 247–258.

Rowton, M. B. "Urban Autonomy in a Nomadic Environment." *Journal of Near Eastern Studies* 32, no. 1/2 (1973): 201–215.

Rowton, M. B. "Enclosed Nomadism." *Journal of the Economic and Social History of the Orient* 17, no. 1 (1974): 1–30.

Ryder, M. L. *Sheep and Man.* London: Duckworth, 1983.

al-Saʿdun, Hamid. *Imarat al-Muntafiq wa Atharuha fi Tarikh al-Iraq wa al-Mantiqa al-Iqlimiyya, 1546–1918.* Amman: Dar Waʾil, 1999.

Saey, Tina Hesman. "The Road to Tameness." *Science News* 191, no. 13 (2017): 20–27.

Şahin, Canay. "The Economic Power of Anatolian *Ayan*s in the Late Eighteenth Century: The Case of the Caniklizades." *International Journal of Turkish Studies* 11, no. 1/2 (2005): 28–47.

al-Saʿidi, Hammud. *Dirasat an Ashaʾir al-Iraq: Al-Khazaʿil.* Najaf: Matbaʿat al-Adab, 1974.

Sağırlı, Abdurrahman. "Cezayir-i Irak-ı Arab veya Şattüʾl-Arabʾın Fethi—Ulyanoğlu Seferi—1565–1571." *Tarih Dergisi* 41 (2005): 43–94.

Saleh, Zaki. *Mesopotamia (Iraq), 1600–1914: A Study in British Foreign Affairs.* Baghdad: Al-Maʿaref Press, 1957.

Salibi, Kamal S. "Middle Eastern Parallels: Syria-Iraq-Arabia in Ottoman Times." *Middle Eastern Studies* 15, no. 1 (1979): 70–81.

Salim, S. M. *Marsh Dwellers of the Euphrates Delta.* London: Athlone Press, 1962.

Salzman, Philip Carl. "Pastoral Nomads: Some General Observations Based on Research in Iran." *Journal of Anthropological Research* 58, no. 2 (2002): 245–264.

Salzmann, Ariel. "An Ancien Régime Revisited: 'Privatization' and Political Economy in the Eighteenth-Century Ottoman Empire." *Politics and Society* 21, no. 4 (1993): 393–423.

Salzmann, Ariel. *Tocqueville in the Ottoman Empire: Rival Paths to the Modern State*. Brill: Leiden, 2004.

Schumm, Stanley A. *The Fluvial System*. New York: John Wiley, 1977.

Scott, James C. *Seeing Like a State: How Certain Schemes to Improve the Human Condition Have Failed*. New Haven, CT: Yale University Press, 1998.

Scott, James C. *The Art of Not Being Governed: An Anarchist History of Upland Southeast Asia*. New Haven, CT: Yale University Press, 2009.

Selivanov, Andrei O. "Global Climate Changes and Humidity Variations over East Europe and Asia by Historical Data." In *Global Precipitations and Climate Change*, edited by Michel Desbois and Françoise Désalmand, 77–104. New York: Springer-Verlag, 1994.

al-Sharqi, Ali. "Al-Baṭaʾih al-Haliyya." *Lughat al-Arab* 7 (January 1927): 375–384.

Simon, Hilda. *The Date Palm: Bread of the Desert*. New York: Dodd, Mead, 1978.

Sinclair, Tom. "The Ottoman Arrangements for the Tribal Principalities of the Lake Van Region of the Sixteenth Century." *International Journal of Turkish Studies* 9, no. 1/2 (2003): 119–143.

Singer, Amy. "*Tapu Tahrir Defterleri* and *Kadı Sicilleri*: A Happy Marriage of Sources." *Tarih* 1 (1990): 95–125.

Singer, Amy. *Palestinian Peasants and Ottoman Officials: Rural Administration around Sixteenth-Century Jerusalem*. New York: Cambridge University Press, 1994.

al-Shabibi, Muhammad Rida. "Al-Rumahiyya." *Lughat al-Arab* 1 (1913): 461–465.

Shamout, M. Nouar, and Glada Lahn. *The Euphrates in Crisis: Channels of Cooperation for a Threatened River*. London: Chatham House, 2015.

Snow, Charles Henry. *The Principal Species of Wood: Their Characteristic Properties*. 2nd ed. New York: John Wiley, 1910.

Soucek, Svat. "Certain Types of Ships in Ottoman-Turkish Terminology." *Turcica* 7 (1975): 233–249.

Springfield, Harry Wayne. *Forage Problems and Resources of Iraq*. Washington, DC: International Cooperation Administration, 1957.

Squatriti, Paolo. *Water and Society in Early Medieval Italy: AD 400–1000*. New York: Cambridge University Press, 1998.

Starr, Chester G. *The Roman Imperial Navy, 31 B.C.–A.D. 325*. 3rd ed. Chicago: Ares, 1993.

Stein, Mark L. *Guarding the Frontier: Ottoman Border Forts and Garrisons in Europe*. New York: I. B. Tauris, 2007.

Steinkeller, Piotr. "City and Countryside in Third-Millennium Southern Babylonia." In *Settlement and Society: Essays Dedicated to Robert McCormick Adams*, edited by Elizabeth C. Stone, 185–211. Los Angeles: Cotsen Institute of Archaeology, 2007.

Strange, Guy Le. *Baghdad during the Abbasid Caliphate from Contemporary Arabic and Persian Sources*. Oxford: Oxford University Press, 1900.

Strange, Guy Le. *The Lands of the Eastern Caliphate: Mesopotamia, Persia, and Central Asia from the Moslem Conquest to the Time of Timur.* Cambridge: Cambridge University Press, 1905.

Stroud, Ellen. *Nature Next Door: Cities and Trees in the American Northeast.* Seattle: University of Washington Press, 2012.

Subrahmanyam, Sanjay. *The Portuguese Empire in Asia, 1500–1700.* 2nd ed. Malden, MA: Wiley-Blackwell, 2012.

Susa, Ahmad. *Atlas al-Iraq al-Hadith.* Baghdad: Mudiriyyat al-Masaha al-Amma, 1953.

Susa, Ahmad. *Tarikh Hadarat Wadi al-Rafidayn.* 2 vols. Baghdad: Wizarat al-Ray, 1986.

Tabak, Faruk. *The Waning of the Mediterranean, 1550–1870: A Geohistorical Approach.* Baltimore: Johns Hopkins University Press, 2008.

Taş, Kenan Ziya. "Osmanlı'nın Son Döneminde Fırat ve Dicle Nehirlerinde Kelek ile Ulaşım." In *Osmanlı Devleti'nde Nehirler ve Göller*, edited by Şakir Batmaz and Özen Tok, 1:413–428. Kayseri: Not Yayınları, 2015.

Teclaff, Ludwik A. *The River Basin in History and Law.* The Hague: Martinus Nijhoff, 1967.

Teclaff, Ludwik A. "Evolution of the River Basin Concept in National and International Water Law." *Natural Resources Journal* 36, no. 2 (1996): 359–391.

Tengberg, Margareta. "Fruit-Growing." In *A Companion to the Archaeology of the Ancient Near East*, edited Daniel T. Potts, 1:181–200. Malden, MA: Wiley-Blackwell, 2012.

Terzioğlu, Derin. "How to Conceptualize Ottoman Sunnitization: A Historiographical Discussion." *Turcica* 44 (2012–2013): 301–338.

Tezcan, Baki. *The Second Ottoman Empire: Political and Social Transformation in the Early Modern World.* New York: Cambridge University Press, 2010.

Thalen, D. C. P. *Ecology and Utilization of Desert Shrub Rangelands in Iraq.* The Hague: Dr. W. Junk, 1979.

The Times Atlas of the World. 10th ed. New York: Crown, 1999.

Thesiger, Wilfred. *The Marsh Arabs.* New York: Penguin, 2007.

Thoreau, Henry David. *1851–1852.* Vol. 4 of *Journal*, edited by Leonard N. Neufeldt and Nancy Craig Simmon. Princeton, NJ: Princeton University Press, 1992.

Thoreau, Henry David. *The Portable Thoreau.* Edited by Jeffrey S. Cramer. New York: Penguin, 2012.

Topal, Mehmet. "Bağdadlı Nazmi-Zade Murteza'nın, İcmal-i Sefer-i Nehr-i Ziyab Adlı Risalesine Göre Ziyab Seferi ve Nehirde Yapılan Düzenlemeler." In *Osmanlı Devleti'nde Nehirler ve Göller*, edited by Şakir Batmaz and Özen Tok, 1:71–85. 2 vols. Kayseri: Not Yayınları, 2015.

Torsvik, Trond H., and L. Robin M. Cocks. *Earth History and Palaeogeography.* Cambridge: Cambridge University Press, 2017.

Tübinger Atlas des Vorderen Orients. Wiesbaden: L. Reichert, 1977–1993.

al-Uqayli, Muhammad Husayn. *Tarikh an-Najaf al-Ashraf.* Edited by Abdulrazzaq Hirzuldin. 3 vols. Qum: Dalil-i Ma, 1427/2006–2007.

Vannote, Robin L., G. Wayne Minshall, Kenneth W. Cummins, James R. Sedell, and Colbert E. Cushing. "The River Continuum Concept." *Canadian Journal of Fisheries and Aquatic Sciences* 37, no. 1 (1980): 130–137.

Varlık, Nükhet. "Conquest, Urbanization and Plague Networks in the Ottoman Empire, 1453–1600." In *The Ottoman World*, edited by Christine Woodhead, 251–263. New York: Routledge, 2012.

Varlık, Nükhet. *Plague and Empire in the Early Modern Mediterranean World: The Ottoman Experience, 1347–1600.* New York: Cambridge University Press, 2015.

Vatin, Nicolas. "Un territoire 'bien gardé' du sultan? Les Ottomans dans leur vilayet de Basra, 1565–1568." In *The Ottoman Middle East: Studies in Honor of Amnon Cohen*, edited by Eyal Ginio and Elie Podeh, 63–91. Leiden: Brill, 2014.

Venzke, Margaret L. "The Ottoman Tahrir Defterleri and Agricultural Producitivty." *Osmanlı Araşturmaları* 17 (1997): 1–51.

Verhoeven, Kris. "Geomorphological Research in the Mesopotamian Flood Plain." In *Changing Watercourses in Babylonia: Towards a Reconstruction of the Ancient Environment in Lower Mesopotamia*, edited by Hermann Gasche and Michel Tanret, 159–245. Ghent: University of Ghent and the Oriental Institute of the University of Chicago, 1998.

Verhoeven, Marc. "Ethnoarchaeology, Analogy, and Ancient Society." In *Archaeologies of the Middle East: Critical Perspectives*, edited by Susan Pollock and Reinhard Bernbeck, 251–270. Malden, MA: Blackwell, 2005.

von Oppenheim, Max. *Die Beduinen*. Leipzig: Otto Harrassowitz, 1896–1983.

Vörösmarty, C. J., B. M. Fekete, and B. A. Tucker. *Global River Discharge Database (RivDIS v1.0)*. Paris: UNESCO, 1996.

Waines, David. "The Third Century Internal Crisis of the Abbasids." *Journal of the Economic and Social History of the Orient* 20, no. 3 (1977): 282–306.

Waldbaum, Jane C. "The Coming of Iron in the Eastern Mediterranean: Thirty Years of Archaeological and Technological Research." In *The Archaeometallurgy of the Asian Old World*, edited by Vincent C. Pigott, 27–57. Philadelphia: University Museum, University of Pennsylvania, 1999.

al-Wardi, Ali. *Dirasa fi Tabi'at al-Mujtama al-Iraqi*. London: Dar al-Warraq, 2009.

Watson, Patty Jo. "The Theory and Practice of Ethnoarchaeology with Special Reference to the Near East." *Paléorient* 6 (1980): 55–64.

Webb, James L. A. Jr. *The Long Struggle against Malaria in Tropical Africa*. New York: Cambridge University Press, 2014.

White, Joshua M. "Shifting Winds: Piracy, Diplomacy, and Trade in the Ottoman Mediterranean, 1624–1626." In *Well-Connected Domains: Towards an Entangled Ottoman History*, edited by Pascal W. Firges, Tobias P. Graf, Christian Roth, and Gülay Tulasoğlu, 37–53. Leiden: Brill, 2014.

White, Joshua M. *Piracy and Law in the Ottoman Mediterranean*. Stanford, CA: Stanford University Press, 2018.

White, Richard. *The Organic Machine*. New York: Hill and Wang, 1995.

White, Sam. "Rethinking Disease in Ottoman History." *International Journal of Middle East Studies* 42, no. 4 (2010): 549–567.

White, Sam. *The Climate of Rebellion in the Early Modern Ottoman Empire.* New York: Cambridge University Press, 2011.

Widell, Magnus, Carrie Hritz, Jason A. Ur, and T. J. Wilkinson. "Land Use of the Model Communities." In *Models of Mesopotamian Landscapes: How Small-Scale Processes Contributed to the Growth of Early Civilizations*, edited by Wilkinson, McGuire Gibson, and Widell, 56–80. Oxford: Archaeopress, 2013.

Wilkinson, T. J. *Archaeological Landscapes of the Near East.* Tucson: University of Arizona Press, 2003.

Wilkinson, T. J. "Hydraulic Landscapes and Irrigation Systems of Sumer." In *The Sumerian World*, edited by Harriet E. W. Crawford, 33–54. New York: Routledge, 2013.

Wilkinson, T. J., and Louise Rayne. "Hydraulic Landscapes and Imperial Power in the Near East." *Water History* 2, no. 2 (2010): 115–144.

Wilkinson, T. J., and Carrie Hritz. "Physical Geography, Environmental Change and the Role of Water." In *Models of Mesopotamian Landscapes: How Small-Scale Processes Contributed to the Growth of Early Civilizations*, edited by T. J. Wilkinson, McGuire Gibson, and Magnus Widell, 9–33. Oxford: Archaeopress, 2013.

Wilkinson, T. J., Louise Rayne, and Jaafar Jotheri. "Hydraulic Landscapes in Mesopotamia: The Role of Human Niche Construction." *Water History* 7, no. 4 (2015): 397–418.

Willcocks, William. *The Restoration of the Ancient Irrigation Works on the Tigris, or, The Re-Creation of Chaldea.* Cairo: National Printing Department, 1903.

Williams, William D. "Salinization of Rivers and Streams: An Important Environmental Hazard." *Ambio* 16, no. 4 (1987): 180–185.

Williamson, Grahame. "Iraqi Livestock." *Empire Journal of Experimental Agriculture* 17 (1949): 48–59.

Wilson, Mary Elizabeth. "The Power of Plague." *Epidemiology* 6, no. 4 (1995): 458–460.

Winter, Stefan H. "The Other *Nahdah*: The Bedirxans, the Millis and the Tribal Roots of Kurdish Nationalism in Syria." *Oriente Moderno* 86, no. 3 (2006): 461–474.

Winter, Stefan H. "The Province of Raqqa Under Ottoman Rule, 1535–1800: A Preliminary Study." *Journal of Near Eastern Studies* 68, no. 4 (2009): 253–268.

Wittfogel, Karl A. *Oriental Despotism: A Comparative Study of Total Power.* New Haven, CT: Yale University Press, 1957.

Worster, Donald. *Rivers of Empire: Water, Aridity, and the Growth of the American West.* New York: Oxford University Press, 1992.

Yapp, M. E. "The Euphrates Expedition." In *The Islamic World from Classical to Modern Times: Essays in Honor of Bernard Lewis*, edited by C. E. Bosworth, Charles Issawi, Roger Savory, and A. L. Udovitch, 891–915. Princeton, NJ: The Darwin Press, 1989.

Yaycioglu, Ali. "Provincial Power-Holders and the Empire in the Late Ottoman World: Conflict or Partnership?" In *The Ottoman World*, edited by Christine Woodhead, 436–452. New York: Routledge, 2012.

Yaycioglu, Ali. *Partners of the Empire: The Crisis of the Ottoman Order in the Age of Revolutions*. Stanford, CA: Stanford University Press, 2016.

Yener, Emir. "Ottoman Seapower and Naval Technology during Catherine II's Turkish Wars, 1768–1792." *International Naval Journal* 9, no. 1 (2016): 4–15.

Yenidünya, Süheyla. "XIX. Yüzyıl Başlarında Osmanlı Devleti'nin Bağdat'ta Kölemen Hakimiyetini Kaldırma Teşebbüsleri." *Türk Dünyası Araştırmaları* 182 (2009): 1–46.

Yılmaz, Ali. "16 ve 17 Yüzyıllarda Fırat'ta Nehir Nakliyatı." In *Osmanlı Devleti'nde Nehirler ve Göller*, edited by Şakir Batmaz and Özen Tok, 1:591–611. Kayseri: Not Yayınları, 2015.

Young, Gavin. *Return to the Marshes: Life with the Marsh Arabs of Iraq*. London: Collins, 1977.

Yurdaydın, Hüseyin. "Matrakçı Nasuh'un Hayatı ve Eserleri ile İlgili Yeni Bilgiler." *Belleten* 29, no. 114 (1965): 329–354.

Zahar, A. R. "Review of the Ecology of Malaria Vectors in the WHO Eastern Mediterranean Region." *Bulletin of the World Health Organization* 50 (1974): 427–440.

Zarinebaf, Fariba. *Crime and Punishment in Istanbul, 1700–1800*. Berkeley: University of California Press, 2011.

Zarinebaf, Fariba. *Mediterranean Encounters: Trade and Pluralism in Early Modern Galata*. Oakland: University of California Press, 2018.

Zhang, Ling. *The River, the Plain, and the State: An Environmental Drama in Northern Song China, 1048–1128*. New York: Cambridge University Press, 2016.

Zorlu, Tuncay. *Innovation and Empire in Turkey: Sultan Selim III and the Modernisation of the Ottoman Navy*. New York: Tauris Academic Studies, 2008.

Index

For the benefit of digital users, indexed terms that span two pages (e.g., 52–53) may, on occasion, appear on only one of those pages.

Abbas I, 48, 131–32
Abbasid dynasty, 36–37, 49, 96
agrarianism
    Ottoman Empire and, 19–20, 40, 64–65, 80–81, 89, 93–94, 101–2, 103, 120
    worldwide, 82–83, 93, 95–96, 130–31
    See also irrigation
agriculture. *See* dry farming; irrigation; pastoralism; *and specific crops and animals*
Ahmad Pasha, 135, 136–37
Ahşamat, 89–92, 90t, 105–6, 128–29, 146
Andes, 60
animals
    as means of transport, 28, 31–32, 56, 58, 121
    as sources of energy, 67–68, 102
    as sources of skin, 29–31, 34–35
    domestication of, 79
    See also herders' associations; pastoralism; *and specific animals*
Aleppo
    as transportation hub, 27–28, 33, 138
    conflict in, 25
    Euphrates and, 26–27, 121
    Ottoman navy and, 43, 52–53, 55, 58, 140, 142
alluvial plain
    definition of, 6–7, 59
    geological history of, 9
    shortage of natural resources in, 33–34
    water wide web and, 60
    See also Iraq; irrigation; pastoralism; wetlands
Anatolia
    agricultural investment in, 65
    cadastral surveys of, 18
    climate in, 104, 112–14
    conflict in, 24–25, 48, 126, 127–28
    dams in, 148–49
    early Ottoman history in, 10–12, 13, 45–46
    ecological surplus in, 4, 34
    forests of, 55–56
    geography of, 44, 114–15
    shepherds of, 90, 124–25
    See also specific cities
aridity, 145
    animal herding and, 82, 87–88
    irrigation and, 3–4, 15–16, 62, 79
    river flow and, 114–15
    wetlands and, 96

artillery, 132, 147–48
  ceremonial use of, 40
  fortresses and, 2–3, 33, 35–37, 38
  mounted on ship, 42–43, 50–51
  transport of, 21, 41, 121, 154*t*
artisans
  guilds and, 30–31
  in military camps, 121
  Ottoman navy and, 52–55
avulsion, 126
  definition of, 111
  full, 120
  herders' associations and, 128–29, 143
  Khazaʿil confederation and, 130
  node of, 116–17
  partial, 116–17, 119
  study of, 113
Ayntab, 52–54, 142

Baghdad
  agricultural development in, 63–64, 71–72, 73–76
  appointment of clerics to, 23, 38, 160–61
  arms to, 154*t*
  cadastral surveys of, 18
  drought in, 113–14
  grain to, 137–38, 152*t*
  marshes of, 96–97, 101–2
  militarization of, 34–37
  Ottoman conquest of, 1, 25–26, 43
  plague in, 117–18
  Safavid conquest of, 48
  shipbuilding in, 29–30, 156*t*
  shortage of grain in, 33–34
  Tigris and Euphrates in relation to, 24, 26–27, 31–33, 38–39, 121–22
  today, 149–50
  Pashalik of Baghdad and, 4–5, 20, 138
  See also Iraq; Pashalik of Baghdad; *and specific cities, governors, and tribal groups*

Bahrain, 46–47
Balkans, 127–28
  agricultural investment in, 65–66, 103
  early Ottoman history in, 11, 12, 13, 145
Banu Kaʿb, 138–40
barley
  cultivated by pastoralists, 86–87, 92
  harvest of, 64–65, 65*t*
  in arable lands, 76–77
  in wetlands, 102, 103, 107
  shortage of, 33–34
  transport of, 4, 152*t*
  See also grain
Basra, 43
  agriculture in, 64–65, 65*t*, 69–72
  appointment of clerics to, 37–39
  arms for, 154*t*
  cadastral surveys of, 18
  disease in, 117–18
  grain from, 152*t*
  labor in, 52–55
  marshes of, 7, 49, 97, 99, 101–2
  naval buildup in, 46–47, 156*t*
  Ottoman conquest of, 26, 43, 120
  Ottoman shipyard in, 44–46
  Pashalik of Baghdad and, 20, 138–41, 143
  pastoralists around, 86, 88–90
  Portugal and, 43
  shipbuilding in, 48, 51–52, 142
  Tigris and Euphrates in relation to, 26–27, 31–33
  timber for, 55–57, 134
  See also malaria; Maniʿ, Sheikh; Qurna; Shatt River Fleet
Bayezid II, 24–25, 46
Bebe Süleyman, 119, 120
Birecik, 20, 51, 134, 147
  labor in, 52–55
  natural resources in, 55–58

Ottoman naval disengagement
   from, 141–44
Ottoman shipyard in, 27–28,
   41, 43–46
shipbuilding in, 5, 46, 48, 49,
   147–48, 156t
transport from, 40, 121, 152t, 154t
Black Sea
   forests of, 55
   naval labor in, 52, 53, 55–56
   Ottoman navy in, 41, 50
   pastures of, 82
   watercraft of, 51
bridges, 31–32, 33, 38
Büyük Süleyman Pasha, 140, 143
Byzantine Empire, 10–11, 13–14, 24,
   43–44, 50

cadastral surveys (Tapu Tahrir
   Defterleri), 17–19
canals. *See* irrigation *and specific canals*
Canbulad Bey, 40–41, 134
cannon. *See* artillery
caprids. *See* goat; sheep
Caspian Sea, 47–48, 65, 82
cattle, 80–82, 92, 93
   in statistics, 87–88, 89t, 90t, 91t
Cemmasat, 105–6, 106t, 128–29, 146
Circle of Justice, 63–64
communication. *See* transportation, river
crops. *See* grain *and specific crops*

Daltaban Mustafa Pasha, 119–20, 121–23
dams, 63, 72, 102–3, 147–50
   See also Dhiyab Canal
Danube River, 8t, 21, 34, 41
   Ottoman admiralty in, 134
   Ottoman navy in, 51–52, 55, 142
dendrochronology. *See* tree ring
desert
   Arabs of, 82–83, 85–86, 87, 131
   formation, 96

of Arabia, 7, 38–39
of Iraq, 14–15, 59–60, 87–88, 145
of Syria, 28, 114–15
pastures in, 83–85, 86–87
See also Jazira; truffle
disease, 16, 60, 113
See also malaria; plague
Diyala
   region of, 63, 73–74, 123–25
   river of, 114–15, 137–38
Diyarbakır
   appointment of clerics to, 38, 159
   as transportation hub, 26–27, 33,
      152t, 154t
   conflict in, 24–25
   demography of, 28t
   Dhiyab Canal and, 121
   drought in, 5, 113–14
   geography of, 44
   irrigation and, 76
   Ottoman navy and, 52–54, 55
   Pashalik of Baghdad and, 136
   See also kelek
Dhiyab Canal, 115–17, 118, 119–20,
   121–22
docks, 16, 31–33, 38, 43, 44t
dogana of Foggia, 89, 91
drainage basin
   concept of, 6
   in global perspective, 7–9
   in Ottoman historiography, 5
   Ottoman fortresses in, 14–15, 24
   political unification of, 1–3, 24–26
   See also alluvial plain
drought, 4–5, 97, 113–15, 126
dry farming, 3–4, 59, 116
Dujayl Canal, 73–76, 75t

East India Company
   officials of, 33–34, 84
   Pashalik of Baghdad and, 139–41,
      142, 143–44

endowments, pious, 37
energy
  gunpowder, 2–3
  plant, 89
  river, 7–8, 14–15, 21, 46, 50, 58, 96–97, 116–17, 147
  solar, 111
  wind, 51, 58, 147
  see also under animals *and* water buffalo
Egypt
  as transportation hub, 33
  geography of, 84–85, 126
  grain from, 34, 152*t*
  irrigation in, 62, 65–66, 72, 77
  military band from, 35–36
  Ottoman navy in, 46
  See also Nile River; Mehmed Ali; Napoleon Bonaparte
Erzurum, 24–25, 44, 114–15
Europe
  experts from, 53, 140
  marginal lands in, 102
  maritime empires of, 47–48
  Ottoman expansion into, 11, 45–46
  Ottoman frontier in, 7–8, 24, 51, 52, 127–28, 142
  Ottoman officials in, 147–48
  Ottoman possessions in, 13, 26, 120
  Ottoman relations with, 17, 27–28
  steamboats of, 42
  See also Portugal; Russia
Evliya Çelebi,
  environmental observations of, 7, 61, 62–63, 69–70
  political observations of, 35, 36–37

Falluja, 32–33, 129–30
flood
  agriculture and, 15–16, 76–77, 97, 102–3
  annual, 59, 61, 71–72, 122–23
  cycle of, 66
  diversion of, 148–49
  extraordinary, 5, 60, 114, 116–17, 120, 130
  pastoralism and, 83, 87
  wetland settlement and, 98
flow cycle. *See* flood
forage conservation, 81–82
forests. *See* timber
frigate, 51, 52, 141–42, 156*t*
frontier, Ottoman eastern
  ecological imbalance in, 3–4, 34, 145
  governance in, 13
  integration of, 14–15, 24
  irrigation in, 61–62
  militarization of, 34–35, 38
  navy along, 49
  pastoralism along, 80–81
  political stability of, 7–8, 38, 60
  power projection in, 19–20, 21, 24, 34, 146–47

galley, 33, 42, 49, 50–52
  construction of, 46–47, 156*t*
  forced labor in, 55
galleon, 51, 156*t*
galliot, 45, 46–47, 50, 52, 156*t*
gallivat, 138–40
Gallipoli, 11, 34, 45–46, 50
Georgia, 141
Georgian mamluks, 135, 136
goat
  as sources of skin, 28, 57–58
  grazing of, 92
  herding of, 87–89
  in statistics, 14–15, 89*t*, 91*t*, 125*t*
  versatility of, 93
grain
  access to, 127, 137–38, 143–44
  cultivated by pastoralists, 93
  farmers of, 61, 87
  in cadastral surveys, 18–19, 89

transport of, 5, 34–35, 38–39, 46, 145, 146, 147, 152*t*
See also specific grains
Greece, 46
　artisans from, 53–54
guilds. *See under* artisans
gunboat, 2–3, 19–20, 41, 42
　construction of, 46, 55, 58, 140
gunpowder, 33–34, 38–39, 140
　production of, 34, 36–37
　technology of, 2–3
　transport of, 154*t*
　weapons, 4, 58
　See also artillery

Habsburg Empire, 1, 24, 46–47, 126
Hasaka, 20, 121–22, 129–30, 135
Hasankeyf, 29–31, 148–50, 152*t*
Hasan Pasha, 132–35, 138
herbivores. *See* animals
herders' associations, 79–80, 128–29, 132, 143
　See also specific associations
Hilla, 71*f*, 116–17, 121–22, 130, 135
　dock of, 32–33
　in statistics, 124*t*
　marshes of, 97
　Safavid occupation of, 48
　trees of, 121
Hit, 7, 8*t*, 66–67, 67*f*, 116
　See also tar
Hormuz, 43, 44–45, 46–47
hydrologic cycle, 3, 9, 77–78, 97, 147

Imperial Council (Divan-i Hümayun), 17–18, 118
　domestic travel and, 23
　grain shortage and, 33–34
　herders' associations and, 128–29
　shipbuilding and, 43, 45, 53
　water control and, 64, 74–76, 119

Imperial Naval Arsenal (Tersane-i Amire), 45–46, 138, 141
　labor recruitment by, 52, 53, 55
　ship design by, 51, 52
　provisioning of natural resources by, 55
India, 46, 62, 69, 114, 139
　timber from, 57
　See also East India Company
Indian Ocean, 12–13, 26, 46, 47, 139
　Ottoman navy in, 41
Iraq
　ancient history of, 38, 50, 57, 145
　arable farming in, 61–62, 66–78
　archives of, 19
　experience of disaster in, 126
　geography of, 7, 9, 59
　herders' associations of, 89–91, 105–6, 128–29
　in the nineteenth century, 147–48
　Ottoman pastoral policy in, 89–91
　Ottoman water management in, 62–66
　pastoralism in, 79–81, 83–84, 92–94
　plague in, 117
　regional governance regimes and, 128
　sheep in, 87–89
　shortage of natural resources in, 33–34
　since twentieth century, 148–50
　tribes of, 49, 50, 129–32
　waterways of, 23, 38–39, 56, 114–15
　See also frontier, Ottoman eastern; marshes, Iraqi; Pashalik of Baghdad; *and specific cities*
iron, 29, 31–32
　age of, 2–3
　shipbuilding and, 57, 121, 141, 147–48
　transport of, 154*t*
　See also metal
irrigation, 59, 66–67, 124
　animal herding and, 86–87
　avulsion and, 111–12, 115–16, 117
　streamflow and, 114–15

irrigation (*cont.*)
  Mongols and, 16
  wetland formation and, 96–97
  before fossil fuels, 6, 59
  engineers of, 63
  historiography of, 62
  in wetlands, 102, 103
  lift, 67–68, 71–72
  Ottoman, 60, 61–66
  perennial, 72–76, 77–78
  Sasanian, 15–16, 33–34
  since nineteenth century, 147–49
  Sumerian, 9, 15–16, 100
  see also under aridity *and specific canals*
Ismail I, 24–25

Jalili dynasty, 127–28, 136–38
janissaries
  in battle, 40–41, 130
  of Baghdad, 34–36, 38, 121, 147–48
  origins of, 12
  political role of, 13–14, 24–25, 113
Jazira, 4, 23–24, 34

Karbala, 73–74, 75*t*, 121–22, 129–30, 140
kelek, 28–31, 33
ketch, 139, 140
Khazaʿil confederation, 125–26, 129–32
Khuzistan, 7, 96
  See also Banu Kaʿb
Kirkuk, 24–25, 34–35
Kurds, 89–90, 113, 136, 148–49

levee, 59, 66–67, 77
  cultivation on, 68–70, 76–77
  definition of, 67–68
  irrigation and, 71–74, 115, 116–17, 118
livestock. *See* animals

malaria, 99–101, 148–49
Malatya, 48, 55, 56, 121, 147
malikane, 127–28, 134

Mamluk Empire, 25, 26, 33, 44–45, 46
mamluks. *See* Georgian mamluks
Maniʿ, Sheikh, 117–18, 119, 120, 129–30
Maraş
  artisans from, 52–54
  grain from, 152*t*
  iron from, 57
  timber from, 55, 56, 121, 142, 147
Mardin
  grain from, 147, 152*t*
  labor from, 29–30
  Pashalik of Baghdad and, 20, 136–38, 143–44
marshes, Iraqi, 7, 59–60
  agriculture in, 76, 101–3
  conflict in, 40–41, 49, 120
  drainage of, 101–2
  historiography of, 96
  hydrography of, 96–97
  pastoralism and, 83–84, 87–88
  settlement in, 98–99
  See also Khazaʿil confederation; malaria; water buffalo
meadow transhumance, 84–85
  See also pastoralism
Mehmed Ali, 143
mesta, 89, 91
metal
  deposits of, 3–4, 33–35
  trade with, 139–40
  transport of, 4, 34
  See also iron
Mediterranean Sea
  as communication zone, 47, 65–66
  nautical traditions of, 50, 52, 55–56
  Ottoman navy on, 12–13, 41, 51, 52, 53, 90–91, 142
  Ottoman provinces along, 145
  pastoralism around, 83, 84, 89
  travel from, 7, 23–24, 33, 138
  wetlands around, 99–100
  See also Aleppo

Military Revolution, 42
Mongols
   in early Ottoman history, 10, 12
   irrigation systems and, 16, 96
   Mamluk Empire and, 33
Mosul
   appointment of clerics to, 38, 159–60
   arms from, 154*t*
   as transportation hub, 33
   dam of, 149–50
   drought in, 113
   grain from, 152*t*
   militarization of, 34–35
   soldiers from, 121
   squadron from, 43
   Tigris and, 26–27
   See also Jalili dynasty; kelek
Muntafiq confederation, 132, 138–39
   See also Maniʿ, Sheikh
Murad IV, 35, 48, 131–32

Nadir Shah, 35, 139, 141–42
Najaf, 121–22, 130
   See also Shahi Canal
Napoleon Bonaparte, 141
navy, Ottoman. *See* Imperial Naval
      Arsenal; Shatt River Fleet; *and
      under specific rivers, seas, and
      provinces*
Nazmizade, Murtaza, 114, 117, 120–21,
   122, 129–30
Nile River, 8*t*
   See also Egypt
Nusaybin, 136, 137–38, 143–44, 147, 152*t*

Oman, 139
ore. *See* metal

pastoralism
   as wealth management service, 88–89
   definition of, 79
   in history, 80–81, 147–48

mobility and, 81–82
   Ottoman sponsorship of,
      79–80, 89–91
   political implications of, 117–18,
      120–21, 124–26
   river water and, 82–87
   supplementary pursuits of, 92–94
   See also herders' associations *and
      specific animals*
palm, date, 40–41, 61
   as keystone species, 7, 69–70
   cultivation of, 68, 76
   harvest of, 64–65, 65*t*
   in Ottoman cadastral surveys, 18–19
   irrigation of, 71–72
   uses of, 69, 103, 121
Pashalik of Baghdad, 4–5, 20,
   143, 146–47
   definition of, 109, 127
   fall of, 147–48
   in Ottoman context, 127–28,
      136, 143–44
   Shatt River Fleet and, 135–36, 141–43
   see also under East India Company *and
      specific governors and cities*
Persia
   ancient empires of, 15–16, 65
   communication with, 33, 38, 67–68,
      136–37, 138
   plague in, 117
   war with, 30–31, 49, 146–47, 156*t*
   See also Safavid Empire
Persian Gulf
   conflict in, 5
   drainage into, 1–2, 66–67,
      83–84, 96–97
   mapping of, 6
   Ottoman access to, 23–24, 27
   See also Banu Kaʿb; Basra; East India
      Company; Oman; Portugal; Shatt
      al-Arab
pitch. *See* tar

plague, 4–5, 117–18, 126
Portugal, 12–13, 43, 44–45, 46–47

Qara Ulus, 90–91, 91*t*, 128–29
Qara'ul, 90–92, 91*t*, 128–29
Qizilbash, 48, 52
Quran, 43, 117, 122, 132
Qurna, 33, 34–35, 119–20, 154*t*

Raqqa, 24–25, 26–27, 142, 152*t*
  shipbuilding in, 49, 156*t*
Raşid Efendi, 116, 120
Rauwolff, Leonhard, 82–83, 84, 87, 93
Red Sea, 33, 41, 44–45, 50
Rhodes, 53, 57–58
rice, 16, 40–41
  harvest of, 64–65, 65*t*
  in wetlands, 76, 95–96, 101, 102–3, 107
  See also grain
Roman Empire, 21, 27–28, 43–44, 53, 96
  See also Byzantine Empire
Rumahiyya, 97, 111–12, 115, 118
  history of, 123–26
  See also Khazaʿil confederation
Russia, 24, 114, 127–28, 141, 142
Ridwaniyya, 32–33

Safavid Empire
  army of, 21, 47
  border with, 24, 80–81
  capital cities of, 14–15, 138–39
  collapse of, 29–30
  diplomacy with, 119, 120
  exile to, 131–32
  Mamluk Empire and, 25
  navy of, 47–48
  Ottoman propaganda against, 38, 64
  rise of, 47–48
  trade with, 69
  war with, 1, 24–26, 34–35, 48–49
  See also Abbas I; Persia; Qizilbash; Shahi Canal; Zuhab Treaty

salinization, 68, 77–78, 83–84
Salman, Sheikh, 130, 132
sediment, 111
  clearance of, 74–76, 121–22
  cores of, 13
  deposition of, 7, 59, 61, 66–68, 72, 76–77, 103
  transport of, 5, 77–78, 114, 116–17, 126, 145
  See also sedimentation
sedimentation, 120
  aridity and, 114–15
  canal construction and, 72–73, 77, 116
  in Sumerian irrigation, 9
Selim I, 24–26, 45–46
Selim II, 23, 40, 64
Shammar confederation, 129–30, 132
Shatt al-Arab, 31, 33, 71–72
  See also Shatt River Fleet
Shatt River Fleet
  abandonment of, 142–43
  admiralty of, 134–36
  auxiliaries of, 138–41
  buildup of, 156*t*
  combat role of, 49, 50
  formation of, 41, 47–48
  rivercraft of, 51–52
Şehrizor, 119, 121, 136, 143
Shahi Canal, 73–74, 75*t*
sheep, 12
  as sources of manure, 86
  as sources of skin, 28, 57–58
  herding of, 87–91, 124, 130–31
  in statistics, 14–15, 89*t*, 91*t*, 106*t*, 124, 125*t*
  versatility of, 80–81, 98–99, 124–25
  See also pastoralism
shepherds. *See* pastoralism; sheep
shipbuilding. *See under* Imperial Naval Arsenal; Shatt River Fleet; *and specific sites and construction materials*

silt. *See* sediment
slaves, 12, 48
　See also Georgian mamluks
soft power, 37–38
Sokullu Mehmed Pasha, 27–28, 45
speleothem, 112–14
Süleyman I, 147–48
　conquest of Iraq by, 1, 25–26, 31–32
　Ottoman navy and, 43, 45–46
　water management and, 62–63, 64, 65, 73–74
Süleymani Canal, 73–74, 75*t*
steam technology, 42, 141, 147–49
Suez, 44–45, 46, 134
Syria
　animal skin from, 29–30
　climate and, 44
　conflict in, 25
　domestic travel and, 38, 136–37
　Euphrates and, 26–27, 55–56, 59, 148–50
　death of Süleymanshah in, 12
　grain from, 34
　labor from, 52–53, 55
　see also under desert

Tanzimat reforms, 129
tapu, 64–65, 73
tar, 57–58, 121
Taurus Mountains
　as sources of water, 1–2, 24, 82, 111
　geology of, 9, 44
　mapping of, 6
　pastoralism in, 84
Tikrit, 7, 66–67
timber
　dam building and, 121, 122–23
　shipbuilding and, 44, 50, 55–57, 98, 141
　shortage of, 44–45, 47–48, 142
　transport of, 4, 6, 134, 137–38
　See also kelek

Thoreau, Henry David, 9, 81, 95, 145
Topkapı Palace, 17, 135
traditionalism, 62–63
transhumance. *See* meadow transhumance
transportation, river
　impact of, 34–35, 38
　Ottoman network of, 3–4, 21, 23–24, 31, 33
　Pashalik of Baghdad and, 138
　through Tigris and Euphrates, 16, 147
　See also specific commodities
tree ring, 112–14
truffle, 18, 92–93
Turkey
　archives of, 17–18
　river management in, 148–50
Turkish language, 12, 23, 93, 135

Urfa, 136

Wahhabi movement, 38–39, 132, 140
water wide web, 59–60
water buffalo
　as source of energy, 67–68
　as source of manure, 100, 105
　as source of milk, 105
　herding of, 95–96, 99, 104–5, 124, 130–31
　in statistics, 88, 89*t*, 103–4, 125*t*
　wetlands and, 18–19, 98 99, 104, 124
　See also Cemmasat
waterwheel, 3–4, 9, 18–19, 67–68, 71–72
West Asia
　geology of, 33–34
　history of, 2–3, 63–64
　irrigation in, 19, 62, 96
　Ottoman presence in, 7–8, 11, 26, 42
　pastoralism in, 79–80, 82, 84–86, 87–88
　rivers of, 1, 15–16, 21, 114–15, 147, 148–49

wetlands
  definition of, 95
  early modern drainage of, 106–7
  formation of, 96–97
  misconceptions about, 95–96
  See also marshes, Iraqi
wheat
  cultivated by pastoralists, 86–87, 92
  harvest of, 64–65, 65*t*
  in arable lands, 66–67, 76–77
  in wetlands, 102, 103, 107
  shortage of, 33–34
  transport of, 4, 152*t*
  See also grain
wood. *See* timber

Zagros Mountains, 34–35
  environmental conditions in, 24, 82, 114–15
  geologic history of, 9
  pastoralism and, 83, 84
  transportation through, 21, 136–37
Zuhab Treaty, 48, 119

www.ingramcontent.com/pod-product-compliance
Ingram Content Group UK Ltd.
Pitfield, Milton Keynes, MK11 3LW, UK
UKHW022155230426